Acoustic and Radio EeV Neutrino Detection Activities

Proceedings of the International Workshop (ARENA 2005)

Acoustic and Radio EeV Neutrino Detection Activities

DESY, Zeuthen, Germany 17 – 19 May 2005

editors

Rolf Nahnhauer
Sebastian Böser
DESY, Zeuthen, Germany

World Scientific

NEW JERSEY • LONDON • SINGAPORE • BEIJING • SHANGHAI • HONG KONG • TAIPEI • CHENNAI

Published by

World Scientific Publishing Co. Pte. Ltd.
5 Toh Tuck Link, Singapore 596224
USA office: 27 Warren Street, Suite 401-402, Hackensack, NJ 07601
UK office: 57 Shelton Street, Covent Garden, London WC2H 9HE

British Library Cataloguing-in-Publication Data
A catalogue record for this book is available from the British Library.

Proceedings of the International Workshop (ARENA 2005)
ACOUSTIC AND RADIO EeV NEUTRINO DETECTION ACTIVITIES

ISBN 981-256-755-0

Printed in Singapore by B & JO Enterprise

PREFACE

The intention of the International Workshop on "***Acoustic and Radio EeV Neutrino detection Activities***" – **ARENA 2005** was to display the current efforts towards the detection of neutrinos of highest energies looking for their emission of radio or acoustic signals in dense-matter interactions.

Believing in the benefit of a combination of both techniques, we announced the workshop with an emphasis on possibilities of collaboration and joint activities, following the tradition of previous meetings in Los Angeles and Stanford.

More than 40 years ago Gurgen A. Askar'yan was the first one who proposed the detection of high energy neutrinos by acoustic signals of thermo-elastic origin due to the energy deposit in a dense medium and by coherent radio Cherenkov radiation from the charge excess in a dense particle shower.

In connection with the DUMAND project first experimental checks of the Thermo-Acoustic Model were performed at the end of the 70's of last century, confirming many of its predictions using high intense proton beams from accelerators. Confirmation of the radio-Cherenkov effect came only a few years ago from dumping an intense photon beam in silica sand.

Nevertheless several radio detection experiments started already in the 90's and in the meantime limits could set to cosmic neutrino fluxes. The acoustic technique had a big revival during the last years but is still in a research- and development-phase. A first flux limit from an acoustic array was published last spring. Also ideas about the direct production of radio signals by air showers in the atmosphere got new theoretical and experimental interest recently. So, it was time to try to get a common view to the different fields.

ARENA 2005 brought together ~90 scientists from 10 countries representing nearly all presently known theoretical and experimental research activities for radio and acoustic particle detection. Three days were filled with interesting talks and fruitful discussions about the status and the future of the topics. Finally, in a round table discussion the chances for closer cooperation between different groups were disputed.

The present volume contains most of the workshop contributions. Unfortunately we were not able to collect all of them in the given time frame. However, the slides of all talks can be found at **http://www-zeuthen.desy.de/arena**. We want to thank all of our colleagues who made ARENA 2005 a success.

Zeuthen, December 2005

Sebastian Böser Rolf Nahnhauer

CONFERENCE ORGANIZATION

International Advisory Board

G. Anton, Erlangen
J. Blümer, Karlsruhe
A. Capone, Rome
H. Falcke, Bonn
P. Gorham, Hawaii
G. Gratta, Stanford
F. Halzen, Madison
J. Learned, Hawaii
R. Nahnhauer, Zeuthen
A. Rostovtzev, Moscow
D. Saltzberg, Los Angeles
L. Thompson, Sheffield
F. Vannucci, Paris
I. Zheleznykh, Moscow

Local Organization Committee

Sebastian Böser
Rolf Nahnhauer
Christian Spiering
Michael Walter

CONTENTS

*Contribution not received.

Experimental Results II (Cherenkov Radio)

Experimental Results III (Air Shower Radio)

Future Projects I

Future Projects II

Future Projects III

Round Table Discussion

Conference Summary

EARLY YEARS OF HIGH-ENERGY NEUTRINO PHYSICS IN COSMIC RAYS AND NEUTRINO ASTRONOMY (1957-1962) *

IGOR ZHELEZNYKH

Institute for Nuclear Research of Russian Academy of Sciences
60th October Anniversary Prospect, 7A, Moscow 117312, Russia

Ideas of deep underground and deep underwater detection of high-energy cosmic neutrinos were firstly suggested by Moisey Markov in the end of 50th. Frederic Reines was one of those who first detected high-energy atmospheric neutrinos in underground experiments in the middle of 60th (as well as low energy reactor neutrinos 10 years earlier!). Markov and Reines closely collaborated in 70th – 80th in discussion of alternative techniques for large-scale neutrino telescopes. Some events of 50 – 80 years relating to the development of a new branch of Astronomy – the High-Energy Neutrino Astronomy, in which Markov and Reines took part, were described in my talk at ARENA Workshop. Below the first part of my talk at the Workshop is presented describing discussions and meetings the neutrino physics and astrophysics relating to the period 1957-1962 when I was Markov's student and later post-graduated student.

1. M.A. Markov and High-Energy Neutrino Physics

In the middle of 50th M.A. Markov worked at the Joint Institute of Nuclear Research (Dubna) and also he was a lecturer at Physics Department of Moscow State University. In those years his interests concentrated on the Quantum Field Theory and Elementary Particle Physics. There was a couple of modern in that time problems which especially attracted Markov's attention:

- the classification of elementary particles.

I would mention in this respect Markov's paper, "On classification of elementary particles" (Report USSR Academy of Sciences, 1955). In this paper he suggested composite model based on proton, neutron, hyperons and their antiparticles. This model preceded more economic Sakata's model (1956) based on proton, neutron and Λ-hyperon;

- weak interactions of elementary particles.

---

* This work is supported by grants 05-02-17410 and 05-02-26618 of the Russian Foundation of Basic Researches, grant 1782.2003.2 of the R.F. President, grant RUP2-2624-MO-04 of the U.S. Civilian Research and Development Foundation and by Program "Neutrino Physics" of the Presidium RAS.

I had a good luck to make my student work in Moscow University in 1957-1958 under Markov's leadership. Topic of my work suggested by Markov was: "interactions of the high-energy neutrinos with matter and detection of atmospheric neutrinos in underground experiments".

In fact Markov at that period initiated development of a new branch of the High-Energy Physics - the High-Energy Neutrino Physics – with aim to investigate fundamental problems of the weak interaction theory. These problems were in 1957 the following:

1. Are the weak interactions of the Fermi-type (four-fermion ones [1]) or Yukawa-type with intermediate bosons [2]?

2. How far the quadratic increase of the weak interaction cross-section with energy $\sigma = E_*^2$, where E_* is the energy in the center-of-mass system, continues to hold at very high energies?

This question was raised for the first time by W. Heisenberg in 1936 [3] who supposed that the four-fermion cross-sections could grow with energy up to $E_* \sim 1000$ GeV.

M.A. Markov, Seminar at JINR, Dubna in the middle of the 50th

3. Possible existence of two kinds of neutrinos related to muon and electron. Idea of two neutrinos was first presented by S. Sakata in 1942 [4], but specific quantum numbers for muon and electron neutrinos were introduced by J. Schwinger [5] and K. Nishijima [6] in 1957.

2. Markov's suggestions of experiments with natural and artificial fluxes of high-energy neutrinos

2.1. Possibilities of underground (underwater) detection of atmospheric neutrinos

In September 1957 according to Markov suggestion as a first step I had to estimate a number of atmospheric neutrino interactions with nucleons and electrons in 1 cubic meter of Pb placed deep underground (in order to decrease background from atmospheric muons). So I had to evaluate fluxes of atmospheric neutrinos with energies higher than 1 GeV from the decays of π-mesons produced by cosmic rays in the Earth atmosphere and to calculate cross-sections of the neutrino-nucleon reactions

$\nu + N \rightarrow N' + \mu(e)$ (1)

in the energy interval of 1-100 GeV for different variants (vector, axial vector, tensor, scalar and pseudoscalar ones) of the weak interaction theory (without and with nucleon formfactors) as well as neutrino-electron cross-sections

$\nu + e \rightarrow \nu' + \mu(e)$ (2)

Our first results presented later in my diploma work [7] were the following:

- νN – cross-sections for vector and axial vector variants were $\sigma_V(\nu N) = \sigma_A(\nu N) \sim 2\text{x}\ 10^{-38}\ cm^2$ at $E_\nu = 1$ GeV and grew linearly with energy, if this growth was not cut off by nucleon formfactors;
- $\sigma(\nu e) \sim m_e/M_N \times \sigma(\nu N)$ and its contribution could be neglected;
- if $\sigma(\nu N) \sim E_\nu$, the number of "internal" events produced by atmospheric neutrinos in an underground detector was ~3 times more than in a case, when $\sigma(\nu N)$ were constant above 1 GeV, for example due to nucleon formfactors; thus there appeared a chance to distinguish both alternatives;
- different numbers of muons and electron events induced by neutrino in a detector could give an evidence of the existence of two neutrino types;
- "several neutrino events (in $10^3\ m^3$ Pb target) per a month seemed to be a reasonable estimate"[7].

During a couple of years after 1958 the detection of the neutrino induced muon flux in the ground by "plane" detectors of ~1000 m^2 for which the number of "external" neutrino events is larger than "internal" ones was suggested [8,9, 10]. In [10] Markov reported his idea of deep underwater neutrino detection: "we propose setting up apparatus in an underwater lake or deep in the Ocean to separate charge particle direction by Cherenkov radiation".

G.T. Zatsepin in the 50th

In this period a number of theoretical papers on the weak interactions (V-A theory of Sudarshan and Marshak and Gell-Mann and Feynman), calculations of cross-sections of the neutrino reactions (T.D. Lee et al.) were published.

F. Reines [11] and K. Greisen [12] pointed out also that detectors of the size of kiloton or more sensitive volume were required to study interactions of cosmic ray neutrinos.

After detailed calculations of energy spectrum and angular distributions of the atmospheric neutrinos by G. Zatsepin and V. Kuzmin [13] possibilities of neutrino physics in cosmic rays were considered more carefully in [14].

2.2. Why not high-energy neutrino experiments at accelerators?

During carrying out my diploma work I used to visit Markov at his home every week to discuss results of my work. And I had a good opportunity to ask him various questions. Markov encouraged student's questions, even if they were naïve. One of my first questions in 1957 was related to possible use of an artificial source of high-energy neutrinos: "Why should we make calculations of atmospheric neutrino fluxes only? Why not to consider neutrino experiments at accelerators?" Markov looked at me carefully. As it turned out later Markov had discussed such possibility for the 10 GeV Dubna accelerator and results were pessimistic. My question probably stimulated Markov to come back to consideration of this problem. In a week Markov told me: "I had a talk with Bruno Pontecorvo. I told him that I would like to suggest neutrino experiments at accelerators. Pontecorvo liked such an idea very much".

Our estimations of possible neutrino fluxes and events in accelerator experiments had shown that high-energy neutrino experiments could be perspective for the future accelerators. Markov offered his student Docho Fakirov from Bulgaria to study this problem.

Fakirov defended his diploma work at Moscow University in 1958 [15]. In 1959 Markov had proposed report on the topic "On the High –Energy Neutrino Physics" to the International (Rochester) Conference on High-Energy Physics in Kiev. But after a negative reaction of some colleagues and the leader of the Weak Interactions Section ("Are you serious..?") withdrew the report from the agenda (see details in the Markov talk at 17th International Congress of History of Science [16]).

B. Pontecorvo had made a report "Electron and muon neutrinos" at Kiev Conference. To solve the problem of two neutrinos he suggested to investigate in an accelerator experiment production of electrons in reaction

$$\bar{\nu}_{\mu} + p \rightarrow n + e^{+} \qquad (3)$$

which could be induced by low energy (~35 MeV) $\bar{\nu}_{\mu}$ from decays of μ^{+} (stopped in a target after production in π^{+}-decays) if $\bar{\nu}_{\mu} = \bar{\nu}_{e}$ [17].

As M. Markov wrote in [16]:

"If our proposal of cosmic experiments was to some extent taken into consideration in realization of the neutrino experiments both in the South African mine (by Reines et al.) and in the Indian gold mine by the Indian-Japanese-English collaboration, our proposal concerning neutrino experiments at high energy accelerators (Fakirov et al.) remained practically unknown to wide scientific community. The note by Fakirov ("On spatial distribution of neutrino beam generated by high-energy neutrino collisions") was published in Sofia [18].

....The possibility of experimental separation of two types of neutrinos at high-energy accelerators in the process

$$\nu_\mu + n \to p + \mu\,(e) \qquad (4)$$

was discussed in a footnote in the proof of the book "Hyperonen und K-mesonen" made not later than in late 1959" [19].

2.3. Discussions about Neutrino-Nucleon Cross-Section Growth with Energy Higher than 1 GeV

It was noted by Markov [16] that in 50th an idea of neutrino experiments was "met with strong opposition of the competent scientific community". One of the objections was that neutrino-nucleon cross-sections would grow with neutrino energy only up to 1 GeV, because of cutting role of nucleon formfactors, and detection of atmospheric neutrinos with energies in the region 10-100 GeV would not be possible.

Markov realized that, if somebody brings forward proposal to investigate neutrino-nucleon cross-section at high energies, he must try to find some arguments substantiating its possible growth above 1 GeV. I had a few discussions with Markov on this subject during 1958, in which he suggested some explanations. First of all contribution of different quasi-elastic processes was considered, but Markov stressed, that the important role of inelastic processes would be essential and their contribution into the total ν N-cross-section could be large.

So I was able to write in the end of 1958 in my diploma work [7]:
"It is possible that neutrino-nucleon cross-section increases at high energies because of appearance of the new channels in neutrino-nucleon reactions. ...It is possible to suggest that many new channels will arise due to the strong interaction in the intermediate states. But it is not clear now what their contribution is".

Markov had described the role of the inelastic channels (of deep inelastic processes, in fact!) in [18] (p.p. 292-293), but probably this Markov's idea had appeared too early: it was 10 years before quark-parton model became to be discussed for the neutrino-nucleon processes.

Later in 1964 I tried to come back to discussions of 1958 and speculated about the possible large contribution of the exclusive process in the ν N – collision, namely production of the nucleon isobar with 3/2 spin by neutrino, into the total ν N – cross-section [20]. But in fact only Markov's idea to take into account all inelastic (inclusive) processes turned out to be true.

It is worth reminding that the first evidence of the linear growth of ν N total cross-section with energy at E >> 1 GeV was obtained from the analysis of underground neutrino experiments [21].

3. High-Energy Neutrinos from Outer Space

3.1. From Radio to Neutrino and Gamma Observations of Astrophysical Sources of Cosmic Rays (Crab, Galactic Centre?)

After consideration of the possibilities of detection terrestrial neutrinos (atmospheric and accelerator ones) it became reasonable to consider opportunities of detecting high-energy neutrinos from extraterrestrial (astrophysical) sources. There was a chance, in my opinion, to find significant (higher than atmospheric) neutrino fluxes from some cosmic objects. I asked Markov's advice and he approved such investigation. My problem was to evaluate fluxes of cosmic neutrinos from the Crab nebula and from the galactic centre using energy arguments. Fortunately I was able to use the results of two recent papers of Vitaly L. Ginzburg (now Nobel prize laureate) on the origin of cosmic rays published in 1953 [22] and 1957 [23].

It had been written in [7] (see also [8-10]):

"With the isotropy of the sources and the isotropic distribution of cosmic rays in the galaxy, the neutrino flux in terrestrial conditions is bound to be determined by the flux of neutrinos produced in the atmosphere, as a probability of the meson production in the interstellar space… is very low. Yet observations seem to favor the theory of the production of cosmic particles in the shells of super new and of new stars [22]. According to radio-astronomical data there are many relativistic electrons in the expanding shells of these stars. It is not clear whether these electrons were accelerated or they were produced as a result of nuclear collisions. In [23] there are some arguments in favor of the secondary origin of these electrons. Then 2 antineutrinos (2 neutrinos) and 1 neutrino (antineutrino) have to be produced together with each electron (positron) in π-μ-decays. …Besides π^0-mesons are produced as well as charged π-mesons. It means that 2 gammas are produced …". Taking into account some data from [22] and [23] about the total number and energies of electrons and under some assumptions it was evaluated that "the neutrino flux from Crab could be equal to the atmospheric neutrino flux".

Some speculations were made in [7] about possible neutrino sources in the centre of our Galaxy. "Neutrinos can be produced not only in the shells of the super new stars, but at later stages after cosmic particles go out from the shells.

The cosmic protons are scattered by chance interstellar magnetic fields and disappear because of the collisions with interstellar substance." It was evaluated that, if the attenuation of the protons coming from the galactic centre is not essential the galactic neutrino flux is small. But in the case of strong proton attenuation, the neutrino flux could be large ("hidden" source).

It was also noted that "according to [24] gamma quanta of energy $\sim 10^{12}$ eV have to pass the Galaxy. ...In any case the presence of high energy photons beyond the atmosphere could be an argument in favour of the existence of, at least, the same fluxes of cosmic high-energy neutrinos"[7,9].

At the end of [7] the conclusion was made: "It is worth searching for high energy neutrinos from Outer Space, especially, if the high energy gammas beyond the atmosphere were found".

3.2. Atmospheric, extra-atmospheric neutrinos and a "scientific atmosphere" around of them.

At the period end of 50th - the beginning of 60th I, similarly to what Markov described in [16], encountered in many cases skeptical attitude to the ideas and possibilities of high-energy neutrino physics and astrophysics. Perhaps it was normal that later some critics began to work actively in this area. But at early stages of my work it was important for me to get any sign of support. Below I would like to recollect with gratitude these very first signs of support.

First of all I would like to recall discussions with my university-fellows: Vilik Arutyunyan, Vladimir Voronin, Igor Alekseev, Docho Fakirov. These discussions influenced in the inspiring way on preparation of my diploma work "On interactions of the cosmic ray high-energy neutrinos with substance ". This work was defended in the end of December 1958.

J.B. Zeldovich in the 50th

Next I would mention the talk with academician Jacob B. Zeldovich, which left deep impression on me. In September 1958 Zeldovich came to Moscow University to talk with the students of university chair "Theory of atomic nuclei" (on which I studied) to find young peoples for his research.

At that time Zeldovich worked in Arzamas Nuclear Center and was one of leading persons in Soviet A-and H-bombs project. I did not aspire then to get to Arzamas. When Zeldovich had asked me, what I was working on, I answered, that I was interested in the origin of cosmic rays and possibilities of investigating this problem by means of detecting high energy neutrinos from various astrophysical objects. I considered that these problems were hardly interesting to Zeldovich. In a second I understood, that it was mistaken: Zeldovich had become interested instantly. He asked me in detail on calculations of neutrino cross-sections, estimations of fluxes of cosmic neutrinos and suggested to work on this. His words were: " During the day time we shall work under the plan, and in the evenings we shall be engaged in neutrino astrophysics ". Fast reaction, energy, scientific enthusiasm of Zeldovich were unique and very attractive. But I wished to continue work with M.A. Markov and answered, that I had another plans.

In a couple of days Markov told me, that Zeldovich called him and said that he would like to take me for work. Several months passed and I became Markov's post-graduated student in the P.N. Lebedev Physical Institute of the USSR Academy of Sciences

My research work in Lebedev Institute under direction of Markov and in close contact with Markov's colleagues Aston A. Komar and Jury D. Usachev were supplemented by my participation in two regular seminars led by extremely vigorous persons Igor E. Tamm and Vitaly L. Ginzburg.

V.L. Ginzburg in the 50th

One discussion with V.L. Ginzburg at the end of 1959 was important for me. M.A. Markov and I met V.L. Ginzburg in his cabinet in the Theoretical department of Lebedev Institute and told him in detail about our estimations of the fluxes of high-energy neutrinos from astrophysical sources. I duly emphasized, that these estimations were stimulated by his works [22, 23]. After this talk prospects of high-energy neutrino astrophysics became one of the points of Ginzburg's permanent interest. In years that followed, Ginzburg meeting me often asked, what was new in the neutrino physics and astrophysics.

During this period I was engaged in calculations of cross-sections of various neutrino processes, including ones with an intermediate meson production. The neutrino experiments at accelerators became a branch of the experimental high-energy physics and many theorists all over the world took part in development of neutrino physics.

Discovery of the nonidentity of ν_e and ν_μ in experiments on accelerators took place [25]. Question of experimental search for intermediate mesons was moved forward on the agenda. An interesting opportunity of detecting charged intermediate meson in the resonant reaction

$$\bar{\nu}_e + e^- \rightarrow W^- \rightarrow \mu^- + \bar{\nu}_\mu \quad (5)$$

had been noted by S. Glashow in 1960 [26]. However because of high value of resonant energy of neutrino in electron rest system ($E= M^2_W /2m_e$>240 GeV , if M_W larger than K-meson mass) this reaction could be observed in the planned underground neutrino experiments only at small M_W (~1 GeV). If masses of intermediate mesons are larger than two nucleon masses, it would be perspective in our opinion to search for the charged and neutral mesons at accelerator experiments with antiproton beams in the resonant reactions

$$\bar{p} + n \rightarrow W^- \rightarrow \bar{a} + b \quad (6)$$

$$\bar{p} + p \rightarrow W^0 \rightarrow \bar{a} + a \quad (7)$$

In these proposals, later published in [27], there were no mentioning yet of resonant quark-antiquark processes, in which W- and Z-mesons were really discovered. But it was a step in the right direction.

3.3. Meeting Moisey **Markov** *with Frederic Reines - 1962*

I cite below the words in which Markov described the beginning of his long (thirty-year) dialogue with Reines during which warm friendly relations between them were established [28]:

" At Geneva conference 1962 I for the first time met Professor F. Rejnes. He approached me, holding in hands a reprint of some paper. It appeared, that it was the paper by I.M. Zheleznykh and me which had been published in 1961 in Journal "Nuclear Physics" under the title "On high-energy neutrino physics in cosmic rays" . The paper contained basically a material of diploma work of the student of Moscow State University I.M. Zheleznykh on the theme offered by me:

F. Reines, Seminar at the Baksan Neutrino Observatory, INR, 1977

" On interactions of high-energy neutrinos in cosmic rays with substance ". The result of the diploma work defended in 1958, was summarized in a phrase:

" Experiment with high-energy neutrinos born in an atmosphere, is difficult, but not hopeless. Anyway, discussion of opportunities of such experiment is meaningful. Favorable circumstance is that experiment could be performed at

any depths under the ground, sufficient for elimination of a background … There is a sense to bring an attention to the question space neutrino detection ".
As I remember, Professor Reines has asked a question, whether there are in Soviet Union any attempts to organize the experiment offered by us? It is natural, that in the country the opportunity of realization of the given experiment was discussed…"

With this meeting the "prehistoric" stage of development of the high-energy neutrino physics in cosmic rays came to the end, race in underground experimental physics had begun and teams of the different countries prepared for start: a team of the USA (F. Reines, et al.), the joint team of the Great Britain, Japan and India (A. Wolfendale, S. Miyake, M.G.K. Menon, P.V. Ramana Murthy, V.S. Narasimham, B.V. Sreekantan, et al.) and a team of the Soviet Union (G.T. Zatsepin, A.E. Chudakov, et al.).

4. *Conclusion*

Reviewing my diploma work, M.A.. Markov, in particular, wrote: " The initiative belongs to Zheleznykh to consider a possible role of space neutrinos, i.e. a neutrino flux, arising not in an atmosphere of the Earth, … but in specific processes in depths of the Universe. Here Zheleznykh has made a number of interesting estimations and proposals of experiments with high-energy gamma quanta, coming to an atmosphere of the Earth from Space, which could check the existing hypothesis of origin of cosmic rays. It is very probable, that such a possibility could be closed after more detailed analysis of this question, based on a larger amount of experimental data ". Review was signed on December 29th, 1958.

After more than 40 years of the theoretical studies of high-energy astrophysics problems and continuous development of alternative methods of detecting high-energy cosmic neutrinos V.L. Ginzburg was able to write in 2002 [29]: " At last, we are literally on the eve of the appearance of the high-energy neutrino astronomy with $E>10^{12}$ eV ".

And I am quite sure that ARENA. Workshop is an important stage for development of cooperation in this new branch of the Astronomy.

Acknowledgments

I am deeply indebted to Rolf Nahnhauer and other members of the Organizing Committee of ARENA Workshop for hospitality during my stay in Zeuthen and to Aston A. Komar for careful reading of this manuscript and valuable discussions.

References

1. E. Fermi, Zs.*Phys.* **88**, 161 (1934).
2. H. Yukawa, *Proc. Phys. Math. Sos. Japan* **17**, 48 (1935).
3. W. Heisenberg, Zs.*Phys.* **101**, 533 (1936).
4. S. Sakata, *Progr. Theor. Phys.* **1**, 43 (1942), in Japanese.
5. J. Schwinger, *Ann. of Phys.* **2**, 407 (1957).
6. K. Nishijima, *Phys. Rev.* **108**, 907 (1957).
7. I.M. Zheleznykh, *Diploma paper,* Dep. of Phys., Moscow St. Univ. (1958).
8. I.M. Zheleznykh and M.A. Markov, in: *High-energy neutrino physics,* D-577, Dubna (1960).
9. M.A. Markov and I.M. Zheleznykh, *Nucl. Phys.* 27, 385 (1961).
10. M.A. Markov, *Proc. 10th Int. Conf. on High-Energy Physics at Rochester,* 579 (1960).
11. F. Reines, *Ann. Rev. Nucl. Sci.* **10**, 1 (1960).
12. K. Greisen, *Ann. Rev. Nucl. Sci.* **10**, 63 (1960).
13. G.T. Zatsepin and V.A. Kuzmin, *JETP* **41**, 1818 (1961).
14. V.A. Kuzmin, M.A. Markov, G.T. Zatsepin and I.M. Zheleznykh, *J. Phys. Soc. Jap.* **17**, Suppl. A-III, 353 (1962).
15. D. Fakirov, *Diploma paper,* Dep. of Phys., Moscow St. University (1958).
16. M.A. Markov, *"Early Development of Weak Interactions in the USSR"*, Nauka Publishers, Central Depart. of Oriental Literature, Moscow (1985).
17. B. Pontecorvo, *JETP* **37,** 1751 (1959).
18. D. Fakirov, *Fac. Sci. Sofia* **2**, 53 (1958/1959).
19. M.A. Markov, *Hyperonen und K-mesonen,* Verl. Wissensch., 292 (1960).
20. I.M. Zheleznykh, *Phys. Letts.,***11,** 251 (1964).
21. L.V. Volkova and G.T. Zatsepin, *J. Nucl. Phys. (Yad. Fiz.)* **14**, 211 (1971).
22. V.L. Ginzburg, *Usp. Fiz. Nauk* **51**, 343 (1953).
23. V.L. Ginzburg, *Usp. Fiz. Nauk* **62**, 37 (1957).
24. N. Klepikov, *JETP* 35,316 (1958).
25. G. Danby et al., *Phys. Rev. Lett.* **9**, 36 (1962).
26. S. Glashow, *Phys. Rev.* **118**, 316 (1960).
27. I.M. Zheleznykh and M.A. Markov, *J. Nucl. Phys. (Yad. Fiz.)* **1**, 303 (1965).
28. M.A. Markov, *"Reflecting on physicists .. physics .. world "*, Moscow, Nauka, p.p. 76-83 (1993).
29. V.L. Ginzburg, *"About science, myself and others"*, M., Fizmatlit (2003).

EXTREMELY ENERGETIC COSMIC NEUTRINOS: OPPORTUNITIES FOR ASTROPHYSICS, PARTICLE PHYSICS, AND COSMOLOGY

ANDREAS RINGWALD
Deutsches Elektronen-Synchrotron DESY,
Notkestraße 85,
D-22607 Hamburg, Germany
E-mail: andreas.ringwald@desy.de

Existing and planned observatories for cosmic neutrinos open up a huge window in energy from 10^7 to 10^{17} GeV. Here, we discuss in particular the possibilities to use extremely energetic cosmic neutrinos as a diagnostic of astrophysical processes, as a tool for particle physics beyond the Standard Model, and as a probe of cosmology.

1. Introduction

We are living in exciting times for extremely high energy cosmic neutrinos (EHECν's). Existing observatories, such as AMANDA[1], ANITA-lite[2], BAIKAL[3], FORTE[4], GLUE[5], and RICE[6] have recently put restrictive upper limits on the neutrino flux in the energy region from 10^7 to 10^{17} GeV (cf. Fig. 1). Furthermore, recent proposals for larger EHECν detectors, such as ANITA[7], EUSO[8], IceCube[9], LOFAR[10], OWL[11], PAO[12], SalSA[13], WSRT[10], together with conservative neutrino flux predictions from astrophysical sources of the observed cosmic rays (CR's), such as active galactic nuclei, offer credible hope that the collection of a huge event sample above 10^7 GeV may be realized within this decade (cf. Fig. 1). This will provide not only important information on the astrophysical processes associated with the acceleration of CR's, but also an opportunity for particle physics beyond the reach of the Large Hadron Collider (LHC). There is even the possibility of a sizeable event sample above 10^{11} GeV, with important consequences for cosmology. The corresponding neutrino fluxes may arise from the decomposition of topological defects – relics of phase transitions in the very early universe – into their particle constituents. Moreover, it may be possible to detect the cosmic neutrino background via absorption features in these neutrino spectra. In this contribution, we will have a closer look

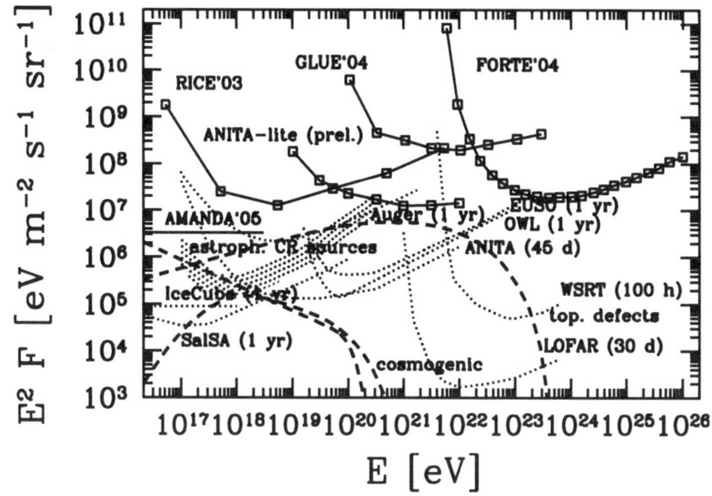

Figure 1. Current status and next decade prospects for EHECν physics, expressed in terms of diffuse neutrino fluxes per flavor, $F_{\nu_\alpha} + F_{\bar{\nu}_\alpha}$, $\alpha = e, \mu, \tau$. Upper limits from AMANDA[1], ANITA-lite[2], FORTE[4], GLUE[5], and RICE[6]. Also shown are projected sensitivities of ANITA[7], EUSO[8], IceCube[9], LOFAR[10], OWL[11], the Pierre Auger Observatory in ν_e, ν_μ modes and in ν_τ mode (bottom swath)[12], SalSA[13], and WSRT[10], corresponding to one event per energy decade and indicated duration. Also shown are predictions from astrophysical CR sources[14], from inelastic interactions of CR's with the cosmic microwave background (CMB) photons (cosmogenic neutrinos)[14,15], and from topological defects[16].

at these exciting opportunities.

2. EHECν's as a diagnostic of astrophysical processes

Neutrinos with energies $\lesssim 10^{12}$ GeV propagate essentially without interaction between their source and Earth. Hence, they are a powerful probe of high energy astrophysics, in particular of the conjectured acceleration sites of the CR's, notably active galactic nuclei (AGN). A paradigm for the acceleration mechanism in the jets of these AGN's is shock acceleration. Protons and heavier nuclei are confined by magnetic fields and accelerated through repeated scattering by plasma shock fronts. Inelastic collisions of the trapped protons with the ambient plasma produces pions and neutrons, the former decaying into neutrinos and photons, the latter eventually diffusing from the source and decaying into CR protons (cf. Fig. 2 (left)).

A quite conservative estimate of the flux of neutrinos from such astrophysical sources can be made as follows[14]. Assuming that the sources are

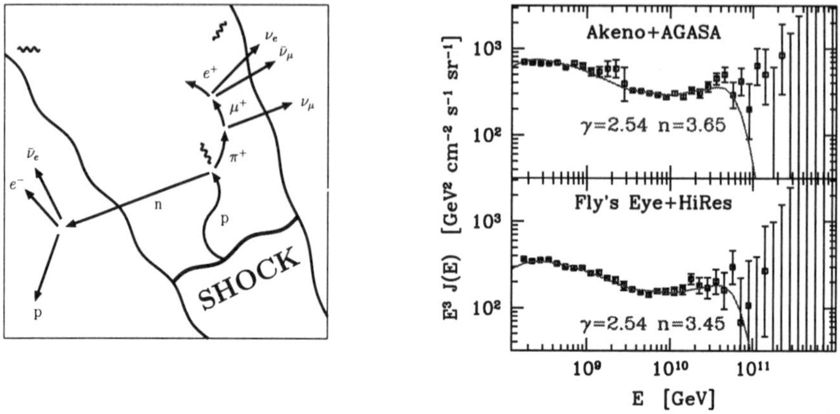

Figure 2. *Left:* Illustration of shock acceleration in the jet of an active galaxy[14]. *Right:* Best fits[14] to the ultra-high energy cosmic ray spectrum in the energy interval $[10^{8.6}, 10^{11}]$ GeV as observed by Akeno[18]+AGASA[19] and Fly's Eye[20]+HiRes[21]. The dip from e^+e^- pair production[22,23] and the bump from Greisen-Zatsepin-Kuzmin[24] (GZK) accumulation are clearly visible in the data and support the simple power law ansatz for the emissivity of the extragalactic sources, in which we have set $z_{\rm min} = 0.012$, $z_{\rm max} = 2$, and $E_{i,\rm max} = 10^{12.5}$ GeV, for the boundaries in redshift and injection energy, respectively. Apparently, this fit undershoots the data for the few highest energy events.

optically thin, i.e. the neutrons can escape, one may determine the neutron emissivity at the sources from the observed CR spectra[17], taking into account propagation effects, in particular e^+e^- and pion production through inelastic scattering off the CMB photons. Figure 2 (right) illustrates that both the AGASA and the HiRes data in the $10^{8.6\div 11}$ GeV range can be fitted nicely under the assumption of a simple power law neutron injection emissivity, $\propto E_i^{-2.5}(1+z)^{3.5}$, of the extragalactic sources, supporting the recent proposal towards a low transition energy, $\sim 10^{8.6}$ GeV, between galactic and extragalactic cosmic rays[22], which is also sustained by chemical composition studies of HiRes data[25]. The neutron injection emissivity is simply related to the neutrino emissivity, and the latter can be translated easily into an expected neutrino flux at Earth. It should be detected very soon, if not already with AMANDA-II, then at least with IceCube (cf. Fig. 1), which therefore can provide significant clues in demarcating the cosmic ray galactic/extragalactic crossover energy[14]. Although the cosmogenic neutrino flux from the inelastic interactions with the CMB photons starts to dominate over the neutrino flux from optically thin cosmic ray sources at energies above a few EeV, it appears to be hard to detect with the EHECν detectors operating in the next decade (cf. Fig. 1).

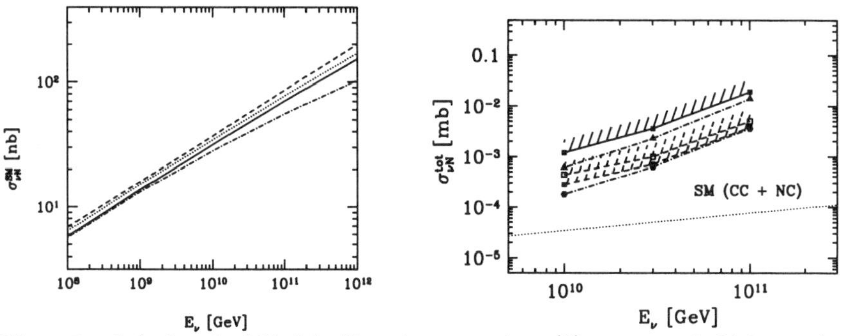

Figure 3. *Left:* Standard Model νN total cross section $\sigma_{\nu N}^{\rm tot}$ at extremely high neutrino energies E_ν obtained by various perturbative QCD resummation techniques: from a unified BFKL-DGLAP approach[26] (solid), based on CTEQ parton distributions[27] (dotted), based on GRV dynamical partons[28] (dashed), and from a unified BFKL-DGLAP approach supplemented by saturation effects[29] (dashed-dotted). *Right:* Model-independent upper bounds on the neutrino-nucleon inelastic cross section[30] derived from the RICE Collaboration search results[6], by exploiting different cosmogenic neutrino flux estimates, by Fodor *et al.* (FKRT[15]) (solid line) and Protheroe and Johnson (PJ[31]) (dashed line joining solid squares). The dashed line joining the open squares (PJ) indicates the upper bound for inelasticity $\langle y\rangle = 0.5$. The dashed-dotted lines indicate the sensitivity (95% CL, for $\sigma_{\nu N}^{\rm tot} < 4$ mb) of PAO in 10 yr of operation assuming zero events observed above SM background (circles PJ, triangles FKRT). For comparison, also shown is the SM total (charged current and neutral current) νN inelastic cross section[27].

3. EHECν's and physics beyond the Standard Model

Cosmic neutrinos with energies E_ν above 10^8 GeV probe neutrino-nucleon scattering at center-of-mass (c.m.) energies above

$$\sqrt{s_{\nu N}} \equiv \sqrt{2 m_N E_\nu} \simeq 14\ \mathrm{TeV}\ \left(E_\nu/10^8\ \mathrm{GeV}\right)^{1/2}\ , \tag{1}$$

beyond the proton-proton c.m. energy $\sqrt{s_{pp}} = 14$ TeV of the LHC, and Bjorken $x \equiv Q^2/(y\, s_{\nu N})$ values below

$$x \simeq 2\times 10^{-4}\ \left(Q^2/m_W^2\right)(0.2/y)\left(10^8\ \mathrm{GeV}/E_\nu\right)\ , \tag{2}$$

where Q^2 is the momentum transfer squared, $m_W \simeq 80$ GeV the W-boson mass, and y the inelasticity parameter. Under these kinematical conditions, the predictions for νN scattering from the perturbative Standard Model (SM) are quite safely under control (cf. Fig. 3 (left)), notably thanks to the input from measurements of deep-inelastic ep scattering at HERA[32,33]. This makes it possible to search for enhancements in the νN cross section due to physics beyond the (perturbative) SM, such as electroweak sphaleron (non-perturbative $B + L$ violation) or Kaluza-Klein, black hole, p-brane or string ball production in TeV scale gravity models.

Since the rate of neutrino-initiated showers is proportional to integrated flux times cross section, the non-observation of quasi-horizontal or deeply-penetrating neutrino-induced air showers as reported by, e.g., Fly's Eye[34], AGASA[35], and RICE[6] can be turned into an upper bound on the neutrino nucleon cross section if a certain prediction for the neutrino flux is exploited[36,37]. This is exemplified in Fig. 3 (right), which displays the limits on $\sigma_{\nu N}$ from the RICE search results on contained showers[30], for two different assumptions about the EHECν flux. These bounds are considerably higher than the SM cross section, albeit in the post-LHC energy region. PAO will be able to improve these limits by one order of magnitude[30].

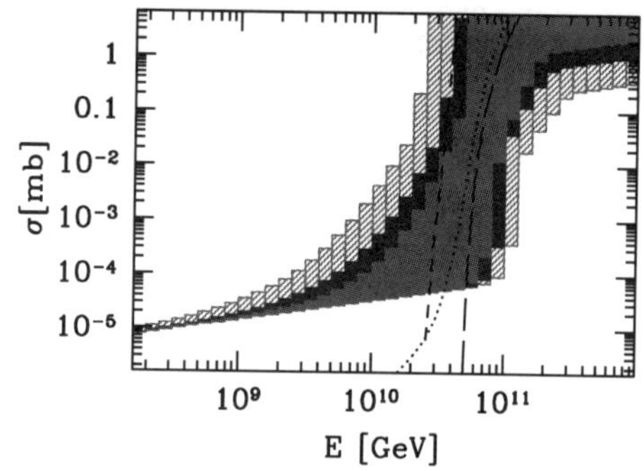

Figure 4. The range of the cross section within the 99%, 95% and 90% CL required for a successful strongly interacting neutrino scenario[39]. The lines are theoretical predictions of an enhancement of the neutrino-nucleon cross-section by electroweak sphalerons[40,41] (short-dashed), p-branes[42] (long-dashed) and string excitations[43] (dotted).

The bounds exploiting searches for deeply-penetrating particles are typically applicable as long as $\sigma_{\nu N} \lesssim 0.5 \div 1$ mb. Models with even higher and more speculative cross sections, $\gtrsim 1 \div 10$ mb, such as electroweak sphaleron production, brane production, or string resonance production, qualify as strongly interacting neutrino scenarios[38,23], according to which the mysterious EHECR beyond the predicted GZK cutoff [24] at $E_{\rm GZK} \simeq 4 \times 10^{10}$ GeV (cf. Fig. 2 (right)) are initiated by cosmogenic neutrinos. Figure 4 illustrates that a combined fit of the existent data on vertical showers by AGASA and HiRes, as well as of the search results on weakly interacting particles of AGASA and RICE, requires a steep increase within one energy decade around $E_{\rm GZK}$ by four orders of magnitude[39] – an enhancement which has indeed been proposed within some extensions of the (perturbative) SM.

We have emphasized here the current constraints from EHECν on physics beyond the SM. A more detailed account of the particle physics reach of the planned EHECν observatories can be found elsewhere[44,45].

4. EHECν's as a tool to study big bang relics

The existence of topological stable solutions of the field equations (topological defects) is a generic prediction of symmetry breaking (SB) in Grand Unified Theories (GUT's) and occurs even at the fundamental level in String Theory in the form of F- and D-strings. Specifically, G $\to$ H $\times$ U(1) SB leads to monopoles, U(1) SB to ordinary or superconducting strings, and G $\to$ H $\times$ U(1) $\to$ H $\times$ Z_N SB to monopoles connected by strings, e.g. necklaces in case of $N = 2$. Such topological defects may be produced through non-thermal phase transitions during preheating after inflation[46]. Their superheavy constituents X, often gauge or Higgs bosons with masses $m_X \sim 10^{12\div16}$ GeV, may be liberated on various occasions[47], e.g. through repeated self-intersections of strings, through annihilation of monopole anti-monopole pairs etc., and rapidly decay into stable SM particles, under which we readily find[48] EHECν's with energies up to $\sim 0.05\,m_X$. The corresponding fragmentation spectra are meanwhile worked out very accurately[49] via Monte Carlo generators or via DGLAP evolution from experimentally determined initial distributions at the scale m_Z to the ones at m_X. The injection rate, which determines in particular the overall normalization of the neutrino flux, depends on cosmic time t in the form $\dot{n}_X = \kappa m_X^p t^{-4+p}$, where κ and p are dimensionless constants depending on the specific scenario[48].

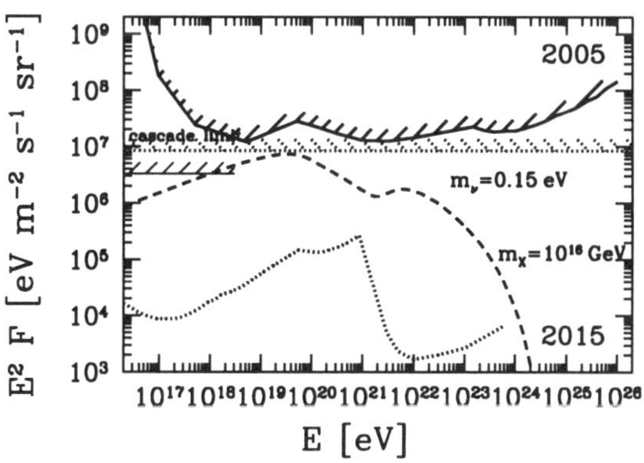

Figure 5. Present (2005) limits on the neutrino flux and projected sensitivity in ten years from now (2015), together with a prediction from topological defects[16] ($m_X = 10^{16}$ GeV, $p = 0$). The absorption dip arising from resonant annihilation of the EHECν's with big bang relic neutrinos of mass $m_\nu = 0.15$ eV into Z-bosons is clearly visible.

For a wide range of overall flux normalizations, the upcoming EHECν

observatories seem to be sensitive enough to obtain, within the next decade, sizeable event rates from topological defects[16] (cf. Fig. 5). Note, that, for the first time in cosmic particle physics, the GUT energy scale can be directly probed. Clearly, a precise measurement of the neutrino spectrum from topological defects would have a strong impact on particle physics and cosmology. Its mere existence would signal the existence of topological defects as relics from early phase transitions after inflation. The high end of the spectrum directly reveals the mass of the X particles, and its shape entails detailed information on the particle content of the desert, on the Hubble expansion rate, and on the big bang relic neutrino background. Indeed, as illustrated in Fig. 5, the resonant annihilation of the neutrinos from X particle decays with big bang relic neutrinos would leave its imprints as absorption dips in the measured spectrum[50]. Such a measurement would not only shed light on the existence and the spatial distribution of the cosmic neutrino background, but would also give important information on the neutrino masses[51], since the dips occur around the resonance energies $E_{\nu_i}^{\rm res} = 4 \times 10^{21}$ eV $(1\ {\rm eV}/m_{\nu_i})$. Note, that, along with a prediction of absorption dips, there goes a prediction of emission features – protons and photons from hadronic Z-decay ("Z-bursts") – which may appear as a CR flux recovery beyond $E_{\rm GZK}$ and be measured by EUSO, OWL, or LOFAR[16].

5. Conclusions

The future seems bright in extremely high energetic neutrinos. There are many observatories under construction, whose combined sensitivity ranges from 10^7 to 10^{17} GeV, the energy scale of Grand Unification. In the likely case that appreciable event samples are collected in this energy range, we can expect a strong impact on astrophysics, particle physics, and cosmology.

References

1. M. Ackermann *et al.* [AMANDA Collab.], Astropart. Phys. **22** (2005) 339.
2. S. Barwick *et al.* [ANITA Collaboration], these proceedings and to appear.
3. R. Wischnewski *et al.* [Baikal Collaboration], arXiv:astro-ph/0507698.
4. N. G. Lehtinen *et al.*, Phys. Rev. D **69** (2004) 013008.
5. P. W. Gorham *et al.*, Phys. Rev. Lett. **93** (2004) 041101.
6. I. Kravchenko, arXiv:astro-ph/0306408.
7. P. Gorham *et al.* [ANITA Collaboration], NASA Proposal SMEX03-0004-0019.
8. S. Bottai and S. Giurgola [EUSO Collaboration], in: *Proc. 28th International Cosmic Ray Conference*, Tsukuba, Japan, 2003, pp. 1113-1116.
9. J. Ahrens *et al.* [IceCube Collab.], Nucl. Phys. Proc. Suppl. **118** (2003) 388.

10. O. Scholten *et al.*, arXiv:astro-ph/0508580.
11. F. W. Stecker *et al.* [OWL Coll.], Nucl. Phys. Proc. Suppl. **136C** (2004) 433.
12. X. Bertou *et al.*, Astropart. Phys. **17** (2002) 183.
13. P. Gorham *et al.*, Nucl. Instrum. Meth. A **490** (2002) 476; private commun.
14. M. Ahlers *et al.*, Phys. Rev. D **72** (2005) 023001.
15. Z. Fodor, S. D. Katz, A. Ringwald and H. Tu, JCAP **0311** (2003) 015.
16. B. Eberle, A. Ringwald, T. J. Weiler and Y. Y. Y. Wong, DESY 05-165.
17. E. Waxman and J. N. Bahcall, Phys. Rev. D **59** (1999) 023002.
18. M. Nagano *et al.*, J. Phys. G **18** (1992) 423.
19. M. Takeda *et al.* [AGASA Collab.], Phys. Rev. Lett. **81** (1998) 1163; Astropart. Phys. **19** (2003) 447; http://www-akeno.icrr.u-tokyo.ac.jp/AGASA/
20. D. J. Bird *et al.* [Fly's Eye Collaboration], Phys. Rev. Lett. **71** 3401 (1993); Astrophys. J. **424**, 491 (1994); Astrophys. J. **441**, 144 (1995).
21. T. Abu-Zayyad *et al.* [HiRes Collaboration], Astropart. Phys. **23** (2005) 157.
22. V. Berezinsky, A. Z. Gazizov and S. I. Grigorieva, arXiv:hep-ph/0204357.
23. Z. Fodor, S. D. Katz, A. Ringwald and H. Tu, Phys. Lett. B **561** (2003) 191.
24. K. Greisen, Phys. Rev. Lett. **16** (1966) 748; G. T. Zatsepin and V. A. Kuzmin, JETP Lett. **4** (1966) 78.
25. D. R. Bergman [HiRes Collab.], Nucl. Phys. Proc. Suppl. **136** (2004) 40.
26. J. Kwiecinski *et al.*, Phys. Rev. D **59** (1999) 093002.
27. R. Gandhi *et al.*, Phys. Rev. D **58** (1998) 093009.
28. M. Glück, S. Kretzer and E. Reya, Astropart. Phys. **11** (1999) 327.
29. K. Kutak and J. Kwiecinski, Eur. Phys. J. C **29** (2003) 521.
30. L. A. Anchordoqui *et al.*, JCAP **0506** (2005) 013.
31. R. J. Protheroe and P. A. Johnson, Astropart. Phys. **4** (1996) 253.
32. C. Adloff *et al.* [H1 Collaboration], Eur. Phys. J. C **30** (2003) 1.
33. S. Chekanov *et al.* [ZEUS Collaboration], Eur. Phys. J. C **32** (2003) 1.
34. R. M. Baltrusaitis *et al.*, Phys. Rev. D **31** (1985) 2192.
35. S. Yoshida *et al.* [AGASA Collaboration], in: *Proc. 27th International Cosmic Ray Conference*, Hamburg, Germany, 2001, p. 1142
36. V. S. Berezinsky and A. Y. Smirnov, Phys. Lett. B **48** (1974) 269.
37. D. A. Morris and A. Ringwald, Astropart. Phys. **2** (1994) 43.
38. V. S. Berezinsky and G. T. Zatsepin, Phys. Lett. B **28** (1969) 423.
39. M. Ahlers, A. Ringwald and H. Tu, arXiv:astro-ph/0506698.
40. A. Ringwald, JHEP **0310** (2003) 008.
41. T. Han and D. Hooper, Phys. Lett. B **582** (2004) 21.
42. L. A. Anchordoqui *et al.*, Phys. Lett. B **535** (2002) 302.
43. W. S. Burgett *et al.*, Nucl. Phys. Proc. Suppl. **136** (2004) 327.
44. T. Han and D. Hooper, New J. Phys. **6** (2004) 150.
45. L. Anchordoqui, T. Han, D. Hooper and S. Sarkar, arXiv:hep-ph/0508312.
46. I. Tkachev *et al.*, Phys. Lett. B **440** (1998) 262.
47. P. Bhattacharjee and G. Sigl, Phys. Rept. **327** (2000) 109.
48. P. Bhattacharjee *et al.*, Phys. Rev. Lett. **69** (1992) 567.
49. R. Aloisio, V. Berezinsky and M. Kachelriess, Phys. Rev. D **69** (2004) 094023.
50. T. J. Weiler, Phys. Rev. Lett. **49** (1982) 234.
51. B. Eberle *et al.*, Phys. Rev. D **70** (2004) 023007.

INVESTIGATION OF EVENT RATES FOR DIFFERENT DETECTOR ARRAYS AND VARIOUS EXTREMELY HIGH ENERGY MODELS

J. K. BECKER AND W. RHODE

Institut für Physik,
Universität Dortmund, Dortmund, Germany
E-mail: julia@physik.uni-dortmund.de, rhode@physik.uni-dortmund.de

New detection methods for extremely high energy (EHE) neutrinos are being discussed. In this paper, the comparison of different detection methods at energies $E > 10^{7.5}$ GeV are examined, using various neutrino flux predictions. Arrays for acoustic and radio signals from neutrino induced electromagnetic cascades as well as the IceCube array with additional strings ("IceCube Plus") are investigated with effective volumes as given in [5,6]. The depth of the detector below the Earth's surface are examined with respect to the absorption of a potential neutrino signal by the Earth. It can be shown that absorption plays an important role and that an array of acoustic and radio antennas should preferably be put at shallow depths of ~ 500 m depth. The detection potential at this depth ranges from several up to tens of events at EHEs depending on the source of the neutrino flux.

1. Introduction

Current neutrino experiments are able to measure the atmospheric neutrino spectrum up to 100 TeV without observing a significant contribution from extragalactic sources [4]. Successor experiments like IceCube aim the detection of neutrinos up to 100 PeV. The detection of a signal at even higher energies is restrictedly also possible with IceCube, but to achieve a good detection possibility of the so called cosmogenic neutrino flux, new methods are being developed which are complementary to optical detection. Acoustic and radio neutrino detection aims at the measurement of neutrinos at Extremely High Energies (EHEs), i.e. $E > 10^8$ GeV. In this paper, different neutrino flux models will be discussed with respect to their detection probability at EHEs. Figure 1 shows neutrino flux predictions with a relatively high contribution at EHEs. The models presented are the following:
Mannheim, Protheroe and Rachen (**MPR**) [2] predict a maximum flux from

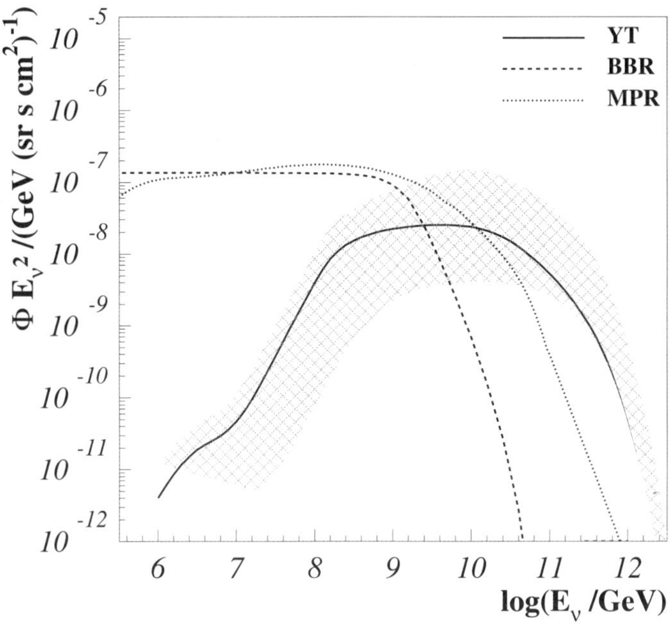

Figure 1. Various diffuse neutrino flux models (see text for detailed explanation).

blazar sources for $p\gamma$ interactions.
Becker, Biermann and Rhode (**BBR**) [1] calculate the number of neutrinos from AGN proton photon interactions considering steep and flat spectrum sources.
Yoshida and Teshima (**YT**) [7] calculate the neutrino flux resulting from CR interactions with the CMB. The spectrum is normalized to the CR spectrum, based on the GZK prediction. The source distribution ρ is assumed to be $\rho \propto (1+z)^m \cdot \Theta(z_{\max} - z) + (1 + z_{\max})^m \cdot \Theta(z - z_{\max})$. The shaded region represents the uncertainties in the model due to the evolution function, using $(m, z_{\max}) = (4, 5)$ as an upper limit and $(m, z_{\max}) = (2, 2)$ as a lower one. The evolution model should match the measured distribution of AGN in space, since AGN follow the Star Formation Rate (SFR) and are believed to be responsible for a significant fraction of the EHE proton flux. Comparing the models with the AGN distribution function given in [3] shows that model $(m, z_{\max}) = (4, 4)$ (solid line in Fig. 1) seems to fit the data most effectively of all given scenarios, and thus it will be used as the standard

GZK prediction with errors given according to the maximum prediction with $(m, z_{\text{max}}) = (4, 5)$ and the minimum expectation of $(m, z_{\text{max}}) = (2, 2)$. Note that $(m, z_{\text{max}}) = (2, 2)$ is a very pessimistic evolution scenario which is far from evolution as it is observed, even at relatively low redshifts.

2. Event rate calculation

The number of neutrinos per time unit is given as the convolution of the initial flux $\Phi_\nu(E_\nu, \theta)$ with the probability of detecting a neutrino within a given detection volume V_{eff},

$$R(E^{\text{min}}, \theta) = \int_{E^{\text{min}}} \frac{dP_{\nu_l \to X}}{dr}(E_\nu, E^{\text{min}}) \cdot P^{\nu}_{shadow} \cdot V_{eff} \cdot \Phi_\nu(E_\nu, \theta)\, dE_\nu \,. \quad (1)$$

This probability is the product of the Earth shadow factor P^{ν}_{shadow} that considers neutrino absorption of the Earth and of the probability $dP_{\nu_l \to X}/dr$ that the neutrino induces a cascade at a certain detection range. The shadow factor depends on the absorption length of the neutrinos in the Earth and in the atmosphere. At energies above the EeV range, the Earth filters most of the neutrino signal and thus, the rates will be calculated assuming a $2\,\pi$ detection field above the horizon.

The probability of inducing a cascade at a certain range r depends on the total cross section of neutrino nucleon interactions including charged and neutral current interactions, σ, and the density of the detection material in units of water density, $\rho' := \rho/\rho_{H_2O}$. This factor is of the order unity for ice and water detectors, for an array in a salt dome, it is $\rho' \sim 2.2$. The effective volume of the detection array varies with the geometric volume and the detection method. One possible scenario is the deployment of acoustic and radio sensors on totally 91 strings with 1000 m separation in the Antarctic ice and use this array in combination with the IceCube optical detection array. Effective volumes for these three detection methods will be used as presented in [5,6].

3. Results and Conclusions

Figure 2 shows the depth dependence of the detection rate at the example of the standard cosmogenic flux YT(4,4) as discussed in section 1. The number of observed events decreases significantly with the depth of the arrays below the Earth surface which is why a shallow depth (< 500 m) of such detection arrays is suggested. Figure 3 shows that the rate is about constant with respect to the lower energy integration limit E_{min} for acoustic

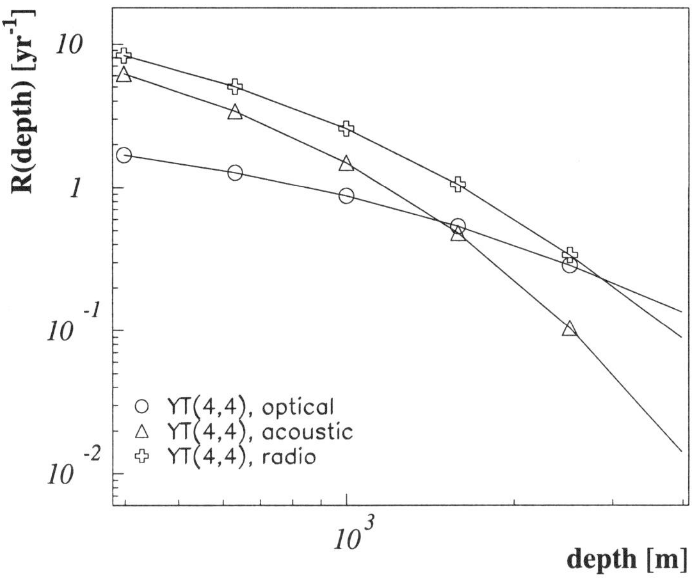

Figure 2. Depth dependence of a possible signal shown at the example of YT(4,4).

detection, since the effective volume does not allow any significant detection below $\sim 10^9$ GeV. Between $10^{9.5} - 10^{10.5}$ GeV, however, the rate decreases about an order of magnitude with the threshold energy for the three models being discussed.

Rate/yr	MPR	YT	BBR
optical	38	$1.4^{+4.6}_{-1.3}$	0.1
acoustic	10	$4.7^{+24.3}_{-4.0}$	0.1
radio	26	$6.6^{+25.4}_{-5.7}$	0.2

The table summarizes the rates for the discussed flux models with a threshold energy of $10^{7.5}$ GeV. BBR does not yield a significant contribution at these high energies and is a better candidate at lower energies. The maximum expected neutrino flux from blazars (MPR) gives the best results with several tens of events per year. It should, however, be kept in mind that this is an upper limit on the flux of these sources. The contribution from the GZK neutrinos (YT) is more guaranteed and leads to a significant rate of several events[a]. Therefore, we conclude that the detection possibilities for acoustic and radio methods are very promising.

[a]Lower and upper errors are given by the YT(2,2) resp. YT(4,5) parametrization.

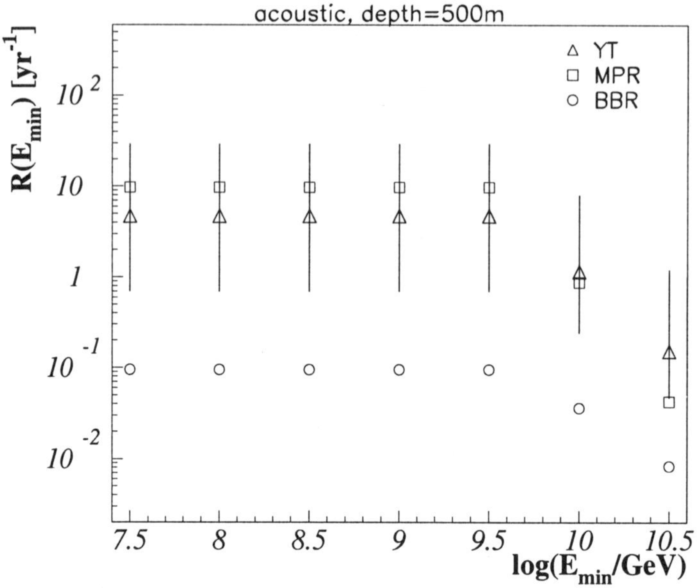

Figure 3. Threshold energy dependence of the total rate for an acoustic array and for various flux models.

Acknowledgments

The authors would like to thank Shigeru Yoshida for helpful discussions. We would also like to thank Justin Vandenbroucke, David Besson, Sebastian Böser, Rolf Nahnauer and Buford Price for their help with this work. This work has been supported by the *Deutsche Forschungsgemeinschaft* DFG.

References

1. J. K. Becker, P. L. Biermann, and W. Rhode. *Astropart. Physics*, 23(4):355, 2005.
2. K. Mannheim, R. J. Protheroe, and J. P. Rachen. *Phys. Rev. D*, 63:23003, 2001.
3. T. Miyaji, G. Hasinger, and M. Schmidt. *Astron. & Astrophys.*, 353:25, 2000.
4. K. Münich et al. In *29th ICRC Proceedings*, 2005.
5. J. Vandenbroucke et al. 2005. these proceedings.
6. J. Vandenbroucke et al. In *29th ICRC Proceedings*, 2005.
7. S. Yoshida and M. Teshima. *Progress of Theoretical Physics*, 89:833, 1993.

MEASUREMENT OF ATTENUATION LENGTH FOR RADIO WAVE IN NATURAL ROCK SALT SAMPLES CONCERNING ULTRA HIGH ENERGY NEUTRINO DETECTION *

MASAMI CHIBA, YUSUKE WATANABE, OSAMU YASUDA

Department of Physics, Tokyo Metropolitan University,
1-1 Minamiohsawa, Hachioji, Tokyo, 192-0397, Japan

TOSHIO KAMIJO

Department of Electrical and Electric Engineering, Tokyo Metropolitan University,
1-1 Minamiohsawa, Hachioji, Tokyo, 192-0397, Japan

YUICHI CHIKASHIGE, TADASHI KON, AKIO AMANO, YOSITO TAKEOKA, YUTAKA SHIMIZU, SATOSHI MORI, SOSUKE NINOMIYA

Faculty of Science and Technology, Seikei University,
3-3-1 Kichijyoji Kitamachi, Musashino-shi, Tokyo, 180-8633, Japan

Ultra high energy (UHE) neutrinos with the energy larger than 10^{15} eV, surely arrive at the earth with Greisen, Zatsepin, Kuz'min (GZK) effect, though the rate is very few. The rare call requires us to utilize a large mass (>10 Gton) of detection medium. UHE neutrino generates a huge number of unpaired electrons in rock salt. They would emit sensible radio wave by coherent Cherenkov (Askar'yan) effect. The longer attenuation length of radio wave in rock salt reduces the number of antennas required. Several rock salt samples including synthesized one are measured in attenuation length for radio wave transmission at 0.3 and 1.0 GHz. Some show attenuation length larger than 300 m, which indicate a possibility for constructing a salt neutrino detector.

1. Introduction

Several models concerning high-energy phenomena in the cosmos predict generation of ultra-high energy (UHE) cosmic neutrinos with the energies larger than 10^{15} eV in astrophysical systems, e.g. active galactic nuclei. UHE protons lose energy traveling 163 M ly (50Mpc) due to collision with 2.7 K cosmic microwave background. The process is called Greisen, Zatsepin and Kuz'min

* Work partially supported by a Grant in Aid for Scientific Research for Ministry of Education, Science, Technology and Sports and Culture of Japan, and Funds of Tokubetsu Kenkyuhi, at Seikei University

(GZK) effect [1]. UHE protons are observed whose energy exceed production threshold of Δ resonance at the collision. The resonance decays to a charged π, which decays to UHE neutrinos (GZK neutrinos). The existence of GZK neutrinos is reliable and we aim to detect them at first [2]. The energy ranges over 10^{15} eV-10^{20} eV, and the flux is as low as 1 ($km^{-2}day^{-1}$). A huge detector larger than the mass of 20 Gton or the volume of $(2km)^3$ in case of rock salt is suitable being sensitive to the energy, the direction, the time and the flavor.

The huge detection medium requires long-range transmission wave with a large attenuation length, which carries the information of the neutrino interaction. Radio wave promisingly transmits the information through rock salt in a long range. G. A. Askar'yan [3] has proposed detection of radio wave emission with coherent amplification produced by excess electrons in an electro-magnetic shower in dense materials. Askar'yan effect was confirmed using a bunched electron beam at SLAC [4]. While for low-density medium, radio emission was calculated in an atmospheric shower by M. Fujii and J. Nishimura and recently confirmed experimentally [5].

Rock salt domes are distributed widely and there will be the suitable sites [6]. We have been studied a rock salt formation having long radio wave transparency [2, 7]. A rock salt dome seems to have a long attenuation length since it does not allow water penetration and it is covered with soil and rock, which prevent surface radio wave to penetrate. Coming in cosmic ray is only μ under the overburden of a few hundred meters, which could not generate often a concentrated high-energy shower emitting radio wave by Askar'yan effect. If the attenuation length of radio wave propagation is long enough in a rock salt dome, a reasonable number of radio wave antennas could detect neutrino interactions in a massive rock salt.

2. Measurement of attenuation length for radio wave in rock salt

We have measured complex permittivity in rock salt samples by a perturbation method [8] using cylindrical cavity resonators of 0.3 GHz (749 mmϕ × 100 mm) and 1.0 GHz (225 mmϕ × 30 mm). Both have the Q of about 10,000. The cavity height is designed as low as possible keeping a moderate value of Q so that the low absorption of the rock salt could be measured. Since long thin rod of rock salt sample is difficult to carve out. They are made of oxygen free copper with sample insertion holes at the center of the cavity. A sample piece having the height of the cavity is inserted and at the same time the holes are closed by copper caps and form a complete cavity.

We obtained the real part (the square of the refractive index) and the imaginary part (absorption in a medium) of the complex permittivity, by measuring the decease of the resonance frequency and the widening of the resonance width after the insertion of the samples, respectively. The method is proper to measure an imaginary part of the permittitivity for low-loss material like rock salt in a laboratory. The attenuation length L is calculated by Eq. (1):

$$\tan\delta = \frac{\varepsilon''}{\varepsilon'}, \qquad L = \frac{\lambda}{\pi\sqrt{\varepsilon'}\tan\delta}. \tag{1}$$

Where $\varepsilon', \varepsilon''$, $\tan\delta$ and λ are real permittivity, imaginary permittivity, loss tangent and wavelength of the radio wave, respectively. At the traveling distance of L, the electric field strength is diminished to 1/e.

Up to now, a few measurements of attenuation length or $\tan\delta$ in rock salt exist in the frequency range of our interest. A. R. Hippel [9] gave data at 10 MHz and 25 GHz, which were the lower limits of the attenuation length for rock salt. Three data at 150, 300 and 750 MHz were given by P.Gorham et al. at United Salt's Hockley mine located in a salt dome near Houston, Texas [10]. The attenuation lengths were larger than 250 m, which were measured in situ. We had measured the rest [2, 7].

We present recent measurements at 0.3 GHz in Table 1 and at 1.0 GHz in Table 2. Most of the sample shape was a cylinder. The synthetic rock salt samples are manufactured by OHYO KOKEN KOGYO Co. Ltd., Japan. Rock salt samples of Hockley, Zuidwending, Asse, Heilbronn and Lugansk are produced at salt domes in USA, Netherlands, Germany, Germany, Ukraine, respectively. At the columns of the tables, diameter (ϕ/mm), real permittivity (ε'), the attenuation length and number of measurements (Meas.#) of the samples are presented.

Table 1. Attenuation length at 0.3 GHz. Height of samples: 30 mm

	ϕ/mm	ε'	L/m	Meas.#
Synthetic	25, 30	5.89 ± 0.03	1000 ± 640	25
Hockley	10×11, 28, 29	5.82 ± 0.32	156 ± 112	48
Zuidwending	28	6.05 ± 0.03	22 ± 2	24
Asse	25, 28	5.94 ± 0.03	405 ± 166	41
Heilbronn	29	5.26 ± 0.03	41 ± 3	18

Table 2. Attenuation length at 1.0 GHz. Height of samples: 100 mm

	ϕ/mm	ε'	L/m	Meas.#
Synthetic	5, 6, 7, 8, 9	5.87 ± 0.14	538 ± 171	32
Hockley	6× 6, 8, 9, 9	6.07 ± 0.18	275 ± 234	59
Zuidwending	8	6.23 ± 0.06	77 ± 11	21
Asse	9, 10	6.04 ± 0.06	60 ± 25	39
Lugansk	9, 9	6.03 ± 0.03	517 ± 339	36

The error value shows a standard deviation of a distribution in the measured values. The deviation comes mainly from imperfect shape of the cylinder, mechanical setting at the center of the cavity with caps and the moisture at the surface of the samples. The humidity under 50 % is required at the measurement and the preservation of the samples. Long-term storage in the higher humidity increases the radio wave absorption and some samples showed a recovery of the attenuation length keeping low humidity for quite a while. The sample of Zuidwending had been stored for a long period therefore the shortening of the attenuation length could be happened. Synthetic samples have a tendency not to be affected so much by the humidity due to their smooth surface. Lugansk sample was carved out from an almost single crystal block showing a long attenuation length as long as synthetic one. The attenuation lengths at 0.3 GHz of synthetic (1000 ± 640 m) and Asse (405 ± 166 m) rock salts are longer than those of 1.0 GHz (538 ± 171 m and 60 ± 25 m), respectively. The tendency is consistent with a hypothesis as $\tan\delta$ being constant with the frequency. On the contrary, the attenuation length of Hockley becomes longer from 156 ± 112 m at 0.3GHz to 275 ± 234 m at 1GHz which is the same tendency consistent with in the situ measurement [10].

3. Summary

We have measured complex permittivity in rock salt samples by a perturbation method using cylindrical cavity resonators of 0.3 GHz and 1.0 GHz. The preservation rock salt samples should be kept in low humidity as well as at the measurement. The attenuation lengths at 0.3 GHz of Asse (405 ± 166 m) and Hockley (275 ± 234 m) at 1GHz indicate the realization the detector. A UHE neutrino detector with economical antenna spacing could detect 10 GZK neutrinos neutrinos/year if we select a suitable site of a salt dome with a diameter of 3 km and the depth of 3 km.

References

1. K. Greisen, Phys. Rev. Lett. **16**, 748 (1966); G.T. Zatsepin, V.A. Kuz'min, Zh. Eksp. Teor. Fiz., Pis' ma Red. **4**, 114 (1966) [Sov. Phys.-JETP Lett. **4**, 78 (1966)].
2. M. Chiba, T. Kamijo, O. Yasuda, Y. Chikashige, T. Kon, Y. Takeoka and R. Yoshida, Physics of Atomic Nuclei **67**, 2050-2053(2004); P.W.Gorham et al., arXiv:astro-ph/0412128 v2 17 Dec 2004; D.Saltzberg, D.Besson, P.Gorham, A.Odian, R.Milincic, and D.Williams, in Proc. of SPIE **4858** Particle Astrophysics Instrumentation, edited by Peter W. Gorham, (SPIE, Bellingham, WA, p.191 (2003).
3. G.A. Askar'yan, Zh. Eksp. & Teor. Fiz. **41**, 616 (1961) [Sov. Phys. JETP **14**, 441 (1962)]; G.A. Askar'yan, Sov. Phys. JETP **48**, 988 (1965) [**21**, 658 (1965)].
4. D. Saltzberg, P. Gorham, D. Walz et al., Phys. Rev. Lett. **86**, 2802 (2001).
5. M. Fujii and J. Nishimura, Proc. 11th Int. Conf. On Cosmic Rays, Butapest, p.709 (1969); H. Falke et al., Nature **435**, 313 (2005).
6. J. L. Stanley, "*Handbook of World Salt Resources*", Plenum Press, New York (1969); T. H. Michel, "*Salt Domes*", Gulf Publishing Company, Houston (1979).
7. M. Chiba, T. Kamijo, M. Kawaki, H. Athar, M. Inuzuka, M. Ikeda, O. Yasuda, Proc. 1st Int. Workshop for Radio Detection of High Energy Particles [RADHEP-2000], UCLA, AIP Conf. Proc. **579**, p.204 (2000); T. Kamijo and M. Chiba, Memoirs of Faculty of Tech., Tokyo Metropolitan University, No.**51** **2001**, 139 (2002) ;M. Chiba et. al., Proc. of the First NCTS Workshop Astroparticle Physics, Taiwan, World Scientific Publishing Co. Ltd. p.99 (2002); Toshio Kamijo and Masami Chiba, in Proc. of SPIE **4858** Particle Astrophysics Instrumentation, edited by Peter W. Gorham, (SPIE, Bellingham, WA) p.151 (2003).
8. H. A. Bethe and J. Schwinger, NDRC Report D1-117 (1943); R.L.Sproull and E.G.Linder, Proc. Of I.R.E., **34**, 305 (1946); 1.J.C. Slater, Review of Modern Physics, **18**, 441 (1946); G.Birnbaum and J. Franeau, J.Appl.Phys., **20**, 817 (1949); N.Ogasawara, J. Inst. Elect Eng., Japan, **74**, 1486 (1954); R. Ueno and T. Kamijo, IEICE Trans. Commun. E83B, 1554 (2000).
9. A.R.von Hippel ed., Dielectric Materials and Applications, P.302, 361, John Wiley&Sons, INC, (1954); Landolt-Boernstein, Zahlenwerte und Function aus Physik, Chemie, Astronomie, Giophysik und Technik, Eigenshaften der Materie in Ihre Aggeregatzustaenden, 6.Teil, Elektrische Eigenshaften I, Herausgegeben von K.H.Hellwege und A.M. Hellwege, P.456, 505, Springer-Verlarg (1959); R.G.Breckenbridge, J. Chem.Phys. **16** (10) p.959 (1948).
10. P. Gorham, D. Saltzberg, A. Odian, D. Williams, D. Besson, G.Fichter and S. Tantawi, Nucl. Instrum. & Methods. **A490**, 476 (2002).

EXPERIENCE ON ACOUSTIC WAVE PROPAGATION IN ROCK SALT IN THE FREQUENCY RANGE 1-100 kHz AND CONCLUSIONS WITH RESPECT TO THE FEASIBILITY OF A ROCK SALT DOME AS NEUTRINO DETECTOR

GERD MANTHEI & JÜRGEN EISENBLÄTTER

Gesellschaft für Materialprüfung und Geophysik mbH
Ober-Mörlen,D-61239, Germany

THOMAS SPIES

Federal Institute for Geosciences and Natural Resources
Hannover,D-30655, Germany

Rock salt is a promising material for the detection of acoustic waves generated by interactions of high energy neutrinos. The economical feasibility of an acoustic neutrino detector strongly depends on the spacing between the acoustic sensors. In this paper we report on our experience on acoustic wave propagation and wave attenuation in rock salt in the frequency range of 1 to 100 kHz and some conclusions with respect to the usefulness of rock salt as a neutrino detector. The experience bases on long-term acoustic emission measurements in a salt mine.

1. Introduction

The attenuation of seismic waves plays a major role for the use of rock salt as a neutrino detector material. To estimate the attenuation of ultrasonic signals during their propagation through the rock salt, we describe a method which is successfully applied since many years during long-term acoustic emission (AE) measurements in salt mines. This method uses the maximum amplitudes of the signals and the location of the events to calculate an event magnitude analogous to the magnitude in seismology and the damping coefficient of AE signals in rock salt.

The following examples originate from a segment of a salt mine in northern Germany which is monitored by a network of 24 AE sensors since 1995. The signals are recorded in the frequency range from 1 to 100 kHz. The network was recently updated to 33 channels and covers a rock volume of about 200 m x 200 m x 100 m. The sensors are distributed at three excavation levels and installed in 3 to 20 m deep boreholes. The average depth of the monitored volume is 400 m. Mining in this area continued until the 60's, but most of the rooms in the rock salt were mined 60 to 70 years ago.

In general, deformation of large rock salt formations occurs for the most part without the formation of macrocracking. Microcracking can occur, however, near cavities and at rock boundaries. The cavities are mined mostly in ductile rock salt, which has a high tendency to creep. It is not always possible to avoid excavating cavities near anhydrite layers. The brittle anhydrite is much more rigid and has a higher strength than rock salt. The redistribution of stresses around cavities leads to deviatoric stresses near rock boundaries. If these stresses exceed a certain level, microcracks form.

For this study, we investigated wave attenuation in rock salt under high deviatoric stress conditions accompanied by high acoustic emission activity. For this purpose, AE events which have been located in a time period of 9 months (November 15, 2004 to August 23, 2005) were considered in our analysis. Half a year before, in this mine segment one cavity was backfilled which showed persisting high AE activity because of stress redistribution and high humidity [1].

2. Location of AE events

The sites of the AE activity between the three excavation levels are shown in a top view in Figure 1. The extension in vertical direction amounts

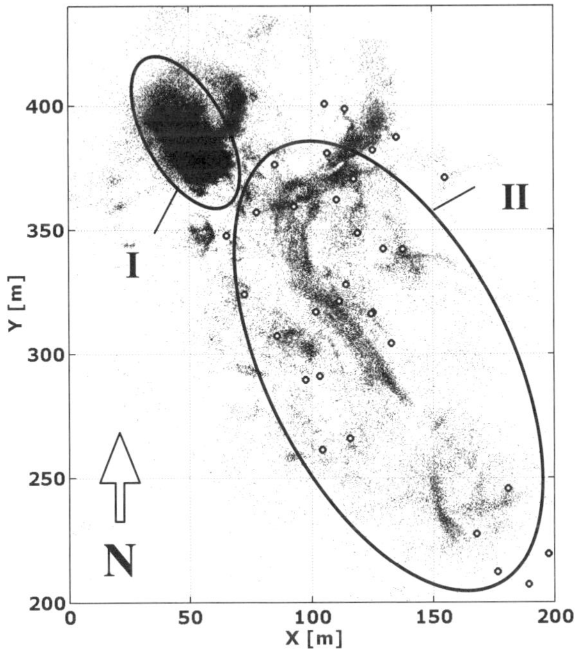

Figure 1: Locations of AE events between November 15, 2004 and August 23, 2005 (696,278 events) in top view.

approximately 120 m. Each AE event is plotted as a point. Only strong events (696,278 events) which were precisely located using at least 16 P- and S-wave arrival times are included in this figure. The locations of the AE borehole sensors are plotted as open circles. The events can be roughly separated into two Regions I and II (marked by ovals). The highest density of AE events was observed in Region I outside the AE network above the backfilled cavity. The AE network is only able to monitor the roof of the cavity not the floor and the walls because all sensors are located at levels higher than the cavity. The events in Region II were preferably located along walls of open cavities which will be backfilled in the future. Compared with Region I, the events of Region II occurred at a 50 to 100 m higher level where the sensor network was placed. The AE activity in this area is interpreted as ongoing damage in the immediate vicinity of mine cavities and rock boundaries due to dilatancy under deviatoric stress conditions.

3. Determination of AE magnitude and damping coefficient

The maximum amplitudes *A(r)* as a function of the distances *r* between the AE source and the sensors are used to determine the magnitude of the event and the damping coefficient. According to the well-known exponential law, the amplitude is expressed by:

$$A(r) = \frac{A_0}{r} \cdot \exp(-\alpha' \cdot r) \tag{1}$$

where α' means the damping coefficient which is covering the effects of intrinsic

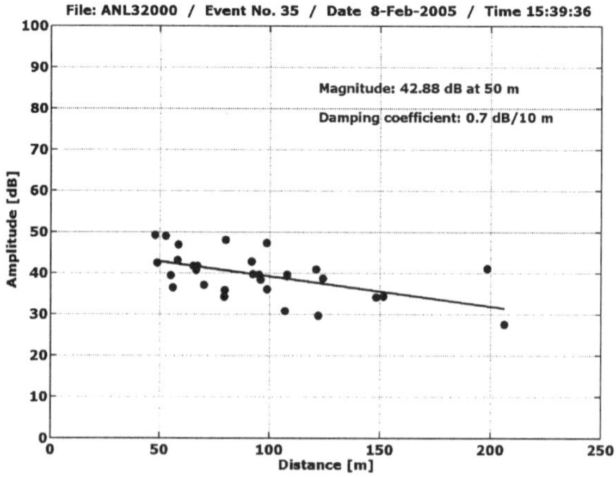

Figure 2: Amplitude (corrected for geometric wave dispersion) versus distance.

absorption and scattering. The amplitudes are specified in dB. Figure 2 shows a semi-logarithmic plot of the product $A \cdot r/r_0$, i.e. the amplitude corrected for geometric wave dispersion, versus r of an AE event.

It can be seen that travel paths of the signals range from 50 to 200 m. A linear relationship is obtained as expected and a straight line is fitted to the data. The slope of the line corresponds to the damping coefficient of 0.7 dB per 10 m. The value of this straight line corresponds to a magnitude of 42.88 dB at a reference distance r_0 of 50 m. In a next step the spatial distribution of mean damping coefficients in the monitored region is evaluated.

4. Results

Figure 3 displays the distribution of the damping coefficient of the events in a gray-scale density plot. The plot shows the mean damping coefficient within horizontal cells of 5 m x 5 m; cells containing less than 10 events are displayed as white areas. Again as in Figure 1, Region I and II are marked by ovals.

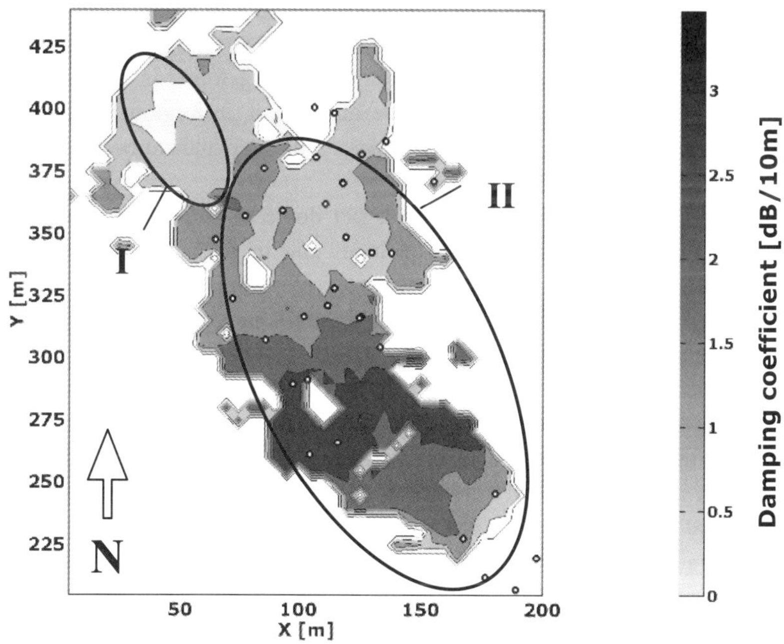

Figure 3: Density plot of damping coefficient obtained from AE events between November 15, 2004 and August 23, 2005 (696,278 events) in top view.

The highest damping coefficient of about 3.47 dB per 10 m occurred in Region II in a small zone between y = 250 m and y = 290 m (black cells). In this area microcracking still takes place at the contours of closely spaced cavities even a long time after excavation. On the other hand, the lowest damping of about 0.17 dB per 10 m was obtained in Region I outside the sensor network (northwest direction) and in the southeastern corner of Region II. The attenuation length, i.e. the distance in which the peak amplitude corrected for geometric wave dispersion, $A \cdot r/r_0$, is reduced to 37% ($1/e$) of the peak amplitude at distance 0, is a reciprocal measure of the damping coefficient. The highest and lowest found damping coefficients, 3.47 dB per 10 m and 0.17 dB per 10 m, correspond to attenuation lengths of 25 m and 510 m, respectively.

5. Conclusions

Apart from geometrical attenuation, intrinsic absorption, and shadowing effects by cavities and drifts, high-frequency acoustic waves are attenuated by scattering at small inclusions of anhydrite, clay, gas, or water which are embedded in most salt rock formations. Scattering at microcracks occurs in regions of deviatoric stress e.g. in the so-called excavation disturbed zone (EDZ) with thickness of a few meters at the contour of underground cavities.

That may be the reason for the high damping found in regions of closely spaced cavities like Region II. On the other hand, in Region I with the lowest attenuation, the AE signals mainly propagate through undisturbed rock salt to the AE sensors.

With regard to a utilization of a salt dome as neutrino detector, the here presented investigations let us conclude, that in pure rock salt, without any boundaries and large inclusions, distances of several hundred meters between sensors for detection of acoustic waves in the frequency range of few tens of kHz are attainable. But, difficulties are to be presumed at boundaries between different rock materials like rock salt, anhydrite, or clay because of reflection, refraction, and mode conversion of acoustic waves. Because of the competent and brittle properties of anhydrite in ductile rock salt formations, anhydrite often strongly determines the tectonic form of salt formations subjected to salt tectonics or halokinesis. Even under such very slow creep conditions, microcracking and consequently AE activity may induced at these geologic boundaries. These interfering signals are to be possibly discriminated from neutrino generated events maybe by careful source analysis.

References

1. J. Hesser and T. Spies, Proc. EUROCK 2004 & 53th Geomechanics Colloquium, Schubert (ed.), ISBN 3-7739-5995-8, 261 (2004).

RADIO SIGNALS FROM PHOTON BEAMS IN SAND AND SALT

D. WILLIAMS*

for

P.GORHAM, E. GUILLIAN, R. MILINCIC, P.MIOCINOVIC†

D.SALTZBERG,D.WILLIAMS‡

R.C. FIELD, R. IVERSON, A. ODIAN, D.WALZ§

G. RESCH¶

P. SCHOESSOW||

In this paper I describe the setup and results of two beamtests which demonstrated the existence and properties of the Askar'yan effect in sand and salt. We observed coherence, 100% linear polarization, and field strength in agreement with simulations. We also demonstrated the possibility of tracking the shower direction with polarization information.

1. Introduction

Observing the GZK neutrino flux from cosmic rays interacting with the cosmic microwave background [2] is of great importance for understanding the highest-energy cosmic rays. The GZK spectrum peaks near 10^{18} eV[3]. G. Askar'yan predicted that neutrino-induced showers in solid dielectrics would develop a charge excess which would then emit Cherenkov radiation [1]. The Cherenkov radiation would be coherent at wavelengths longer than the Moliere radius of the shower, so that the electric field strength would increase linearly with the shower energy. At energies $\geq 10^{18}$ eV, radio emission would dominate the shower output, making the Askar'yan effect a promising method of detection for GZK neutrinos.

*Penn State University

†Univ. of Hawaii at Manoa

‡Univ. of California, Los Angeles

§Stanford Linear Accelerator Center

¶deceased,Jet Propulsion Laboratory

||Argonne National Laboratory

Askar'yan proposed ice, the lunar regolith (similar to sand) and salt as possible detectors. Sand, salt and ice are transparent in the coherent regime, which corresponds to frequencies below 1-10 GHz.

Attempts to detect the Askar'yan effect radiation in air showers were complicated by competing geomagnetic effects [4]. Several experiments attempted to detect neutrinos via the Askar'yan effect in ice [5,6], and in the lunar regolith [7,8,9]. As of 2000, these experiments had not seen a signal.

A beamtest at Argonne [10] attempted to measure coherent Cherenkov radiation from a shower induced by an electron bunch going into a sand target. However, the Cherenkov radiation was obscured by transition radiation (TR) as the electron bunch crossed the interface between the beampipe and the target. Two beamtests were performed at the Stanford Linear Accelerator Center (SLAC) using photon bunches, which would not produce TR. This paper describes the setup and results of the beamtests at SLAC: one with a sand target in 2000, and one with a salt target in 2002.

The following sections summarize the setup and results of the beamtests. For full details, see the published papers from each test [11,12].

2. Setup

Both beamtests were performed at the Final Focus Test Beam (FFTB) facility at SLAC. The FFTB provides bunches of 28.5 GeV electrons which go through thin bremsstrahlung radiators. The resulting gamma rays go to the target, 30 m downstream, while the electrons are diverted into a beampipe going below the target to avoid TR. There are two radiators which can be used separately or in tandem, producing gamma ray bunches with energies from $(0.06 - 1.1) \times 10^{19}$ eV. The typical beam current is 10^{10} electrons per bunch, and this can be lowered by several orders of magnitude to produce bunch energies down to 10^{15} eV.

Figure 1 shows the target geometry for both beamtests. Both targets have an outer surface slanted at 10° with respect to the beam axis (as seen from above), to avoid total internal reflection (TIR), as the TIR angle is complementary to the Cherenkov angle.

The sand target in the 2000 beamtest was a plywood box containing 3200 kg of dry silica sand. The signal was measured from outside the box, through a polypropylene wall, with standard gain pyramidal horns (1.7-2.6 GHz and 4.4-5.6 GHz). We also buried dipoles in the sand for internal measurements.

The salt target in the 2002 beamtest was a stack of 1.8 kg salt bricks

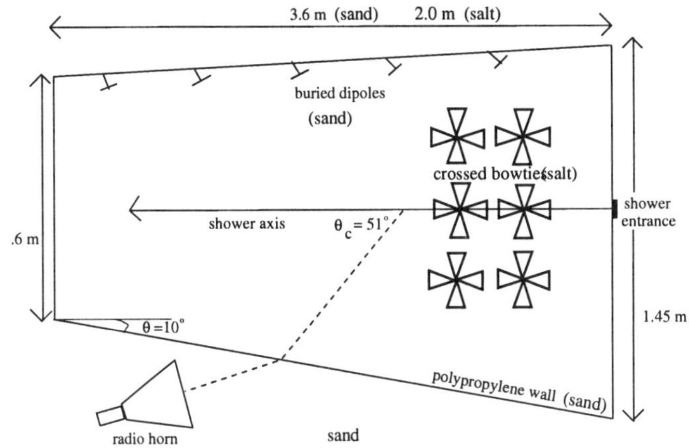

Figure 1. Target geometry for the SLAC beamtests, as seen from above. Note that the buried dipoles along the far wall were used only in the sand target (2000) and the crossed bowties along the beam axis were used only in the salt target (2002). The actual bowtie array is 3x7, but only 2 rows of antennas are shown along the beam axis.

from Morton Salt Inc. with a total mass of 4 metric tons. An array of dual-polarization crossed bowtie antennas was embedded in the stack, and the external horns were used as in 2000.

3. Results

The key property of the Askar'yan effect is coherence at low frequencies; that is, the electric field is linear in shower energy. Figure 2 shows electric field strength as a function of shower energy for both beamtests, with very clear linear dependence. Coherence was measured up to 5 GHz in 2000 over two orders of magnitude in shower energy, and up to 14 GHz in 2002 over four orders of magnitude in shower energy.

Cherenkov radiation is 100% linearly polarized in the plane formed by the direction of observation and the shower direction [13]. In 2000, we used a linearly polarized pyramidal horn to measure the linear polarization, by rotating the horn with respect to the plane of polarization. When the horn was rotated at 90° with respect to the plane of polarization, the signal was almost totally suppressed until late reflections arrived, as seen in Figure 3 (left). In 2002, we reconstructed the shower direction using the relative amplitudes in both polarizations of the crossed bowtie antennas. Figure 3 (right) shows the reconstructed plane of polarization along with the known direction; they are in excellent agreement.

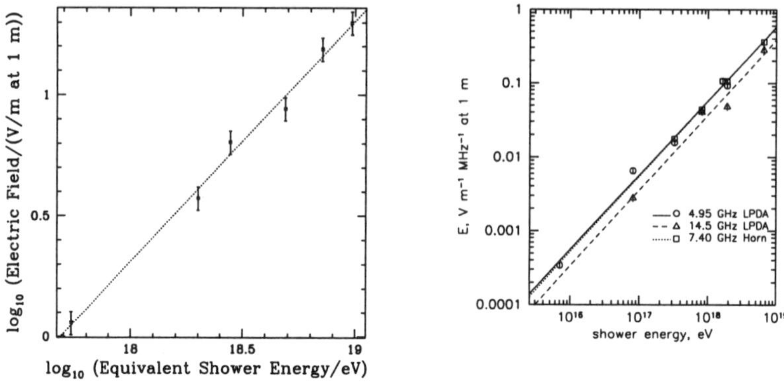

Figure 2. Left: Electric field vs. shower energy at 2.2 GHz from SLAC 2000 (sand). Right: electric field vs. shower energy at several frequencies, from SLAC 2002 (salt).

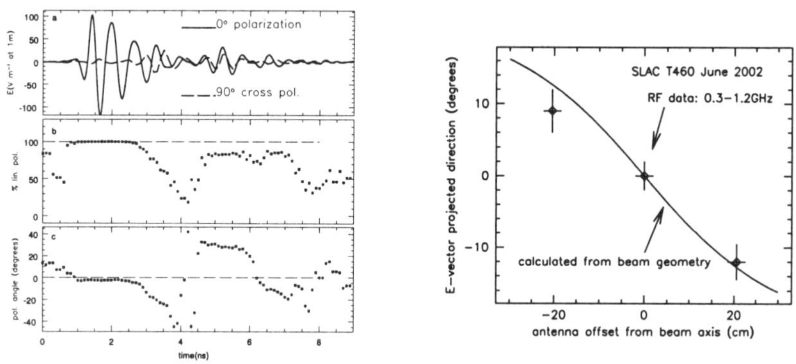

Figure 3. Left: Polarization measurements from SLAC 2000 (sand). Right: plane of polarization vs. transverse distance of antenna from beam axis, from SLAC 2002 (salt).

The absolute field strength was measured in the frequency range of the various antennas at both beamtests. The measured field strength is in good agreement with the field strength predicted by simulations [14,15]. Field strength vs. frequency for both beamtests is shown in Figure 4.

4. Conclusions

The Askar'yan effect has been observed in two beamtests with the expected properties. Coherence is observed over four orders of magnitude in shower energy across a range of frequencies. The polarization information

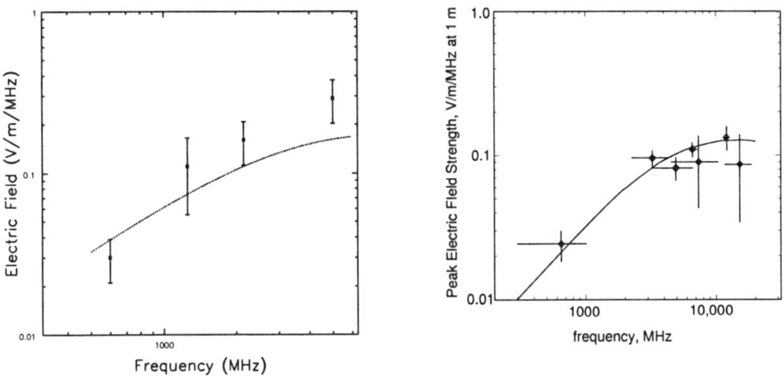

Figure 4. Left: Field strength vs. frequency from SLAC 2000(sand). Right: Field strength vs. frequency from SLAC 2002(salt).

can be used to track the shower direction. The field strength agrees with simulations. The existence of large radio-transparent volumes in nature makes this a promising avenue of detection for ultra-high energy neutrinos, especially GZK neutrinos.

References

1. G. Askar'yan, Soviet Physics JETP 14, 441 (1962); G. Askar'yan, Soviet Physics JETP 21, 658 (1965).
2. K. Greisen, Phys. Rev. Lett. 16, 748 (1966); G. Zatsepin and V. Kuzmin, JETP Lett. 4, 78 (1966).
3. R. Engel, D. Seckel and T. Stanev, Phys.Rev. D64, 093010 (2001).
4. H. Allen, Prog. in Elem. Part. and Cosmic Ray Physics 10, 171 (1971).
5. RICE collaboration, astro-ph/9709223.
6. N. G. Lehtinen *et al.*, Phys.Rev. D69 (2004) 013008.
7. I. M. Zheleznykh, 1988, Proc. Neutrino '88, 528;R. D. Dagkesamanskii, & I. M. Zheleznykh, 1989, JETP 50, 233.
8. T. Hankins, R. Ekers & J. O'Sullivan, 1996, MNRAS 283, 1027.
9. P. W. Gorham *et al.*, Phys.Rev.Lett. 93 (2004) 041101.
10. P. Gorham, D. Saltzberg, & P. Schoessow *et al.*,Phys. Rev. E62, 8590 (2000).
11. D. Saltzberg, P. Gorham, D. Walz, *et al.* 2001, Phys. Rev. Lett., 86, 2802.
12. P. W. Gorham, D. Saltzberg *et al.*, Phys.Rev. D72 (2005) 023002.
13. J. Jackson, Classical Electrodynamics, third edition.
14. E. Zas, F. Halzen,, & T. Stanev, 1992, Phys Rev D 45, 362.
15. J. Alvarez–Muñiz, & E. Zas, 1997, Phys. Lett. B, 411, 218;J. Alvarez-Muñiz, E. Zas Phys.Lett. B434 (1998) 396-406.

BROADBAND ANALYSIS OF ASKARYAN PULSES

PREDRAG MIOČINOVIĆ

Department of Physics and Astronomy,
University of Hawaii at Manoa,
2505 Correa Rd,
Honolulu, HI 96822, USA
E-mail: predrag@phys.hawaii.edu

A coherent radio emission from particle shower in a dielectric medium, so-called Askaryan pulse, has been recorded in a laboratory setting. In this report the analysis of measurement made with a broadband (1-18 GHz) log periodic dipole antenna (LPDA) is described. The properties of the observed radio pulses are compared with the current theoretical model. The radiation intensity as a function of frequency was found to agree well with the expectation, while the phase characteristics of pulses disagree with the theoretical work.

1. Introduction

G. Askaryan proposed in 1962 that a compact particle shower in a dielectric medium will produce a coherent radio Cherenkov emission.[1] Subsequent theoretical work supported this prediction.[2,3,4] The experimental verification came in 2001,[5] with follow up measurements confirming the frequency and the polarization properties of the emitted radiation.[6]

The emission of coherent radio signal comes from the charge asymmetry in particle shower development. The asymmetry is due to combined effects of positron annihilation and Compton scattering of atomic electrons. There is ~20% excess of electrons over positrons in a particle shower, which moves as a compact bunch, few cm wide and ~1 cm thick, at the velocity above the speed of light in the medium. The frequency dependence of Cherenkov radiation emitted is $dP \propto \nu d\nu$. For radiation with wavelength $\lambda \gg l$, where l is the scale of the particle bunch, the radiated signal will add coherently and thus be proportional to the square of shower energy. A radio signal emitted by a particle shower in material such as ice or salt is coherent up to few GHz, is linearly polarized, and lasts about a nanosecond. A shower with the total particle energy of 10^{19} eV interacting in the ice will produce a radio pulse with peak strength of ~ 1 V/m/MHz at the distance of 1 m.

This report describes broadband measurements of frequency and phase characteristics of Askaryan pulses recorded with log periodic dipole antenna (LPDA) in the experiment (SLAC T460) performed at the Final Focus Test Beam (FFTB) facility at SLAC in June 2002.[6]

2. Measurements

A detailed experimental setup is described by Gorham[6] and is illustrated in Fig. 1. Bremsstrahlung photon bunches of varying total energy were directed into a salt-block target. The resulting radio signals were collected by bowtie antennas embedded in the salt and by horn and LPDA antennas outside the salt. The LPDA antenna is Electro-Metrics model EM-6952 with bandwidth from 1-18 GHz. The antenna was connected by two pieces of 75-foot heliax cable, Andrew LDF4-50A, and by three pieces of 12-inch semi-rigid Haverhill cable, to CSA8000 sampling scope with 20 GHz bandwidth and up to 1000 GSa/s sampling rate. During each run, to improve SNR, the recorded waveforms were averaged over many pulses, using ultra-stable microwave transition-radiation trigger from an upstream location. This work will concentrate on two runs, runs 35 and 109 in which no microwave filters were placed between the LPDA and the scope, providing the full bandwidth data. The photon bunch energies were estimated to be 0.66

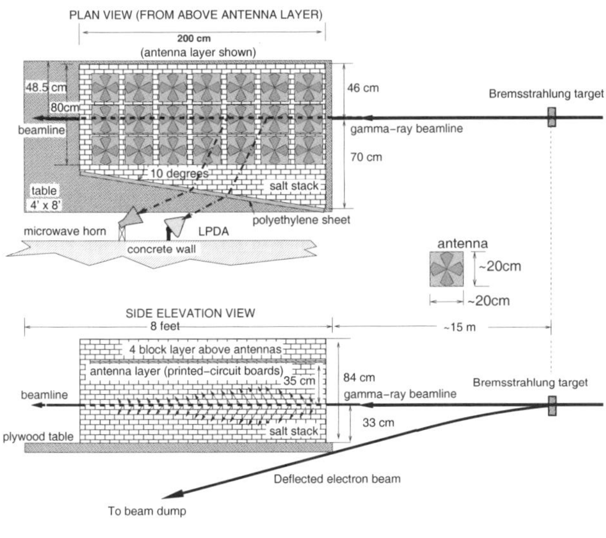

Figure 1. Geometry of salt-block target and receiving antennas.

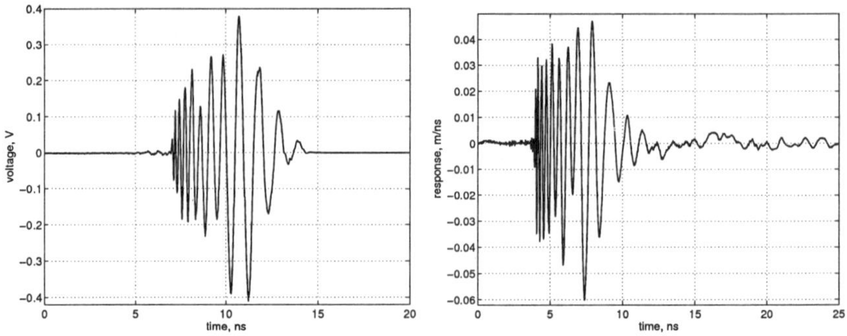

Figure 2. Left: Raw voltage recorded through LPDA in run 35. Right: Impulse response of LPDA-cable system used in measurements.

and 1.9 EeV in two runs, respectively, with ~20% uncertainty. The raw signal recorded in run 35 is shown in Fig. 2.

The calibration of LDPA gain and phase delay was performed in an anechoic chamber at University of Hawaii by a reciprocal S12 method. Two identical antennas were mounted about 60-in apart, which ensured they were in each others far-field region.[a] The transmitting antenna was stimulated by 200 mV step generated by HP 54121A logging head. The receiving signal was amplified by Agilent 83017A broadband amplifier and recorded by HP 54120B digitizing scope with 20 GHz bandwidth at 100 GSa/s. The frequency and phase response of LPDA were extracted by subtracting the responses of the amplifier, cables and the scope which were measured as a reference. The numerical operations involved are described in the next section. The heliax cable response was measured in a similar way, by stimulating it on one side by 200 mV step and recording the resulting pulse at the other end with the digitizing scope. For semi-rigid Haverhill cable, only attenuation was measured as a function of frequency with a network analyzer. The phase response was ignored due to a short relative length of the cable used. The resulting time-domain response of LPDA-cable system as used in the experiment is shown in Fig. 2.

3. Analysis

The voltage recorded by the scope can be expressed as

$$V(t) = E_{Ask}(t) \circ h_{LPDA}(t) \circ T_c(t), \tag{1}$$

[a]For LPDA, far-field requirement reduces to $d \geq \lambda/4$.

where E_{Ask} is a magnitude of the electric field at antenna due to an Askaryan pulse, h_{LPDA} is the effective height of LPDA,[7] T_c is the response of the cable to a delta-like impulse, i.e. the cable transfer function expressed in the time domain, and $\circ$ is a convolution operator. The reciprocity theorem states that the transmitting and the receiving effective heights of an antenna are related by $h^t(t) = \partial_t h^r(t)$, and thus the voltage recorded in our LPDA calibration measurement, after correcting for the transfer function of amplifier and cables, can be expressed as[7,8]

$$V^r(t) = \frac{1}{2\pi rc} h^r(t) \circ h^t(t) \circ V^t(t) = \frac{1}{2\pi rc} h^r(t) \circ h^r(t) \circ \partial_t V^t(t), \qquad (2)$$

where $h^t(t)$ and $h^r(t)$ are re-normalized to the impedance mismatch between the radiation resistance of an antenna and the free-space impedance. In the last step, the advantage was taken of commutativity between convolution and differentiation operators. The straight forward way to extract the receiving effective height of an antenna is to switch to Fourier domain where,

$$\tilde{h}^r(\omega) = \sqrt{\frac{2\pi rc}{i\omega} \frac{\tilde{V}^r(\omega)}{\tilde{V}^t(\omega)}}. \qquad (3)$$

As per Eq. 1, to extract the incident electric field from the measured voltage and the known response of LPDA-cable system, deconvolution has to be performed. Again, by converting to Fourier domain,

$$\tilde{E}(\omega) = \frac{\tilde{V}(\omega)}{\tilde{h}_{LPDA}(\omega)\tilde{T}_c(\omega)}. \qquad (4)$$

However, this approach is very sensitive to high NSR over frequencies where no signal is expected ($\omega \geq 14$ GHz) due to attenuation in the cable. In order to reduce this "out-of-band" NSR, deconvolution algorithm with noise reduction can be applied. In this work, Weiner filter was chosen, and the level of NSR was chosen such to maximize out-of-band noise rejection while fully preserving "in-band" signal. The same approach was taken to subtract amplifier and cable effects from antenna S12 measurements.

4. Results

The resulting Askaryan pulses and averaged frequency and phase properties of the pulses are shown in Fig. 3. The frequency spectrum of pulses has been corrected for field-line divergence at the air-salt interface and for the

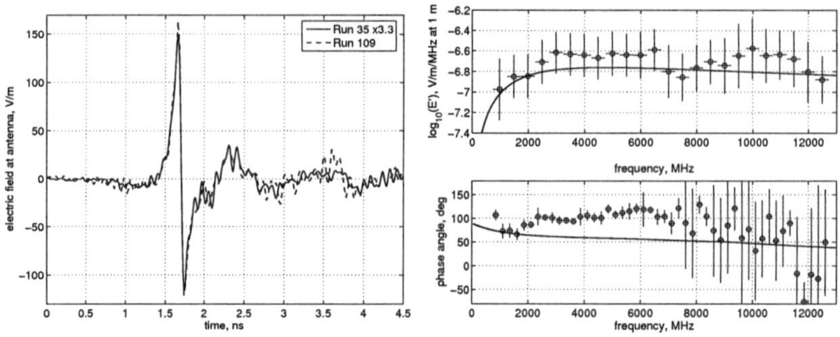

Figure 3. Left: Electric field at antenna due to Askaryan pulses from two runs, with run 35 rescaled by factor 3.3 for comparison. Right top: Average electric field magnitude vs. frequency rescaled to 1 TeV shower with corrections described in the text, compared to the theoretical expectation. Right bottom: Average electric field phase vs. frequency compared to the theoretical expectation for an Askaryan pulse in ice. Large errors at higher frequencies indicate loss of SNR due to cable attenuation.

near-field effects.[b] The theoretical frequency expectation is given by[3,6]

$$|E| = 2.53 \cdot 10^{-7} \cdot 0.52 \left[\frac{E_{sh}}{1\text{TeV}}\right] \left[\frac{\nu}{\nu_0}\right] \left[\frac{1}{1+(\nu/\nu_1)^{1.44}}\right] \text{V MHz}^{-1}\ \text{m}^{-1}, \quad (5)$$

where $\nu_0 = 1.15$ GHz and $\nu_1 = 2.86$ GHz. There is a good agreement between the theoretical expectation and the measured spectrum. The measured phase angles are compared with the phases expected from an Askaryan pulse in ice.[2] The agreement is not good and warrants further theoretical investigation.

This work was supported by NASA ROSS - UH grant NAG5-5387.

References

1. G. A. Askaryan, JETP **14**, 441 (1962); JETP **21**, 658 (1965)
2. E. Zas, F. Halzen, and T. Stanev, Phys. Rev. **D45**, 362 (1992).
3. J. Alvarez-Muñiz, R. A. Vázquez, and E. Zas, Phys. Rev. **D62** 063001 (2000).
4. S. Razzaque *et al.*, Phys. Rev.**D65**, 103002 (2002); *errata* Phys. Rev.**D69**, 047101 (2004).
5. D. Saltzberg *et al.*, PRL **86**, 2802 (2001).
6. P. Gorham *et al.*, Phys. Rev. **D72**, 023002 (2005).
7. A. Shlivinski, E. Heyman, and R. Kastner, IEEE AP **45** 1140 (1997).
8. E.G. Farr and C.E. Baum, http://www.farr-research.com/Papers/ssn426.pdf

[b]While pulse generating shower is in far-field of the antenna, the antenna is in near-field region of the shower at frequencies above 800 MHz, and thus it only samples a portion of the full pulse strength. The near-field correction is of the form $E' \rightarrow (\nu/800\ \text{MHz})E$.

HYBRID SCHEME OF SIMULATION OF ELECTRON-PHOTON AND ELECTRON-HADRON CASCADES IN DENSE MEDIUM AT ULTRA-HIGH ENERGIES*

DEDENKO L.G.†
Moscow State University, Faculty of Physics

Moscow,119992, Leninskie Gory, Russia

MIRONOVICH A.A., ZHELEZNYKH I.M.
Institute for Nuclear Research of Russian Academy of Sciences

Moscow, 117312, 60th Anniversary of October Prospect, 7a, Russia

The multilevel scheme to simulate electron-photon and electron-hadron cascades has been suggested. The Monte Carlo approach should be used to take into account fluctuations in a development of an individual cascade. Then transport equations should be exploited. The low energy particles should be treated again by the Monte Carlo approach.

1. Introduction

It is remarkable that a discovery of the most elusive particle by C.Cowan and F.Reines [1] and a call to develop high energy neutrino physics by M.A. Markov [2] coincide practically with suggestions of acoustical and radio methods of detections by G.A. Askaryan [3-5]. As a result of the growing interest in the high energy neutrino physics [6-7] these methods of neutrino detection have been found perspective [8-15]. The thermo-acoustic mechanism of sound generation suggested by G.A. Askaryan has been confirmed [16]. The Askaryan effect, the Cherenkov radio emission by negative charge excess in dense materials have been also confirmed in accelerator experiment with a silica sand target [17] and

* This work is supported by US Civilian Research and Development Foundation RUP2-2624-MO-04 and the RFBR (grant 05-02-17410).
† This work is partially supported by INTAS (grant 03-51-5112) and RFBR (grant 03-02-16290).

with a salt target [18]. Thus such experiments as GLUE [19], RICE [20], ANITA [21] and suggestions [8,9,22] and LORD [23] have firm grounds.
Estimates [24-25] of possible signals are of importance. At the same time some alternative approach allows to interpret data more confidently. In this paper the hybrid multilevel scheme has been discussed.

2. The hybrid multilevel scheme

In interactions of ultra-high energy (UHE) neutrinos with atomic nuclei hadrons and leptons may be produced. Thus to estimate any (radio, acoustical) signal produced by a neutrino-induced interaction both the electron-hadron (eh) and electron-photon (eγ) cascades should be taken into account. In a dense medium (sea water, ice, rock salt domes, the lunar regolith) a development of a cascade of secondary particles at ultra-high energies differs from that in the atmosphere. At energies above E_{LPM}~10^{14} – 10^{16} eV due to the Landau – Pomeranchuk – Migdal (LPM) effect [26-27] the cross sections for the pair production and the bremsstruhlung process decrease considerably.
Let us consider the transport of hadrons first. If a few UHE hadrons are produced then the standard Monte Carlo approach may be used to take into account fluctuations in development of the particular cascade. When a number of hadrons increases to save the CPU time the transport equations method may be exploited. The decay products of neutral pions are regarded as a source function $S_\gamma(E,x)$ of gamma quanta which give origins of electron-photon cascades in dense media. At energies above ~10^{18}eV it may be assumed that neutral pions are stable. The rest of mesons may be considered as stable particles at all energies. As a result almost all energy of produced hadrons would be deposited into high energy gammas which are regarded with help of the additional source term in transport equations for electron and photons. Thus in case of hadron production the standard cascade development is expected and the LPM effect is not much of importance.
Contrary to this the case when a large fraction of neutrino energy is transferred to an electron may be very peculiar. If the energy of an electron will be above ~10^{19}eV then due to the LPM effect the cascade length would increase considerably. Besides fluctuations in a cascade development will change a shower profile so dramatically that a cascade curve would have a stochastic picture. The fact is that the mean free paths λ for both the pair production and the bremsstruhlung process at UHE become much larger than the length of partial subcascades developed due to relatively small energy depositions. Thus only the Monte Carlo approach should be used to simulate the real individual cascade development.

Then a number of secondary particles increase and fluctuations will come to some balance. Therefore transport equations may provide sufficient accuracy till the lateral spread of particles become important at energies below E_s which may be chosen as 10 GeV.

The low energy electrons and photons (in the energy range of $E_{\min} \leq E \leq E_s$) are produced by higher energy particles. This production may be estimated by the source functions $SL_e(E,x)$ and $SL_\gamma(E,x)$ calculated with help of transport equations. For these low energy particles the 3-dimentional cascades should be simulated by Monte Carlo method with help of the GEANT4 code [28] in advance. The 3-dimensional cascade profiles and signal profiles $SE(E,x,r)$ and $SG(E,x,r)$ should be estimated and stored in some library as the data base.

Finally the signals $s(x_0,r)$ may be estimated straightforward at any depth x_0:

$$s(x_0,r) = \int_{x_b}^{x_0} d\xi \int_{E_{\min}}^{E_{th}} dE(SL_\gamma(E,\xi)SG(E,x_0-\xi,r) + SL_E(E,\xi)SG(E,x_0-\xi,r)) \quad (1)$$

here $E_{th} = 10$ GeV, $E_{\min} = 1$ MeV

The equation of the type (1) enables to estimate any signal of interest.

3. Example

This rather complicated scheme may be simplified. As some approximation it is possible to use the NKG formulae [29] to represent the 3-dimentional development of cascades at energies below E_{LPM}. The 3-dimensional shower profile may be approximated by the product of the NKG formula for the longitudinal development and the modified NKG formula [30] for the lateral spread of secondary particles.

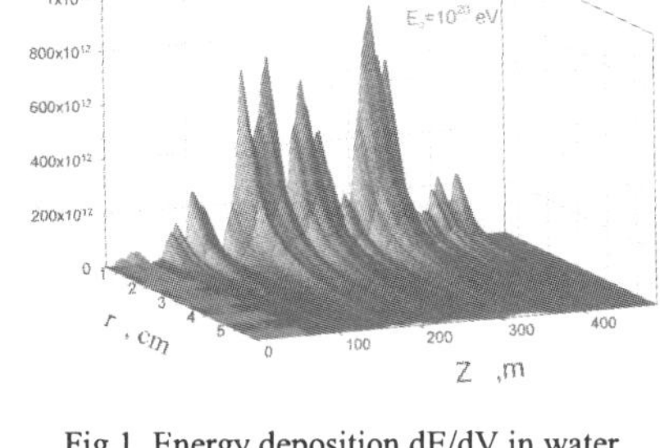

Fig.1. Energy deposition dE/dV in water for $E_0 = 10^{20}$ eV

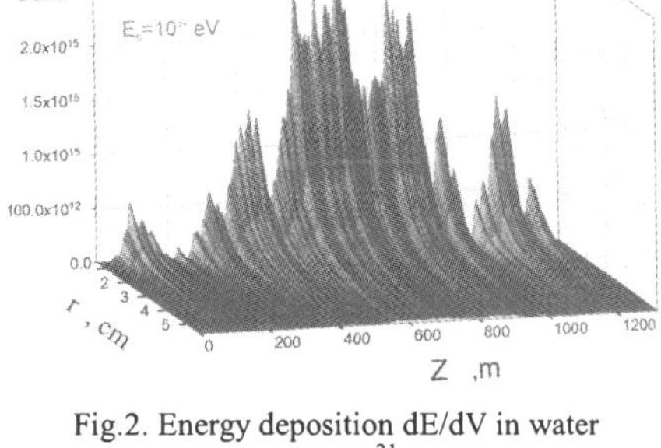

Fig.2. Energy deposition dE/dV in water for $E_0 = 10^{21}$ eV

As test examples Figures 1-2 display the influence of the LPM effect on a cascade development in dense media. Distributions of energy deposition in water are shown for the primary electrons with energies 10^{20} and 10^{21}eV accordingly. The stochastic nature of cascades is clearly seen. Due to the LPM effect the mean interaction path of electrons and photons in a dense medium at very high energies exceed considerably the length of a subcascade with energies below the LPM threshold energy. So if an energy deposition to a secondary particle is small then a subcascade develops and decays at a standard rate. But particles with very high energies have much enlarged mean paths for interactions and can penetrate very deep into the medium. Thus fluctuations in a cascade development are very important and only the Monte Carlo approach is the just method to simulate individual cascades. Just to the same increase of mean interaction paths the total cascade length enlarges very much. As energy of the primary electron changes from 10^{18}eV up to 10^{21}eV the total length of a cascade increases from ~50 m up to nearly 1 km. The cascade length *l* increases proportionally to $\sim E^{0.395}$. Not large changes are expected if the NKG approximation would be replaced by results simulated with the GEANT4 code.

Conclusion

The hybrid multilevel scheme has been suggested to estimate acoustical (radio) signals produced by eγ and eh cascades in a dense medium. The Monte Carlo approach for UHE particles should be used to take fluctuations into account. Then transport equations are exploited to estimate source functions of low energy particles which are regarded in advance by Monte Carlo method with help of the GEANT4 code.

Acknowledgments

We thank G.T. Zatsepin for useful discussions, US Civilian Research and development Foundation RUP2-2624-MO-04, RFBR (grants 05-02-1740 and 03-02-16290) and INTAS (grant 03-51-5112).

References

1. C. L. Cowen, F. Reines, F.B. Harrison et al. Science 124, 103 (1956)
2. M.A. Markov, Proc.of the X Intern. Conf. on high-energy physics. Rochester, p. 579 (1960),
   M.A. Markov and I.M. Zheleznykh, *Nucl. Phys.* **27**, 5 (1961).
3. G.A. Askaryan, Atomnaya Energiya, 3, 8 (1957) (Sov. At. Energy 3, 921 (1957))

4. G.A. Askaryan, Soviet Physics JETF, v.41, №2(8), p. 616 (1961)
5. G.A. Askaryan, Soviet Physics JETF, v.48, №3, p. 985 (1965)
6. V.S. Berezinsky and G.T.Zatsepin, Uspekhi Fiz. Nauk 5 (1977)
7. H.Blood, J. Learned, F. Reines and A. Roberts, Proc. 1981 Int. Conf. Neutrino Physics and Astrophysics, Vol.2, eds., R. Cence, E. Ma and A. Roberts, Maui, Hawaii, (1981).
8. R.D. Dagkesamanskii and I. M. Zheleznykh, *JETP Lett.* **50,** 259 (1989). A.R. Beresnyak, R.D. Dagkesamanskii, I. M. Zheleznykh et al., *Astronomy Reports* **4,** No. 2, 127 (2005).
9. G.A. Gusev and I.M. Zheleznykh, *Pis'ma Zh. Eksp. Teor. Fiz. [JETP Lett.]* **38**, 505 (1983).
10. L.G. Dedenko et al., *Proc. 24th ICRC, Rome,* **1,** 797 (1995).
11. A. Capone, L.G. Dedenko, A.V. Furduev et al., *Proc. 27th ICRC, Hamburg* 1264, (2001).
12. G.A.Askaryan, B.A.Dolgoshien, Soviet Physics JETF pisma, v.25, №5, p.232 (1977)
13. G.A.Askaryan et al., Nucl. Instr. and Meth., v.164, №2, p. 267 (1979)
14. J.G.Learned, Phys.Rev., D19, 11 (1979)
15. T.Bowen, Proc. 15th ICRC, Plovdiv, v.6, p. 277 (1977)
16. L.R.Sulak et al. Nucl. Instr. and Meth., 161, 203 (1979)
17. D.Saltzbery et al. Phys. Rev. Letters, 86, 2802 (2001)
18. P.W.Gorham et al. Phys. Rev. D, in press (2005), astro-ph/0412128.
19. P.W.Gorham et al. Phys. Rev. Letters, 93, 0411101 (2004).
20. I.Kravchenko et.al., Astropart. Phys., 20 , 145 (2003)
21. P.Miosenovic et al. 22nd Texas Symposium on relativistic Astrophysics, Stanford (2004)
22. M.A.Markov AND I.M.Zheleznykh, Nucl.Instr. and Methods A 248, 242 (1986)
23. V.A.Tsarev, International ARENA Workshop, Zeuthen (2005)
24. E.Zas, F.Halzen and T.Stanev, Phys.Rev. D45, 362, (1992)
25. J.Alvarez-Muniz, and E.Zas, RADHEP 2000, UCLA (2001)
26. L.D.Landau and I.Pomeranchuk, Dokl. Akad. Nauk SSSR, 92, 535 (1953); 92, 735 (1953)
27. A.B.Migdal, Phys.Rev. 103, 1811,1956. Zh. Eksp. Teor. Fiz., 32, 633 (1957) (Sov. Phys.JETP 5, 527 (1957)
28. The GEANT4 Collaborations, http:///wwwinfo.cern.ch/asd/geant4.html
29 K.Greisen in progress in Cosmic Ray Physics, III, ed by J.G.Wilson (Interscience,New York), p. 3 (1956)
30. L.G.Dedenko et al., Proc. 14th ICRC, Munich, 8, 2731 (1975)

STRUCTURE FUNCTION OF EXCESS CHARGE IN ROCK SALT†

YUSUKE WATANABE, MASAMI CHIBA, OSAMU YASUDA

Department of physics, Tokyo Metropolitan University, 1-1 Minami-Osawa Hachioji-shi, Tokyo, 192-0397, Japan

TOSHIO KAMIJO

Department of Electrical Engineering, Tokyo Metropolitan University, 1-1 Minami-Osawa Hachioji-shi, Tokyo, 192-0397, Japan

YUICHI CHIKASHIGE, TADASHI KON, YUTAKA SHIMIZU, AKIO AMANO, YOSITO TAKEOKA, SATOSHI MORI, SOUSUKE NINOMIYA

Faculty of Science and Technology, Seikei University, 3-3-1 Kichijyoji Kitamachi Musasino-shi, Tokyo, 180-8633, Japan

Ultra high-energy (UHE) neutrino (10^{15} eV <) detection in natural huge rock salt formation has been studied. Radio wave is generated by Askar'yan effect in rock salt. That is caused by the number asymmetry of electrons and positrons (excess charge) in electromagnetic shower from an UHE neutrino. An Electromagnetic shower in rock salt is generated by a computer code of Geant4.5. A modified Greisen parameterization in longitudinal distribution of the shower and a modified Nishimura, Kamata, and Greisen function in lateral distribution of the shower (structure function) are fitted to a space distribution of the excess electrons. The structure function would assist to calculate radio wave emission without using an electromagnetic shower simulation.

1. Introduction

There are many kinds of models to generate ultra high-energy (UHE) neutrino in the universe. The observations of UHE cosmic rays and cosmic microwave background predict existence of UHE neutrinos (Greisen, Zatsepin, Kuz'min effect) [1]. An UHE neutrino interacts with rock salt yielding a hedonic shower in which π^0s are included, and π^0s generates a gigantic electromagnetic (EM) shower. The radio wave radiated from EM shower would be detected. The radio wave is generated by the number asymmetry of electrons and positrons (excess electron) [2]. The excess electron is the number of electrons minus positrons.

We simulated space distribution of EM shower with a computer code of Geant4.5 [3]. We divided the EM shower into small parts in longitudinal and lateral directions, and counted the number of excess electrons in them. We have modified Greisen parameterization [4] and modified Nishimura, Kamata, and Greisen function (NKG function) [5] (structure functions) to be applied in UHE region. Halzen, E.Zas, and T.Stanev calculated the electric field strength generated from neutrino interaction having the energy up to 10^{15}eV in ice [6]. Our aim is to calculate electric field strength at antennas in rock salt generated by neutrino energy up to 10^{21}eV. The structure function of excess electrons is indispensable to calculate that without using EM shower simulation.

2. Electromagnetic shower simulation with Geant4.5

We simulated the EM shower generated by an electron of incident energy from 10^{12} eV to 10^{16} eV by using Geant4.5 in rock salt (NaCl). The radiation length (r.l.), critical energy, and density of rock salt are 9.9 cm (X_0), 43 MeV (ε_0), and 2.2 g/cm^3 (ρ), respectively. The simulation includes all the EM processes provided by Geant4.5. The EM shower development was traced to minimum 1 MeV for e^+, e^-, and γ. The Longitudinal length of UHE EM shower becomes longer than Greisen parameterization. According to Landau, Pomeranchuk, and Migdal, the formation zone of bremsstrahlung extends to the foreword atomic nuclei. That causes destructive coherent effect (LPM effect) [7, 8]. Geant4.5 includes LPM effect of bremsstrahlung, but not gamma conversion. We simulated EM shower with LPM, and without LPM effect (Fig, 1).

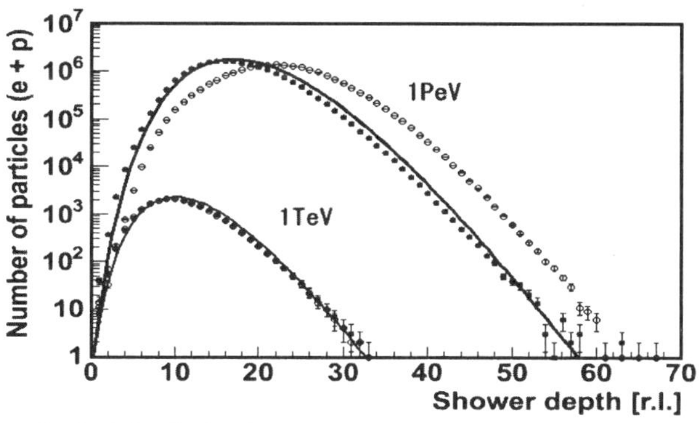

Figure 1. Longitudinal distribution for 1 TeV and 1 PeV. Open circle represent with LPM. Closed circle represent without LPM. Solid line is Greisen parameterization. Vertical axis presents the total number of electrons (e) and positron (p). Horizontal axis presents shower depth.

Figure 1 shows the total number of electrons and positrons crossing planes perpendicular to the shower axis inside the rock salt. The shower shape in longitudinal direction with LPM is longer than that of without LPM. The shower shape without LPM effect agrees well with Greisen parameterization. The shower maximum with LPM effect moves from 16 to 22 radiation length for 1PeV. Thus, the EM shower of higher incident energy could not ignore LPM effect. The shower length of 1 PeV shower with LPM is 60 radiation lengths (about 6 m).

3. Longitudinal distribution of excess charge

Greisen and Rossi have dicussed to the longitudinal distribution of electrons and positrons in EM shower [9]. But, The distribution of excess electrons has been tried to parameterize yet. The shower shape of excess electrons is similar to figure 1. We have modified Greisen parameterization to represent the number of excess electron crossing planes perpendicular to the shower axis inside the rock salt (N(t)):

$$N(t) = \frac{\alpha_L \cdot 0.22}{\sqrt{\ln\left(\frac{E_0}{\varepsilon_0}\right)}} \exp[t(1 - \beta_L \ln\left(\frac{\gamma_L t}{t + \chi_L \ln\left(\frac{E_0}{\varepsilon_0}\right)}\right))]. \qquad (1)$$

Where t is shower depth per radiation length. E_0 is incident energy of an electron. The rate of excess electrons is 0.22 against to total electron and positron for 1PeV. α_L, β_L, γ_L, and χ_L are the parameters to be fitted by least-squares method. α_L corrects the rate of excess electron, and depends on the incident energy. β_L, γ_L, and χ_L are introduced to correcting LPM effect. Those are constant (~2., ~2.2, ~1.6) up to incident energy 100TeV. In this energy range, LPM effect is not important. But, above 1PeV, LPM effect expands the shower length and, β_L, γ_L, and χ_L decrease. The longitudinal structure function fitted by equation (1) is shown in Fig 2. The structure function agrees well with Geant 4.5.

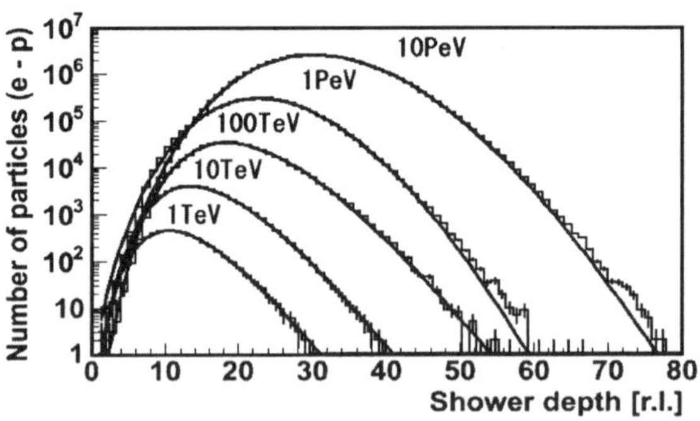

Figure 2. Longitudinal distribution of excess electron for incident electron energy from 1 TeV to 10PeV. The histogram is Geant4.5. Solid line is the structure function.

4. Lateral distribution of excess charge

The lateral development of the EM shower is determined by average angular deflection due to multiple Coulomb scattering. The average angular deflection is presented as:

$$\langle \delta\theta^2 \rangle = \left(\frac{E_s}{\varepsilon_0} \right)^2 \delta t. \tag{2}$$

Where $E_s=(4\pi/\alpha)^{1/2}m_ec^2=21.2$ MeV; δt is very little length per radiation length. The lateral development of the shower is well described by a unit called the Moliere length (m.l.). Due to Rossi's definition, the Moliere unit (t_M= 4.6 cm in rock salt) is related to the critical energy and radiation length of the material:

$$t_M = \frac{E_s}{\varepsilon_0} X_0 . \tag{3}$$

We have modified NKG function to represent the density of excess electron at a perpendicular distance t_R from shower axis (N_{Lat}):

$$N_{Lat}(t, t_R) = \alpha_T \frac{N_L(t)}{t_M^2} t_R^{\ s-\beta_T} (1+t_R)^{s-\gamma_T} . \tag{4}$$

Where t_R is the lateral distance per Moliere unit from shower axis. N_L is the number of generated excess electrons in EM shower at t radiationlengths. The age parameters defined to be s=3t/(t+2y) in case of without LPM effect. The parameter y is defined as ln (E_0/ε_0). α_T, β_T, and, γ_T are introduced to include LPM effect, and β_T, γ_T have also related to extension of age parameter range. As a result, α_T, β_T, and γ_T have become to have incident energy dependence. NKG function holds with age parameter (0.8-1.6). We expanded the effective region to 0-1.6. The lateral distribution and the fitted curved by equation (4) is shown in Fig 3.The equation (4) represents the lateral distribution with LPM effect very well.

5. Summary

Our aim is to calculate electric field strength at antennas without an EM shower simulation e.g. Geant4. In this study, we simulated EM shower with Geant4.5. The longitudinal and lateral distributions are fitted by the structure functions of the modified Greisen parameterization and the modified NKG function, respectively. The structure function included LPM effect and the extension age parameter agree with the longitudinal and the lateral distribution very well. The electric field generated by an UHE EM shower will be calculated by using structure functions. And, the calculation need average momentum vector of excess electrons in a unit volume.

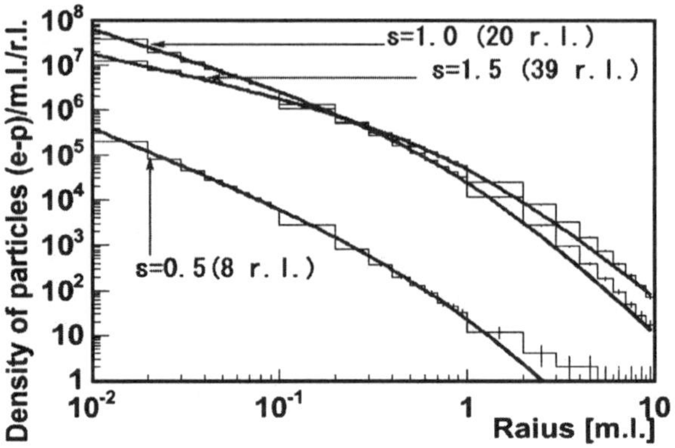

Figure 3. Lateral distribution of excess charge for 10 PeV. The histogram is Geant4.5. Solid line is the structure function. The age parameters are 1.5, 1.0 (shower max), and 0.5. Vertical axis presents the number of electrons (e) minus positron (p). Horizontal axis presents radius.

References

1. K. Greisen, Phys. Rev. Lett. **16**, 748(1966)
2. G.A. Askar'yan, Soviet Physics JETP **14** 441 (1962)
3. S.Agostinelli *et al*, Nuclear Instruments and Methods in Physics Research, NIM **A 506** (2003), 250-303.
4. K. Greisen, Prog. Cosmic Ray Phys. **3**, 1 (1956)
5. J. Nishimura, Handbuch der Physik, Vol. **46**, Iss. 2.
6. E. Zas, F. Halzen, and T. Stanev, Phys. Rev. D **45**,362 (1992)
7. L.Landau and I.Pomeranchuk, Dok.Akad.NaukSSSR **92**,535 (1953)
8. A. B. Migdal, Phys. Rev. **103**, 1811 (1956)
9. B. Rossi, K. Greisen, Rev. Mod. Phys. **13**, 240 (1941)

SIMULATIONS OF RADIO EMISSION FROM ELECTROMAGNETIC SHOWERS IN DENSE MEDIA*

J. ALVAREZ-MUÑIZ, E. MARQUES, R.A. VAZQUEZ AND E. ZAS
Depto. de Física de Partículas, Universidade de Santiago de Compostela, 15782, Santiago de Compostela, SPAIN

By means of GEANT4-based Monte Carlo simulations, we have studied the frequency and angular behavior of Cherenkov radio pulses originated by the excess charge in electromagnetic (EM) showers in different dense media. We have developed a simple model to relate the main characteristics of the electric field spectrum to the longitudinal and lateral development of the EM showers. Using this model the electric field spectrum is shown to have a scaling behavior with a number of medium parameters. We explore the validity of our model by comparing its predictions against full Monte Carlo simulations.

1. Introduction

The detection of ultra high energy neutrinos above ~ 1 TeV is a challenging experimental task due to their small interaction cross section, and to the small expected fluxes at ultra high energy[1]. A very promising detection technique is to search for coherent Cherenkov radio pulses in the MHz-GHz radio frequency range, produced by neutrino induced showers in dielectric, transparent, dense media. When the wavelength of the emitted radiation is larger than the typical dimensions of the shower the emission is coherent, and the contribution to the electric field from positive and negative particles would approximately cancel out were it not for the existence of an excess of electrons over positrons in the shower as pointed out by Askar'yan [2]. This has been experimentally confirmed[3]. For coherent frequencies, the power spectrum scales as the square of shower energy. In dense media the shower dimensions are of the order of meters, and coherence extends to high

*This work is supported by the Xunta de Galicia (PGIDIT02 PXIC 20611PN), and the MCYT (FPA 2001-3837, FPA 2002-01161 and FPA 2004-01198). We thank the "Centro de Supercomputación de Galicia" (CESGA) for computer resources. J.A-M is supported by the Spanish "Ramón y Cajal" program.

frequencies, typically $\sim$ GHz, where more power is available. Besides, large formations of dense, transparent media exist in nature having attenuation lengths in radio frequencies of a few hundred meters. The cost-effectiveness of antennas also adds to the attractiveness of the radio technique.

Several experiments are searching or planning to search for neutrinos exploiting radio emission in ice[4,5], the Moon regolith[6,7,8], and in salt domes[9,10]. Some of them have already placed upper bounds on neutrino fluxes[4,6]. In order to interpret this data, a reliable and well tested Monte Carlo simulation of radio emission in dense media is needed. Also it would be desirable to have a simple model that relates medium properties to the features of the radiopulses, which will allow the quick evaluation of the capabilities of the planned experiments. In this work we show that many of the observables of the electric field spectrum emitted by an EM shower developing in dense media, scale to a few percent level with several parameters of the media, such as density (ρ), radiation length (X_0), Molière radius (R_M), critical energy (E_c), and refraction index (n).

2. A reliable MC simulation

A realistic simulation of showers for radio applications must follow all particles explicitly down to the 10-100 keV kinetic energy range, since these are responsible for the the bulk of the coherent radio emission[11]. This is a very challenging problem that was first approached by a specific package called the ZHS code[11].

The ZHS code[11] is a fast simulation program specifically deviced for calculating radio-pulses from EM showers in ice. It takes into account all low energy processes and accurate timing of the particles in the shower. It calculates the radio emission from finite particle tracks using the standard electrodynamics expression for the electric field produced by a charged particle travelling along a small straight path at constant speed[11]. The calculation of the emission from a shower requires the subdivision of all the individual charged particle tracks in small steps. For each subtrack the average velocity is calculated between the corresponding end points of the track and the electric field is added up. We have also implemented this same algorithm in a GEANT4 simulation[12] of EM showers. GEANT is a well known, well tested and widely used simulation and detection package. In Fig.1 we show the frequency spectrum of the electric field for a 1 TeV shower in ice as predicted by ZHS and GEANT4. The agreement is strikingly good. The differences between both predictions that appear at high

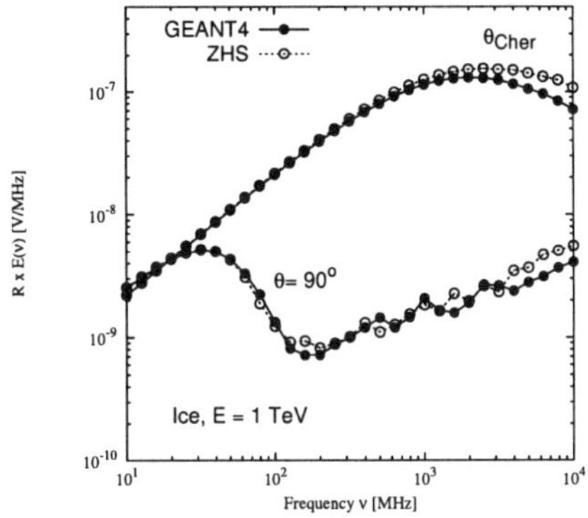

Figure 1. Frequency spectrum of the electric field emitted by a 1 TeV EM shower in ice as predicted by the ZHS and GEANT4 codes for observation at the Cherenkov angle θ_c and at $\theta = 90^\circ$ from the shower axis.

frequencies (above 1 GHz) are due to the slightly different algorithm of subdivsion of particle tracks used in both codes. At high frequencies (small wavelengths) the emission is sensitive to the fine details of the subdivision algorithm.

3. Radiopulses in different dense media

When a photon or electron of energy E incides on a thick absorber it initiates an EM shower. Using the Heitler model[11] it is easy to show that the total tracklength T due to the charged particles in a photon initiated shower can be approximated by $T \sim (E/E_c)(X_0/\rho)$. The particles in the shower also spread in the transverse dimension mainly due to multiple scattering. One can imagine a shower as a thin pancake in which particles travel at the speed of light ($\beta = 1$), parallel to shower axis along a length $L = k_L X_0/\rho$ proportional to the radiation length, and has a width $R = k_R R_M/\rho$ proportional to the Molière radius R_M. Both k_L and k_R are normalization constants of the model to be determined in numerical simulations (see below). This very simple model allows us to establish the relation between the frequency spectrum and magnitude of the electric field, and the parameters of the medium in which it is emitted.

The overall time duration of the radio pulse for a given observation angle from shower axis (θ) is related to the turnover frequency at which the spectrum departs from a linear behavior with frequency (see Fig.1). There are two relevant time scales in our toy model: $\delta t_{\rm L}$ the time delay between two light rays emitted at the two end points in the longitudinal development of the shower, and $\delta t_{\rm R}$ the time delay between two rays emitted from the two ends of the lateral spread of the shower. It is easy to obtain that:

$$\delta t_{\rm L} = \frac{1}{c}\,\frac{k_L X_0}{\rho}\,(1 - n\cos\theta) \text{ and } \delta t_{\rm R} = \frac{n}{c}\,\frac{k_R R_M}{\rho}\sin\theta. \tag{1}$$

For observation at θ_c, $\delta t_{\rm L} \simeq 0$, and only the time delay associated to the lateral spread of the shower remains. The electric field spectrum then increases linearly with frequency up to a turnover frequency given by

$$\nu_{\rm R}(\theta = \theta_c) \simeq \frac{1}{\delta t_{\rm R}} = \frac{\rho}{k_R R_M}\,\frac{c}{\sqrt{n^2 - 1}}. \tag{2}$$

where we have used $\cos\theta_c = 1/n$. The turnover frequency associated to $\delta t_{\rm L}$, and relevant for observation away from θ_c is,

$$\nu_{\rm L}(\theta) \simeq \frac{1}{\delta t_{\rm L}} = \frac{\rho}{k_L X_0}\,\frac{c}{|1 - n\cos\theta|}. \tag{3}$$

The magnitude of the field at low frequencies (when the electric field $\propto \nu$) is known to be determined by the tracklength associated to the excess negative charge after projecting it onto the plane perpendicular to the line of sight [11]. As a result the magnitude of the electric field is expected to scale as,

$$r \times |\vec{E}| \sim k_E\ \nu\ T\sin\theta \simeq k_E\ \frac{E}{E_c}\,\frac{X_0}{\rho}\ \nu\ \sin\theta. \tag{4}$$

The constants k_E, k_L and k_R appearing in the toy model, are obtained by normalizing the scaling relations above to the predictions of simulated EM showers in ice using the GEANT4-based code. These give: $k_E \simeq 4.97 \times 10^{-14}$ V m^{-1} MHz^{-2}, $k_L \simeq 25.6$ and $k_R \simeq 0.6$.

We can now use the scaling relations together with the normalization constants to predict the properties of the radio emission in salt and the regolith. We assume the constants to be medium-independent, and hence all the dependence with the medium is contained in ρ, X_0, R_M, E_c and n. In Table 1 we compare the numerical values of the electric field magnitude and the turnover frequencies as obtained in the GEANT4 simulations and as predicted by the toy model. The turnover frequencies of the electric field spectrum agree with those obtained by GEANT simulations within $10 - 15\%$. However the scaling of the magnitude of the electric field with the excess tracklength projected onto the direction perpendicular to the direction of observation is only approximate, mainly because the toy model implicitely assumes that particles in the shower follow straight lines parallel to the shower axis, while it is well known that they deviate by an average angle which is medium dependent. As a result, the projection of the tracks onto the direction perpendicular to the θ_c-direction is not simply achieved by a factor $\sin\theta_c$, but some dependence on the average deviation angle should also be present. This leads to an overestimate of the electric field magnitude by the scaling model in salt and the Moon regolith by about 30%.

Table 1. Several features of the electric field emitted by a 10 TeV electromagnetic shower in salt and the Moon regolith as obtained in GEANT4 simulations and as predicted by our scaling model (quantities labeled with "model"): Turnover frequency at θ_c; turnover frequency at $\theta = 90$ deg.; electric field at 10 MHz for observation at θ_c.

Medium	$\nu_R^{\rm model}(\theta_C)$ [GHz]	$\nu_{\rm cut}^{\rm MC}(\theta_C)$ [GHz]	$\nu_L^{\rm model}(90^\circ)$ [MHz]	$\nu_{\rm cut}^{\rm MC}(90^\circ)$ [MHz]	$r \times \lvert\vec{E}\rvert^{\rm model}$ [V MHz^{-1}]	$r \times \lvert\vec{E}\rvert^{\rm MC}$ [V MHz^{-1}]
Salt	4.0	4.5	105.5	113.5	1.2×10^{-8}	8.3×10^{-9}
Moon	5.6	6.5	90.9	82.5	1.2×10^{-8}	8.6×10^{-9}

4. Summary

We have shown that the predictions of the ZHS and GEANT4 codes on the electric field spectrum emitted by EM showers are in very good agreement. We have developed a toy model to relate the main characteristics of the electric field spectrum to the longitudinal and lateral development of the EM showers. We have shown that the electric field spectrum has a scaling behavior with a number of medium parameters that works to a few percent level. Further improvements of our scaling model are under development.

References

1. J.G. Learned and K. Mannheim, Ann. Rev. Nucl. Sci. **50**, 679-749 (2000).
2. G.A. Askar'yan, Soviet Physics JETP **14,2**, 441 (1962).
3. D. Saltzberg *et al.*, Phys. Rev. Lett. **86**, 2802 (2001).
4. I. Kravchenko *et al.*, Astropart. Phys. **20**, 195 (2003).
5. P. Miocinovic *et al.* these proceedings
6. P. W. Gorham, *et al.* Phys. Rev. Lett. **93**, 041101 (2003).
7. J. Bacelar *et al.* these proceedings.
8. R. Protheroe et al., http://www.physics.adelaide.edu.au/astrophysics/
9. P.W Gorham *et al.*, Phys. Rev. D **72** 023002 (2005).
10. A. M. van den Berg *et al.* these proceedings.
11. E. Zas, F. Halzen, T. Stanev, Phys. Rev. D **45**, 362 (1992).
12. J. Alvarez-Muñiz *et al.*, Phys. Rev. D **67**, 101303 (2003).

MONTE CARLO SIMULATIONS OF RADIO EMISSION FROM COSMIC RAY AIR SHOWERS

T. HUEGE

Institut für Kernphysik,
Forschungszentrum Karlsruhe,
Postfach 3640,
76021 Karlsruhe, Germany
E-mail: tim.huege@ik.fzk.de

H. FALCKE

ASTRON,
P.O. Box 2,
7990 AA Dwingeloo, The Netherlands
E-mail: falcke@astron.nl

As a basis for the interpretation of data gathered by LOPES and other experiments, we have carried out Monte Carlo simulations of geosynchrotron radio emission from cosmic ray air showers. The simulations, having been verified carefully with analytical calculations, reveal a wealth of information on the characteristics of the radio signal and their dependence on specific air shower parameters. In this article, we review the spatial characteristics of the radio emission, its predicted frequency spectrum and its dependence on important air shower parameters such as the shower zenith angle, the primary particle energy and the depth of the shower maximum, which can in turn be related to the nature of the primary particle.

1. The simulations

Two main mechanisms have been considered to contribute to radio emission from cosmic ray air showers: Askaryan-type Čerenkov radiation arising from a negative charge excess moving through the atmosphere at velocities faster than the speed of light in air, and emission generated as a consequence of the deflection of charged particles in the earth's magnetic field. A number of historical results illustrated that, while the Čerenkov emission mechanism dominates in dense media such as ice, the geomagnetic mechanism is dominant in the atmosphere. Our simulations thus focus on the latter mechanism, interpreting the radio emission as "coherent geosynchrotron radiation" arising from the geomagnetic deflection of highly relativistic electron-positron pairs generated in the air shower cascade[1].

In order to gain a solid understanding of the emission characteristics, in a first step, we have performed analytical calculations of the expected radio signal[2]. In a second step, we have then improved on the analytical results with detailed Monte Carlo simulations, which we have directly compared and thus verified with the analytical results and the available historical data[3]. While these Monte Carlo simulations still incorporate a somewhat simplified air shower model based on analytical parametrizations, they do already take into account the most important air shower characteristics such as longitudinal (arrival time) and lateral particle distributions, energy and track-length distributions, the overall longitudinal development of the air shower and the geometry of the air shower and magnetic field. Our simulations thus for the first time have provided a prediction of the radio emission from realistically modeled air showers. Furthermore, the model is currently being enhanced by substituting the parametrized particle distributions with CORSIKA[4]-generated distributions.

We here present only a subset of the results derived with our Monte Carlo code. A more detailed analysis, including a parametrization of the derived dependences, has been published elsewhere[5].

2. Simulation results

We first consider the very simple scenario of a vertical air shower with primary particle energy of 10^{17} eV developing to its maximum at $\sim 630\,\mathrm{g\,cm^{-2}}$, i.e., at $\sim 4000\,\mathrm{m}$ above ground. The geomagnetic field is adopted as 70° inclined with a strength of 0.5 G, corresponding approximately to the field configuration in central Europe.

Figure 1 shows the ground-level total field strength emission pattern at 10 MHz, visualized as a contour plot, and the frequency spectra derived at various radial distances from the shower center. The emission pattern shows remarkable symmetry and is almost circular. This is not a trivial result, as the emission process itself, i.e., the deflection of electrons and positrons in the geomagnetic field, is a highly directed process. The circularity of the footprint illustrates that most of the emission stems from particles having short track lengths. A slight north-south asymmetry introduced by the inclination of the geomagnetic field is also visible. The frequency spectra shown in the right panel illustrate that the field strength drops quickly to higher observing frequencies. This is a direct consequence of diminishing

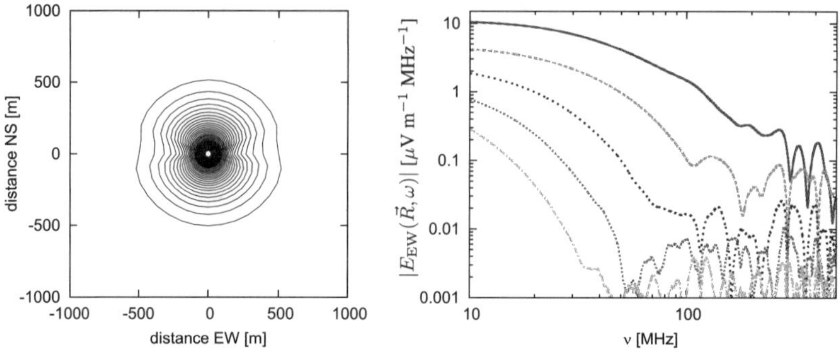

Figure 1. Radio emission from a vertical 10^{17} eV air shower. Left: 10 MHz total field strength emission pattern. Right: Frequency spectra at (from top to bottom) 20 m, 140 m, 260 m, 380 m and 500 m north of the shower center.

coherence as the wavelength becomes shorter and thus comparable to the dimensions of the air shower pancake, in particular its thickness of a few meters. The decrease is stronger at larger distances from the shower center. When one enters the incoherent regime the frequency spectra exhibit unphysical seeming features such as rapidly alternating series of maxima and minima. Realistic calculations of the emission in this regime can only be performed with a better underlying air shower model taking into account inhomogeneities in the shower in very mildly thinned calculations.

Figure 2 demonstrates the changes arising in the transition from a vertical to a 45° inclined air shower (coming from the south). The emission pattern becomes elongated considerably along the shower axis. This is mainly

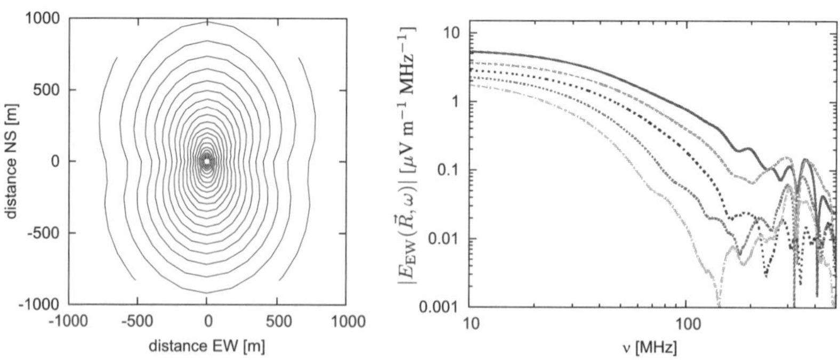

Figure 2. Radio emission from a 45° inclined 10^{17} eV air shower. Left: 10 MHz total field strength emission pattern. Right: Frequency spectra at (from top to bottom) 20 m, 140 m, 260 m, 380 m and 500 m north of the shower center.

a projection effect directly associated with the inclination of the shower axis. On closer look, however, the emission pattern becomes wider (and less peaked) as a whole, even in the direction perpendicular to the shower axis. The reason for this is that the maximum of the inclined shower located at the same (slant) atmospheric depth of $\sim 630\,\mathrm{g\,cm^{-2}}$ now is at much greater geometrical distance from the observer at ground-level. This geometric effect has direct influence on the slope of the radio emission's lateral distribution. A look at the frequency spectra in the right panel shows that coherence is also retained up to higher frequencies in case of inclined showers. Their larger radio footprint combined with the large solid angle associated with medium to high zenith angles thus makes inclined showers a particularly interesting target for radio observations[5,6]. At near horizontal inclination, even neutrino-induced air showers might become observable.

Figure 3 illustrates two additional parameters that have direct influence on the radio signal. The left panel shows the impact of the primary particle's energy. The electric field strength at all distances scales as a power-law with the primary particle energy. The power-law index is very close to unity, i.e., that of the linear relation expected for coherent emission. To larger distances, the slope of the power-law gets flatter due to the effect that more energetic showers on average penetrate deeper into the atmosphere and thus have their shower maximum geometrically closer to the observer. As already discussed in the context of inclined showers and illustrated in the right panel, this directly influences the lateral distribution of the radio emission. Since the depth of the shower maximum can in turn

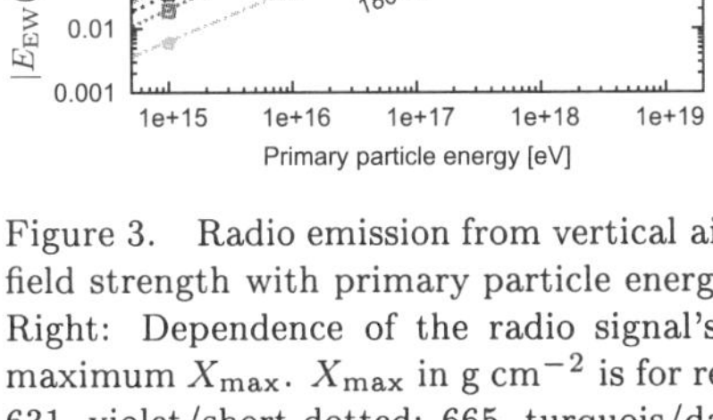

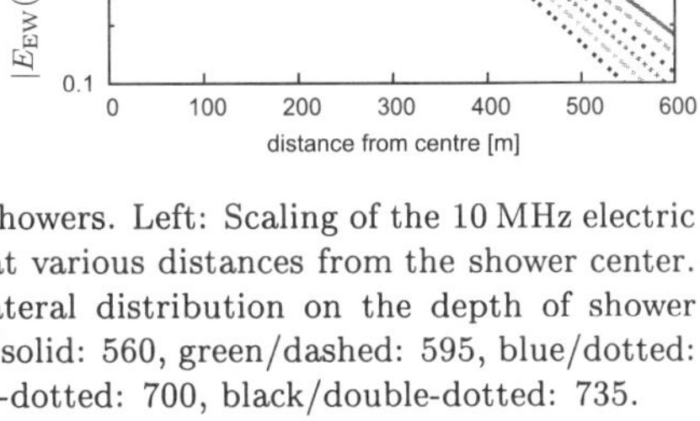

Figure 3. Radio emission from vertical air showers. Left: Scaling of the 10 MHz electric field strength with primary particle energy at various distances from the shower center. Right: Dependence of the radio signal's lateral distribution on the depth of shower maximum $X_{\rm max}$. $X_{\rm max}$ in g cm^{-2} is for red/solid: 560, green/dashed: 595, blue/dotted: 631, violet/short dotted: 665, turquois/dash-dotted: 700, black/double-dotted: 735.

be related to the nature of the primary particle, its influence on the radio signal's lateral distribution can potentially be used to probe the primary particle composition with radio measurements[7].

Another important result of the simulations (not shown here explicitly) are the predicted linear polarization characteristics of the radio signal[5]. They can be used to directly verify the geomagnetic origin of the emission. To make the simulation results available for easy comparison with experimental data, they are also available as a parametrization formula[5].

3. Conclusions

We have carried out elaborate Monte Carlo simulations of geosynchrotron radio emission from cosmic ray air showers. Special care has been taken to verify the Monte Carlo results with analytical calculations and historical data, giving us a good understanding of the emission process and thus solid confidence in the predictions. The simulations predict many important characteristics of the radio emission and their relation to parameters of the associated air shower. The total field strength emission pattern is very regular and symmetric in the coherent regime. The geomagnetic origin of the emission can be directly verified with polarization measurements. The frequency spectra cut off quickly to high frequencies, making low observing frequencies around a few tens of MHz desirable. Inclined air showers have a much wider emission pattern and are thus particularly suitable for radio observations. The slope of the lateral distribution can be directly related to the geometrical distance between observer and shower maximum. It is thus not only sensitive to the shower zenith angle, but also to the nature of the primary particle. As expected for coherent emission, the electric field strength scales approximately linearly with the primary particle energy. These predictions will allow to analyze and interpret experimental data such as those provided by LOPES[8] and other experiments.

References

1. H. Falcke and P. Gorham, *Astropart. Phys.* **19**, 477-494 (2003).
2. T. Huege and H. Falcke, *Astronomy Astroph.* **412**, 19-34 (2003).
3. T. Huege and H. Falcke, *Astronomy Astroph.* **430**, 779-798 (2005).
4. D. Heck et al., *Forschungszentrum Karlsruhe Report* **FZKA 6019** (1998).
5. T. Huege and H. Falcke, *Astropart. Phys.* **in press** (2005), astro-ph/0505180.
6. T. Gousset, O. Ravel and C. Roy, *Astropart. Phys.* **22**, 103-107 (2004).
7. T. Huege et al. - LOPES collaboration, *Proc. 29th ICRC, Pune, India* (2005).
8. H. Falcke et al. - LOPES collaboration, *Nature* **435**, 313-316 (2005).

SIMULATION OF RADIO SIGNALS FROM 1-10 TeV AIR SHOWERS USING EGSNRC

R. Engel[a], N.N. Kalmykov[b], A.A. Konstantinov[b].
(a) Forschungszentrum Karlsruhe, Institut fuer Kernphysik, Postfach 3640, D-76021 Karlsruhe, Germany
(b) Skobeltsyn Institute of Nuclear Physics, Moscow State University, Leninskie Gory 1, 119992, Moscow, Russia

Cherenkov and geosynchrotron radiation are considered as two fundamental mechanisms of the radio emission generated by extensive air showers (EAS). The code EGSnrc is used for Monte-Carlo simulations of the individual shower development. Calculations of the radial dependence and frequency spectrum of the emitted radiation are performed for the LOPES experiment frequency range.

1. *Introduction*

Coherent radio emission generated by extensive air showers was theoretically predicted by Askaryan in 1961 [1] and experimentally discovered by Jelly et al. in 1965 at a frequency of 44 MHz [2]. Over a period of time this phenomenon has been considered as an interesting alternative to traditional methods of detection of high-energy cosmic rays with energy greater than 10^{17} eV. In the 1960th and 1970th the experimental and theoretical efforts in this direction had no actual success [3]. Modern experiments, such as CODALEMA [4] and LOPES [5], aimed at EAS radio emission studies use modern, improved instruments and thus can hope for the final success. But there are still many questions concerning the quantitative radio emission theory.

Several mechanisms of radio emission generation in air have been identified after the pioneering work of Askaryan where the coherent Cherenkov radiation of the charge excess was put forward [1]. This radiation is very strong for showers developing in dense media [6]. In the case of EAS there is also an alternative radiation due to the acceleration of charged shower particles in the Earth's magnetic field. It is called geosynchrotron mechanism and has been recently investigated in detail [7]. However we still have no sufficiently clear understanding what interrelation exists between these two essential mechanisms. So, one needs to perform accurate radio emission calculations for these mechanisms within the framework of a unified approach. In our work we present a model in which Cherenkov and geosynchrotron radiation are combined. In a sense, our work is complementary to [7] where only the geosynchrotron radiation was considered.

2. *Calculations*

To calculate the radio emission of air showers an EGSnrc-based [8] program code has been developed. For reproduction of the Earth's atmosphere we have taken 200 strata of air, with density and optical properties varying from stratum to stratum according to the atmospheric profile. The declination and strength of the Earth's magnetic field correspond to those for Karlsruhe, where the LOPES experiment is being performed. Radio emission characteristics (radial dependence, frequency spectrum, polarization and some others) are calculated taking into account contributions from each charged particle.

There are two different radiation mechanisms adopted in the model and the separation of them is realized as follows. If a charged particle is moving in the magnetic field characterized by the field strength $\vec{B}$ and the refractive index is equal to n, we may present the electric field $\vec{E}$ as the sum of two parts with the following properties

$$\vec{E} = \vec{E}_1 + \vec{E}_2,$$

where $E_1 \to 0$, when $B \to 0$, and $E_2 \to 0$, when $n \to 1$. We accept that $\vec{E}_1$ is the electric field due to the Earth's magnetic field (*geosynchrotron radiation*) and $\vec{E}_2$ is the electric field due to medium (air) properties (*Cherenkov radiation*).

3. *Monte Carlo simulation results*

Vertical showers were simulated for primary photons with the energies 1 and 10 TeV and for energy threshold of 100 keV. The primary particle is injected at 30 km above the ground level. The lateral distributions of radio emission were calculated simultaneously at several frequencies: 10, 30 and 100 MHz. In total 50 ground-level observation points were uniformly distributed over a straight line from the shower axis to the direction of the geographic north in the range of distances up to 500 m.

The mean longitudinal profile of showers with 1 TeV primary photon energy is presented on Fig.1. Such showers have the negative charge excess (ε) of about 20% in the maximum. It should be stressed that electrons and positrons emit Cherenkov radiation if their energy exceeds the Cherenkov threshold (that is equal to 21 MeV at sea level) and thus only ≈ 1/3 of the above mentioned excess particles give a contribution to the observed electric field. This is in contrast to the situation in ice where, due to a rather large refrac-tive index, almost all excess particles emit Cherenkov radiation.

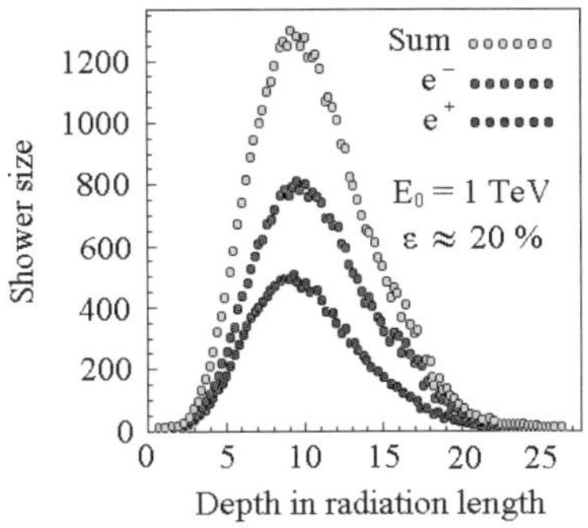

Figure 1. Number of shower particles as a function of depth (averaged over 20 showers).

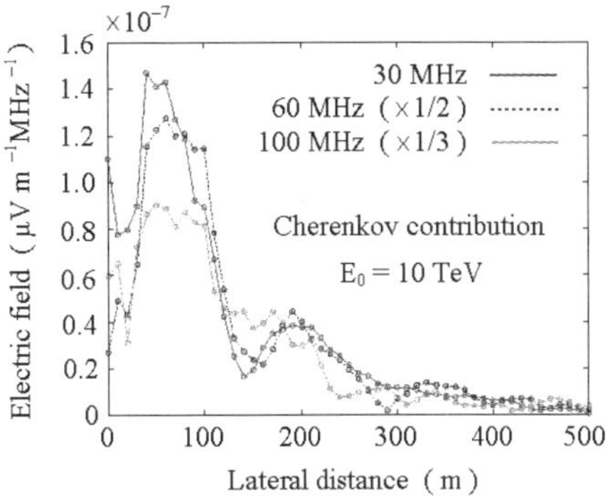

Figure 2. Lateral distribution of Cherenkov radio emis-sion at different frequencies (averaged over 10 showers).

Fig.2 shows the lateral distribution of the electric field produced by Cherenkov radiation of shower particles. The primary energy is 10 TeV and the electric field is normalized at the frequency 10 MHz. We associate this radiation with Askaryan's mechanism (radiation of the negative charge excess). This idea was confirmed by direct calculations: when the excess is zero then we have a decrease of the field by two to three orders of magnitude (depending on the considered frequency). It is also interesting to note that the Cherenkov radiation demonstrates a diffraction pattern.

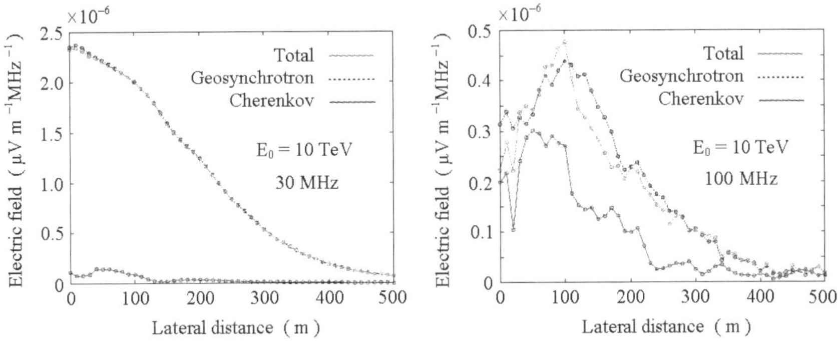

Figure 3. Lateral distribution of geosynchrotron, Cherenkov and total radio emission at different frequencies (averaged over 10 showers).

The full pattern of the radio emission lateral distribution is shown in Fig.3 for 10 TeV showers. Plotted are the Cherenkov, geosynchrotron and total (the sum of Cherenkov and geosynchrotron contributions) radio emission at 30 and

100 MHz. We see that there exists practically full domination of the geosynchrotron radiation in the low frequency part of the radio emission spectrum at all distances. But it is not so for higher frequencies and especially at the main Cherenkov peak.

It seems that we can interpret this behavior as due to the difference in spectral properties of the two types of radiation. This is confirmed by Fig.4 where the spectral distribution of the radio emission at 100 and 300 m from the shower axis of 1 TeV showers are shown. We see (picture for 100 m) that the coherent regime for the Cherenkov emission is maintained up to higher frequencies than in the case of the geosynchrotron emission. It seems that the main reason of this situation is that the effective dimension of the radiation region is smaller for the Cherenkov emission than for the geosynchrotron emission due to the large Cherenkov threshold energy. The situation is similar at larger distances from shower axis (results are given at 300 m).

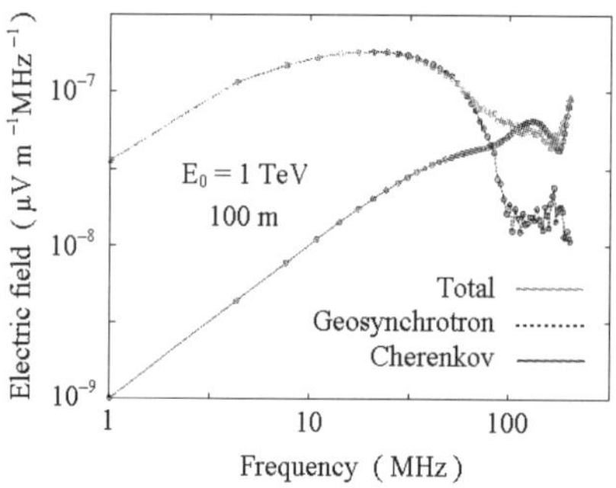

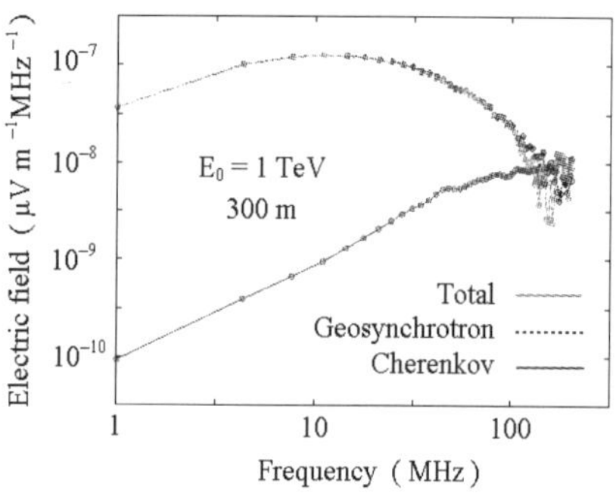

Figure 4. Spectral distribution of geosynchrotron, Cherenkov and total radio emission at different distances from shower axis (averaged over 10 showers).

4. *Conclusions*

Realistic air shower and radio signal simulations for primary energies 1 and 10 TeV have been performed. The calculations show that there is no full domination of one of the two radiation mechanisms in the Earth's atmosphere. It seems that an appropriate radio emission theory needs to take into account the Cherenkov radiation as well as the geosynchrotron mechanism.

The contribution of the Cherenkov radiation to the total field is not identical at different distances from shower axis. At small distances, including the main peak, the role of the Cherenkov component grows with the increase of the observation frequency due to violation of the coherence condition for the geosynchrotron radio emission whereas it is conserved for the Cherenkov radiation.

We also observe the same situation at larger distances from shower axis. However the flow of the geosynchrotron radio emission falls with distance more slowly than for the Cherenkov emission and thus the amplitude of the Cherenkov radiation at these distances is much smaller.

The amplitude of the geosynchrotron mechanism essentially depends on the configuration of the system "shower axis - magnetic field" and there is a need to simulate showers with different arrival directions relative to the local magnetic field. In parallel one certainly needs to push up the primary energy and statistics of the simulations to attain better understanding of radiation processes in air.

Acknowledgments

The authors thank T. Huege for fruitful discussions on the simulation of geosynchrotron radio emission.

References

1. G.A. Askaryan, Soviet Phys. JETP **41**, 616 (1961).
2. J.V. Jelley *et al.*, Nature **205**, 237 (1965).
3. H.R. Allan, Prog. in Element. part. and Cos. Ray Phys. **10**, 171 (1971).
4. O. Ravel *et al.*, CODALEMA Collab., Nucl. Instrum. Meth. **A518**, 213 (2004), astro-ph/0409039.
5. H. Falcke *et al.*, LOPES Collab., Nature, **435**, 313 (2005).
6. E. Zas, F. Halzen and T. Stanev, Phys. Rev. **D45**, 362 (1992).
7. T. Huege and H. Falcke, Astron. Astrophys. 412, 19 (2003); Astron. Astrophys. 430, 779 (2005).
8. http://www.irs.inms.nrc.ca/inms/irs/EGSnrc/EGSnrc.html .

SCALING OF ASKARYAN PULSES

D. SECKEL

Bartol Research Institute and Department of Physics and Astronomy, University of Delaware, Newark, DE 19716, USA
E-mail: seckel@bartol.udel.edu

The theory of the Askaryan process, impulsive RF emission from particle showers, is reviewed. The radiated electric field may be calculated at all angles and frequencies from just two phenomonological functions related to the longitudinal and transverse profiles of the shower. A prescription is given for extracting the relevant profiles from shower Monte Carlo calculations. Results obtained for one shower may be scaled to other energies and environments. A two parameter analytic model for shower profiles is proposed.

Askaryan[1] proposed that particle induced showers in dense medium would be a source for impulsive radio signals. The mechanism depends on medium energy processes in the shower: δ-ray production, Compton scattering, and in flight positron annihilation all favor the production of a negative excess charge as the shower evolves. As a pattern, this charge propagates relativistically through the medium, acting as a source for Cerenkov radiation. This process has been modeled numerically and analytically, going back to the seminal paper by Zas, Halzen, and Stanev[2]. More recently, experimental confirmation[3] of the basic process has provided justification for several experimental efforts to detect ultra-high energy cosmic neutrinos, as summarized in these proceedings.

This contribution reviews theoretical understanding of the Askaryan pulses, and proposes a simple model[4] suitable for inclusion in experiment Monte Carlos. The model gives an acceptable picture in both the frequency and time domains, and is valid at all viewing angles. Input to the model is taken from a limited set of Monte Carlo simulations, and extended using the scaling properties of showers with energy and environment. The scaling concepts discussed here should apply to acoustic signals as well.

Modeling of Askaryan pulses. There are two complementary ways to discuss Askaryan pulses. On the one hand, the pulse arises from the coherent sum of radiation from all particles in the shower. General purpose particle Monte Carlo codes (GEANT) have been used to model Askaryan pulses at shower energies below 1 PeV[5,6,7]. For higher energies, the application specific "ZHS" code[2] is currently maintained by the Santiago group[8,9,10]. In either case, utilizing a shower simulation in a full experiment Monte Carlo is not practical due to the large CPU resources required.

Alternatively, one may model the shower as a smooth profile of an appropriate current density, leading to a compact description of the radiation pattern[5,11]. That current density may be determined directly from numerical simulaton or from a theoretical model. Once a profile is determined for one shower, or a class of showers, scaling properties may be invoked to extend the result to other energies or environments.

Sum over particles. The spectrum of the radiated electric field observed at a distance R from a single charged particle track may be written as

$$R\vec{E}_\omega = \frac{1}{\sqrt{2\pi}}\frac{q}{c}e^{ikR}e^{i\omega(t_1-\hat{n}\cdot\vec{r}_1/c)}\vec{\beta}_\perp\frac{e^{i\omega\delta t(1-\hat{n}\cdot\vec{\beta})}-1}{1-\hat{n}\cdot\vec{\beta}} \quad (1)$$

The track begins at $(t_1, \vec{r}_1)$ and extends for time δt at velocity $\vec{\beta}$. $\vec{\beta}_\perp$ is the component of velocity perpendicular to the line of sight to the observer, and $\hat{n}$ is the direction from the shower to the observer.

For observers near the Cerenkov cone the factor $1 - \hat{n} \cdot \vec{\beta}$ is small, leading to an approximation where E is proportional to the projected track length $\delta t \beta_\perp$. This approximation is not valid for all particles, so there is a loss of signal for particles scattered significantly from the shower axis. Further, the electric fields of such particles are not aligned vectorially. The complex phase is determined by $\phi = kR + \omega(t_1 - \hat{n} \cdot \vec{r}_1/c)$. For a pattern velocity traveling at $v_1 = c$ along the shower axis, this phase vanishes on the Cerenkov cone. Several conditions lead to a loss of phase coherence between different particles. For an observer outside the Cerenkov cone, radiation from the initial/latter part of the shower arrives earlier/later, and vice versa for an observer to the inside. Transverse dislocatation of a particle track yields a shorter or longer path to the detector. Smaller effects are due to the width and curvature of the shower front. A consequence of these considerations is that the effective source function, summed over particles, requires a numerical code for evaluation.

Shower profiles and form factors. For a large number of particles, the sum transitions into a space-time integral over a smooth source distribution which, ignoring the small effects of shower curvature and thickness, reduces to a 2-D integral over the longitudinal and transverse profile of the shower.

$$E(\theta,\nu) = i\frac{\nu}{R}\sin(\theta)\int dydz\rho(y,z)e^{i\phi(\theta,\nu,y,z)} \tag{2}$$

where ρ is an effective source density. The lateral distribution of the excess charge is nearly independent of the age of the shower, so one may separate the shower profile into the product of a normalization, a transverse profile and a longitudinal profile, $\rho = f_0 f_y(y) f_z(z)$. The phase lag also separates: $\phi \simeq \phi_y + \phi_z = a_y y\nu + a_z z\nu$, with $a_y = (2\pi/c)(\cos\delta\tan\theta_c + \sin\delta)$, and $a_z = (2\pi/c)(\sin\delta\tan\theta_c + (1-\cos\delta))$, where δ is the observer's direction relative to the Cerenkov cone. The result is that the electric field can be expressed as the product of transverse G_y and longitudinal G_z form factors.

$$E(\theta,\nu) = i\frac{\nu}{R}\sin(\theta) f_0 G_y(\theta,\nu) G_z(\theta,\nu) \tag{3}$$

where $G_y(\theta,\nu) = G_y(a_y\nu) = \int dy f_y(y) e^{ia_y\nu y}$, and $G_z(\theta,\nu) = G_z(a_z\nu) = \int dz f_z(z) e^{ia_z\nu z}$. In either case, the form factor reduces to a function of a single variable.

Interpretation of f, G. It is useful to choose the normalization f_0 such that $G_y(0) = G_z(0) = 1$, in which case f_0 is given by the total excess projected track length, modulated by a medium dependent efficiency to account for Coulomb scattering.

On the Cerenkov cone, $\delta = 0$ and $G_z = 1$. The electric field spectrum is of the form $E(\nu) \sim i\nu G_y$, which may be recognized as the Fourier transform of df_y/dy. For ice, the spectrum rolls over at about 2 GHz, as determined by the transverse dimensions of the shower. For $\delta = 0$, the time domain waveform is bipolar and anti-symmetric.

For δ larger than a few degrees, G_z dominates the frequency cutoff, and G_y may be approximated as constant. In this case, $E(\nu) \sim i\nu G_z$, and the time domain pulse is proportional to the time derivative of the longitudinal shower profile. The waveform should be strongly bipolar, but not strictly anti-symmetric, depending on the details of f_z. There is a null at t_0, corresponding to shower max. There is a reflection symmetry $E(-\delta,(t-t_0)) = -E(\delta,-(t-t_0))$.

Extracting form factors from MC calculations. The normalization and shower profiles, f_0, f_y and f_z, may be determined directly from a Monte Carlo simulation[5]. Alternatively, the ZHS Monte Carlo has been used[10] to directly parameterize the Askaryan signal, and these results may be inverted to determine the form factors G_y and G_z. For a 1 PeV electromagnetic shower, the spectrum on the Cerenkov cone is given by

$RE_\nu = 2.53 \cdot 10^{-4}\tilde{\nu}/(1+\tilde{\nu}^{1.44})$ V/MHz, with $\tilde{\nu} = \nu/1.15$ GHz. With $G_y(0) = 1$, this yields $f_0 = 2.66 \cdot 10^{-3}$ V/MHz^2 and $G_y(\tilde{\nu}) = 1/(1+\tilde{\nu}^{1.44})$. The width of the Cerenkov cone is described as a Gaussian, which translates to a form factor $G_z(\delta, \nu) = \exp(-1140(a_z\nu/\text{GHz})^2)$.

Scaling of electromagnetic showers. Askaryan and early researches focussed on charged current scattering of electron neutrinos, in which case the shower is initiated by an electron, which typically carries $\sim 80\%$ of the neutrino energy. Such showers are dominated by e^- and γ processes, and are generally referred to as electromagnetic. The transverse profile is primarily determined by the Moliere radius R_M and is nearly independent of energy. The longitudinal evolution is determined by the the radiation length X_{rad}. The position of shower max t_0 increases as $\log(E_s)$. For Askaryan pulses, it is more relevant to consider the "width" of the shower t_w, the range around shower max which dominates the radiation pulse[8]. The shower width grows with energy, but more slowly than t_0. At high energies, X_{rad} increases as $E^{1/2}$ due to the LPM mechanism, widening the showers, narrowing the Cerenkov cone, and reducing the volume over which a single antenna may detect signals. It has been suggested[12,13] that nuclear interactions of γ's may convert EM showers to hadronic showers, reducing the importance of the LPM effect at the highest energies.

Scaling of hadronic showers. Deep inelestic $\nu - N$ scattering produces a recoil quark, which fragments and produces an hadronic shower typically with $\sim 20\%$ of the neutrino energy. The normalization of hadronic showers must account for the efficiency for converting hadronic to electromagnetic energy. This efficiency is greater than 95% for shower energies above a PeV, but decreases for lower energy showers[7,9]. In a dense medium, at energies above ~ 1 PeV π^0's interact before they decay. This postpones the transition from a hadronic shower to an EM shower via $\pi^0 \rightarrow \gamma\gamma$ until particle energies are below the threshold for the LPM effect to be important. The growth of shower width with energy has been studied for showers initiated by protons[9] and quark jet fragments[7]. The results show small differences, but in both cases hadronic showers are relatively immune to the LPM effect and are expected to dominate neutrino detection rates.

An analytic model for f_z. The age evolution of a shower may be modeled as a Γ function, $f_z(t) = Nt^{a-1}e^{-bt}$. Here, $N = \Gamma(a)/b^a$ provides the normalization, and a, b are chosen appropriate to the type and energy of the shower, and the medium in which it develops. This form has several appealing features. The Fourier transform is analytic and causal,

$$G_z = (\frac{b}{\sqrt{b^2+\omega^2}})^a e^{ia\arctan(\omega/b)}. \tag{4}$$

The parameters a, b can be chosen to model shower max $t_0 = (a-1)/b$ and the shower width $t_w = \sqrt{a-1}/b$. At energies below a PeV, the predicted pulse shapes are similar to those produced by Monte Carlo simulation[4].

Scaling to different media. By scaling shower profiles to known properties of the medium, it is possible to take results derived for one medium (e.g. ice) and apply them to another (e.g. salt). The most obvious consideration is the change in the Cerenkov angle due to the different index of refraction.

The transverse shower profile scales by the Moliere radius, which in turn scales as $R_M \sim 1/(\rho dE/dX)$. The longitudinal profile is scaled by the radiation length $x \sim X/\rho$. The width of the Cerenkov cone also depends on the Cerenkov angle, which affects the viewing perspective. Normalization of the signal strength depends on the total track length, which scales as $1/(\rho dE/dX)$, but also on the excess charge and the degree to which the tracks are scattered. These quantities should be roughly independent of energy, but must still be determined once for each medium.

Acknowledgments

This work was supported in part by NASA grant NASA-NAG5-5390.

References

1. G.A. Askaryan, Zh. Eksp. Teor. Fiz. 41, 616 (1961); G.A. Askaryan, Soviet Physics JETP 14, 441 (1962).
2. E. Zas, F. Halzen & T. Stanev, Phys. Rev. D45, 362 (1992).
3. D. Saltzberg, et al., Phys. Rev. Lett. 86, 2802-2805 (2001); P. Gorham, et al. Phys.Rev. D72 (2005) 023002.
4. S. Hussain & D. Seckel, in preparation.
5. S. Razzaque, et al., Phys.Rev. D65, (2002) 103002; Phys.Rev. D69 (2004) 047101.
6. J. Alvarez-Muniz, E. Marques, R.A. Vazquez & E. Zas, Phys.Rev. D68 (2003) 043001.
7. S. Hussain & D. McKay, Phys.Rev. D70 (2004) 103003.
8. J. Alvarez-Muñiz & E. Zas, Phys. Lett. B411, 218-224 (1997).
9. J. Alvarez-Muñiz & E. Zas, Phys. Lett. B434, 396-406 (1998).
10. J. Alvarez-Muñiz, R.A. Vázquez & E. Zas, Phys.Rev. D62, 063001 (2000).
11. R.V. Buniy & J.P. Ralston, Phys. Rev. D65, 016003 (2002).
12. J.P. Ralston, S. Razzaque & P. Jain, astro-ph/0209455 (2002).
13. S. Klein, astro-ph/0412546 (2004).

SIGNAL PROCESSING FOR ACOUSTIC NEUTRINO DETECTION IN WATER, ICE AND SALT *

SEAN DANAHER

for the ACoRNE Collaboration†

School of Computing, Engineering and Information Sciences, Northumbria University, Newcastle upon Tyne, NE1 8ST, United Kingdom

The recovery of neutrino induced acoustic signals, whether it be in water, ice or salt is very challenging, as the amplitude of the signals are small, the event rate low, and there is the presence of a strong and not completely understood acoustic background. There is a need for new and sophisticated techniques, which have not traditionally been used in the area of Astroparticle Physics in order to separate the signal from this background. There exists however a mature area named Signal Processing which is extensively used in the fields of Electronic Engineering and Communications. This paper is an attempt to show how signal-processing techniques can be applied in acoustic neutrino detection.

1. Introduction

Signal processing techniques have a wide application in acoustic neutrino detection and indeed can be used in almost every part of the process, from initial simulation of the shower parameters to improved reconstruction accuracy of the vertex position and arrival direction of the neutrino. The techniques can for example be used in the following cases:

1. Accurate parameterisation of distributions with no amenable analytical form e.g. shower energy distribution profiles;
2. Speed up computation e.g. acoustic integrals by many orders of magnitude;
3. Design, understand, simulate hydrophones, microphones and other acoustic transducers (and amplifiers);
4. Design, understand and simulate filters both analogue and digital (tailored amplitude, phase response, minimise computing time);
5. Estimate and recreate noise and background spectra;
6. Design optimal filters e.g. matched filters;
7. Design optimal classification algorithms (separating neutrino pulses from background);

* This work is supported by PPARC and the MoD.

† http://www.shef.ac.uk/physics/research/pppa/research/acorne.htm

8. Improve reconstruction accuracy.

Due to space limitation it is not possible to cover all these topics in a single paper. Only the first four of these topics will therefore be discussed.

2. Parameterisation of Shower Energy Distribution Profiles

Parameterisations of the longitudinal and transverse shower energy deposition distributions have been made in the past [1,2,3] using conventional methods. We show a method of parameterisation here, which should be much faster than these conventional methods. Fluctuations will take place in both the lateral and longitudinal distribution from shower to shower. As this will directly affect both the shape and angular spread of the acoustic pulse, it is highly desirable that this effect be included in future simulations. Both accurate parameterisation and the inclusion of fluctuations can be achieved using singular value decomposition.

2.1. *Singular Value Decomposition*

Singular value decomposition (SVD) represents an efficient numerical technique for the analysis of multivariate data [4]. SVD is also an extremely effective technique for the reduction of white noise assuming the input data is over-determined. The technique is very similar to Principle Component Analysis used in the area of statistics. If the data is presented in the form of an observation matrix, **O**, with each row an experiment (e.g. a radial distribution) and each column the value for a particular shower, the SVD of **O** is defined as

$$\mathbf{O} = \mathbf{WLV} \tag{1}$$

where **W** and **V** are unitary matrices containing the column eigenvectors of $\mathbf{O}^T\mathbf{O}$ and row eigenvectors $\mathbf{OO}^T$ respectively. The **L** matrix contains the eigenvalues in decreasing order of significance on the leading diagonal.

In many cases it is useful to partition the matrices as follows

$$\mathbf{O} = (\mathbf{W}_1\mathbf{W}_2)\begin{pmatrix} \mathbf{L}_1 & \mathbf{0} \\ \mathbf{0} & \mathbf{L}_2 \end{pmatrix}\begin{pmatrix} \mathbf{V}_1 \\ \mathbf{V}_2 \end{pmatrix} \tag{2}$$

where $\mathbf{W}_1$ contains the first k columns of **W**, $\mathbf{L}_1$ the first k rows and columns of **L** and $\mathbf{V}_1$ the first k rows of **V**. The **O** matrix can then be decomposed into two terms

$$\mathbf{O} = \mathbf{O}_S + \mathbf{O}_N = \mathbf{W}_1\begin{pmatrix} \mathbf{L}_1 \\ \mathbf{0} \end{pmatrix}\mathbf{V}_1 + \mathbf{W}_2\begin{pmatrix} \mathbf{0} \\ \mathbf{L}_2 \end{pmatrix}\mathbf{V}_2 \tag{3}$$

where the subscript $_S$ denotes signal space and $_N$ noise space. As the value of k increases $\mathbf{O}_S$ becomes a closer and closer approximation to **O** and the values in $\mathbf{O}_N$ become smaller and smaller. It is this property that is used for dimensional reduction. It is found in this application that the number of dimensions can be reduced from 600 (one point every 0.5 mm from 0 to 300 mm) to 3 without any significant loss in signal. This is illustrated in Figure 1. The radial distributions have been generated using the hadronic energy produced with GEANT IV for energies between 1 and 10TeV. In this illustration the radial distributions are created by collapsing the energy deposition onto a plane normal to the shower axis. In general the shape of the distribution scales with energy but there are significant fluctuations from shower to shower. The three distributions shown have been chosen to cover the widest variation of shapes within the limited statistics (~100 showers). Graphs *a-c* show the best fit using a single vector. This is equivalent to assuming that all the profiles are scaled versions of the same shape. Graphs *d-f* and *g-i* show the best approximation using a linear combination of two and three vectors respectively. Whereas one vector is sufficient to match the first distribution, three vectors are needed to span the information space. As any valid distribution must be within this information space, a further advantage is that similar shower profiles can be obtained using an appropriate mixture of the three vectors allowing for very accurate and efficient modelling of shower shape variation.

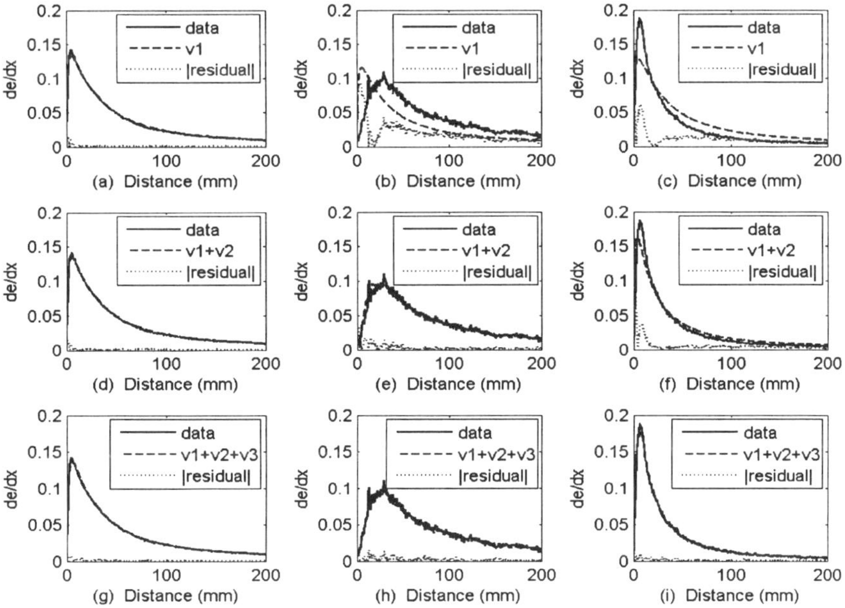

Figure 1 a-i. Comparison between actual radial distributions and fits using 1 (a-c), 2(d-f) and 3 (g-i) eigenvectors

3. Improving Computational Efficiency for Acoustic Integrals

The procedure for the determination of acoustic integrals from neutrino cascades is well established and follows Learned [5]. Analytically the overall pulse shape can be determined by integrating over the volume in which the energy is deposited is given by

$$P(t) = \oint \frac{E_0 \beta \varepsilon}{4\pi C_p r} \frac{d}{dt}\left(\delta(r/c - t)\right) dV \tag{4}$$

where $P(t)$ is the pressure wave created from the cascade, β is the bulk thermal expansion coefficient of seawater, C_p the specific heat capacity, E_0 the total energy deposited and ε is the relative energy deposition at each point in the volume. This can be calculated numerically using the following steps

1. Throw a number of MC points, typically around 10^7, according to the distribution of the neutrino cascade;
2. Assume the energy at each point in the cascade is deposited as a Gaussian distribution, with a width of at least an order of magnitude smaller than the shower cross section. This creates a known bipolar pulse;
3. Propagate the bipolar pulse to the observer using a linear approximation to acoustic attenuation in water as a function of frequency (the pulse remains bipolar but gets wider with distance);
4. Sum over all the points.

This procedure is accurate; however it is very numerically intensive. The critical steps are 3 and 4. If the assumption is made that the pulse be resolved on a time axis with 1024 points, steps 4 and 5 use about 5x1024 flops for each of the 10^7 MC points yielding a total flop count of ~ **5x10^{10}**.

If it were only necessary to do this calculation a few times, this would not be a major problem with modern fast computers, however as this procedure needs to be repeated for every angle, distance and shower distribution it constitutes a major bottleneck to efficient computation. Using Signal Processing techniques it is possible to improve this by many orders of magnitude *as steps 4 and 5 can be done once for the entire shower rather than individually for each point.*

3.1. *Fast Convolution*

One of the fundamental theorems of signal processing is the convolution theorem, which states *"Convolution in the time domain is multiplication in the frequency domain and vice versa."* [6] Multiplication is computationally much more efficient than convolution and the existence of Fast Fourier Transform

(FFT) algorithms allows the conversion from the time domain to the frequency domain and back again with little computational overhead.

By noting that we can change the order of integration in Equation 4, we get

$$P(t) = \frac{E_0\beta}{4\pi C_p}\frac{d}{dt}\oint \frac{\varepsilon}{r}\delta(r/c-t)dV = \frac{E_0\beta}{4\pi C_p}\frac{d}{dt}E_{xyz}(t) \tag{5}$$

where $P(t)$ is the overall pressure, $E_{xyz}(t)$ is the integrated energy deposition over each point in the volume, projected to an arbitrary point in space. By taking the Fourier transform we have

$$P(\omega) = \int_{-\infty}^{\infty}\frac{E_0\beta}{4\pi C_p}\frac{d}{dt}E_{xyz}e^{i\omega t}dt = \frac{E_0\beta}{4\pi C_p}i\omega\int_{-\infty}^{\infty}E_{xyz}e^{i\omega t}dt = \frac{E_0\beta}{4\pi C_p}i\omega E_{xyz}(\omega) \tag{6}$$

using the standard Fourier transform theorem, that taking the derivative in the time domain is identical to multiplying by $i\omega$ in the frequency domain [6]. The procedure is as follows. The distribution $E_{xyz}(t)$ can be determined using a histogram technique requiring ~ $\mathbf{10^3}$ flops. The histogram is scaled by $1/d$: $\mathbf{10^3}$ flops (*only needed in near field*). The distribution $E_{xyz}(t)$ is now converted to the frequency domain using a FFT (requiring $n\log(n)$ steps ~ $\mathbf{10^4}$ flops). The Pressure $P(t)$ can now be evaluated using:

$$P[t] = \sum_{n=-512}^{n=511} E_{xyz}(\omega)w(\omega)D(\omega)A_w(\omega)e^{i\omega nT} \tag{7}$$

where $w(\omega)$ is an optional windowing function (Blackmann) to reduce high frequency noise, $D(\omega)$ is the derivative term ($i\omega$) and $A_w(\omega)$ is the water attenuation. The summation with the complex exponential term produces the inverse Fourier transform. As the multiplication terms can be pre-stored this process takes $\mathbf{10^3}$ flops for the multiplication and $\mathbf{10^4}$ flops for the IFFT. Hence the total computation takes of the order of $\mathbf{2x10^4}$ flops, which is better than 6 orders of magnitude faster than the conventional technique.

A further advantage is that the correct water attenuation is used rather than the linear approximation of the older technique. Figure 2 shows a simulation of the pulse shape at right angles to the shower for various distances. In the near field the pulse is asymmetric with the positive part (Max P) being larger than the negative part (Min P). We note the excellent agreement (worst case 2% and 0.5% typical) between the two methods, even at near field where the classic method has a tendency to be particularly noisy.

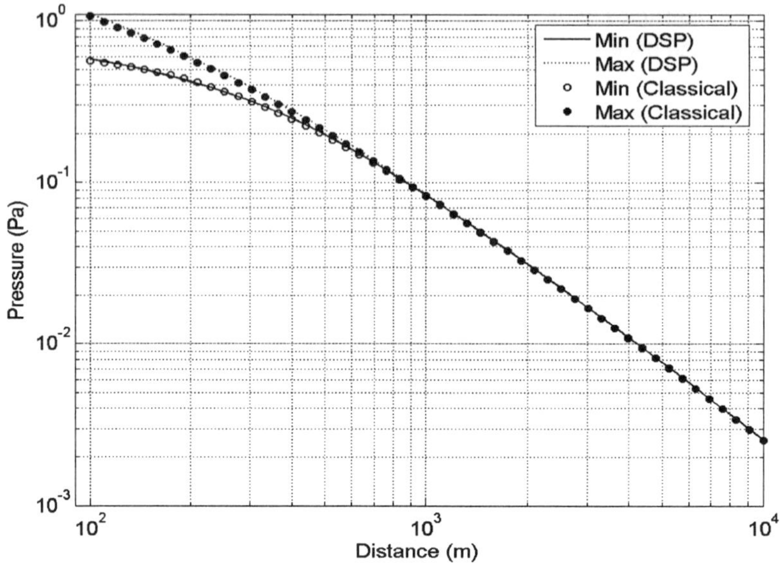

Figure 2.Comparison of the Fast Convolution method (DSP) and classical method.

4. Acoustic Transducer and Electronic Modelling

The understanding of both the behaviour of acoustic transducers and the associated signal conditioning/processing chain (e.g. amplifiers, filters) is fundamental to the recovery of our acoustic signals. Classically these have been modelled using transfer functions (Laplace transform techniques), however in recent years the trend is moving towards using state space analysis, which offers considerable advantages.

4.1. *State Space Analysis*

State space analysis is a powerful technique for the analysis of linear systems, such as amplifiers, filters, rockets and many types of transducers [7]. It is a time domain technique based on matrix representation of systems. It has the advantage of simplicity in combining models from different areas such as electro-mechanical systems, is very computationally robust, and can deal with very high order systems and with multiple inputs and outputs. Once the state space representation is known, standard algorithms can be used to determine the response of the system to arbitrary inputs and initial conditions. The states are degrees of freedom of the system; for example capacitor voltages and inductor currents for electrical systems and positions and velocities of masses for mechanical systems. The state space model of a system is defined as

$$\begin{aligned}\dot{\mathbf{X}} &= \mathbf{AX}+\mathbf{BU}\\ \mathbf{Y} &= \mathbf{CX}+\mathbf{DU}\end{aligned} \tag{8}$$

where $\mathbf{X}$ and $\dot{\mathbf{X}}$ are column vectors containing the states and their time derivatives respectively, $\mathbf{U}$ and $\mathbf{Y}$ are column vectors containing the inputs and outputs and **A**, **B**, **C** and **D** are matrices containing physical parameters such as capacitance and mass.

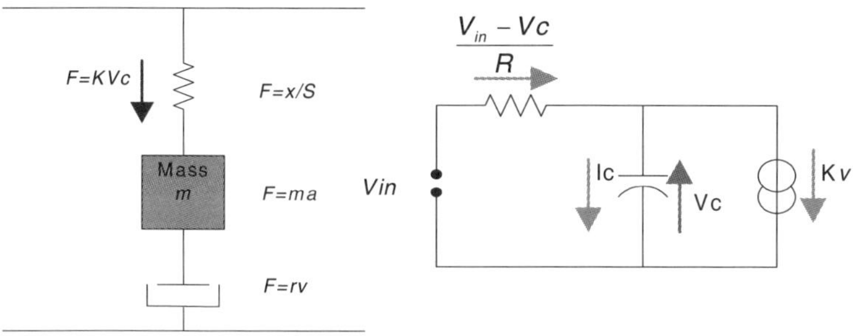

Figure 3 Electrical and Mechanical model of a hydrophone system

4.2. *A Simple Hydrophone Model*

Figure 3 shows a simple model of a hydrophone [8] consisting of a damped mass connected to a spring and a capacitance in parallel with a current generator connected to an input/output via an external series resistor R. This model can be used in both transmitter and detector mode. The three states are taken to be: the capacitor voltage and the velocity and position of the mass; the inputs: the applied voltage, V_{in}, and external force F; and the outputs: velocity of the mass (as the amplitude of sound is proportional to velocity rather than position) and current through the series resistor. This leads to eqn. 9*a*) for the mechanical system and 9*b*) for the electrical system.

$$\begin{aligned}&Kx_1 + u_2 = m\dot{x}_2 + rx_2 + \frac{1}{S}x_3\\ &\dot{x}_2 = \frac{K}{m}x_1 - \frac{r}{m}x_2 - \frac{1}{Sm}x_3 + \frac{1}{m}u_2\\ &\dot{x}_3 = x_2\end{aligned} \qquad 9\,a)$$

$$\begin{aligned}&C\dot{x}_1 = \frac{u_1 - x_1}{R} - Kx_2\\ &\dot{x}_1 = -\frac{1}{RC}x_1 - \frac{K}{C}x_2\end{aligned} \qquad 9\,b)$$

where K is the piezo-electric constant, m mass, r mechanical resistance, S compliance, C capacitance and R the electrical resistance.

These can easily be converted to a state space model yielding:

$$\dot{\mathbf{X}} = \begin{pmatrix} -\frac{1}{RC} & -\frac{K}{C} & 0 \\ \frac{K}{m} & -\frac{r}{m} & -\frac{1}{Sm} \\ 0 & 1 & 0 \end{pmatrix} \mathbf{X} + \begin{pmatrix} \frac{1}{RC} & 0 \\ 0 & \frac{1}{m} \\ 0 & 0 \end{pmatrix} \mathbf{U}, \mathbf{Y} = \begin{pmatrix} -\frac{1}{R} & 0 & 0 \\ 0 & 1 & 0 \end{pmatrix} \mathbf{X} + \begin{pmatrix} \frac{1}{R} & 0 \\ 0 & 0 \end{pmatrix} \mathbf{U} \tag{10}$$

This model can then be used either independently or in conjunction with a MC simulation, such as described in section 3. In figure 4 there is shown the resultant hydrophone signal caused by firing a pulsed laser into a tank of water[9] and the result of a simulation showing good qualitative agreement. To first approximation the pulse shape is the derivative of E_{xyz} convolved with the impulse response of the hydrophone. The large pulse at ~40cm is where the laser enters the tank; the smaller at~30cm where the laser is closest to the hydrophone.

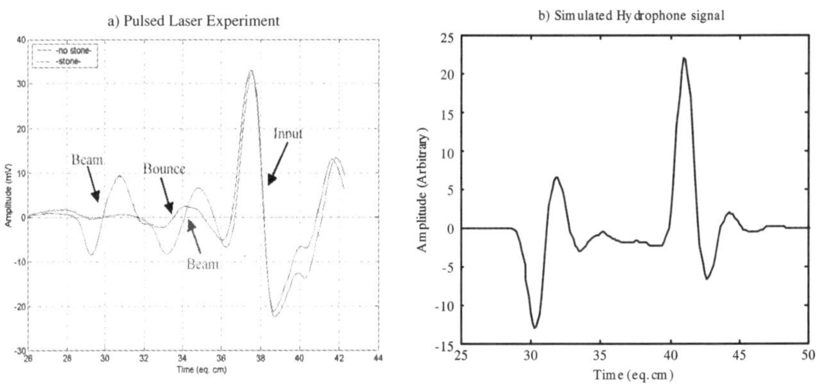

Figure 4 showing a result of the Erlangen Pulsed Laser Experiment (a) and simulation using a MC and state space simulation (b.)

5. The behaviour of Amplifiers and Filters

At acoustic frequencies very high performance amplifiers are cheaply available, causing minimal distortion to acoustic pulses. With filters however it is important to understand both the amplitude and phase response. Indeed the phase response is often most important as it is determined by the time delay as a function of frequency and can cause severe distortion to pulses. As a very simple guide any sharp transition in the frequency domain will cause severe ringing in the time domain. Both the synthesis and analysis of filters are very mature disciplines [10]. It is traditional to start with low pass filters with a cut off of 1 rad s^{-1} and then transform the filter design to the desired type e.g. band pass or high pass and cut-off frequency. One of the most popular filter designs is the Butterworth. A low pass Butterworth filter with a cut off angular velocity of 1 rad s^{-1} has the response

$$|H(\omega)| = \frac{1}{\sqrt{1+\omega^{2n}}} \tag{11}$$

where n is the order of the filter. When $\omega \gg 1$ the fall off approximates to $1/\omega^n$, which is 6 dB per octave or 20dB per decade. At acoustic frequencies three electronic components are used: resistors, capacitors and op-amps. Filters with order greater than two also require feedback, which must be negative or the output will become unstable. For reasons beyond the scope of this paper this stability requirement dictates that there is a 90^0 phase shift per order between low and high frequencies for such filters [10].

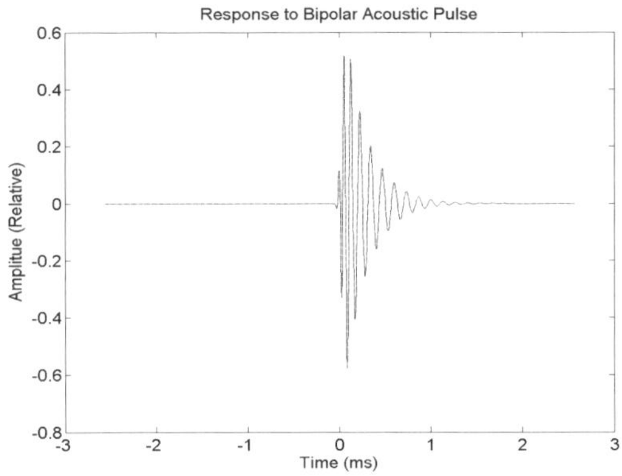

Figure 5. Response of the high pass filter to the anticipated bipolar pulse

The sea is dominated by low frequency noise [11] and high order high pass filters are commonly used for the detection of acoustic signals in the 10-100kHz range. This makes perfect sense if the expected signals are sinusoidal in nature, however such filters will cause considerable distortion to pulses. For example some commercial hydrophone systems have a 7.5kHz Butterworth high pass filter with a roll off of 100dB per octave. The response of such a filter to the anticipated acoustic pulse from a neutrino-induced shower can be determined by signal processing techniques. In Figure 5 the ringing caused by such a filter is evident. Because negative feedback is intrinsic to the operation of analogue filters, positive feedback is needed to reverse the process. As positive feedback is intrinsically unstable, perfect inversion is not possible. It is however possible to remove most of the ringing by recording the signal and sending it again through the filter, but reversed in time. This will create a zero phase delay for all frequencies. It should be noted however that this is non-causal as it uses future knowledge of the signal. It is possible however to design digital filters with no phase distortion. It is strongly recommended that digital techniques be used in this area [12].

5.1. *Digital Filters*

Whereas analogue filters work on continuous signals, digital filters work on sampled signals. Sampling is normally at fixed frequency called the sampling frequency, f_s, and with a precision dictated by the number of bits on the analogue to digital converter (ADC). The sampling frequency needs to be at least twice as high as the highest frequency in the signal, otherwise aliasing will occur. The precision of the ADC will depend on the desired signal to noise ratio and is approximately 6dB per bit. A 12-bit ADC will therefore have a signal to noise ratio of approximately 72dB. Analogue filters are normally modelled using: differential equations, the Fourier and Laplace transforms; whereas digital filters are modelled using: difference equations, the FFT and the z transform. The equation for a causal (the output depends only on past values but not on future values) digital filter is given by

$$\sum_{k=0}^{N} a_k y[n-k] = \sum_{k=0}^{M} b_k u[n-k] \qquad (12)$$

where u is the input sampled signal, y the output signal and a and b filter coefficients.

Digital filters may be subdivided into two classes, finite impulse response filters (FIR), and infinite impulse response filters (IIR). The finite and infinite refer to time not amplitude. A FIR filter will have an impulse response (the response to a sequence starting with 1 and followed by an infinite number of zeros) which will last for a finite duration. Consider the following filters.

$$y[n] = \frac{1}{2}x[n] + \frac{1}{2}x[n-1] \quad (13a), \qquad y[n] - y[n-1] = x[n] \quad (13b)$$

Equation 13*a*) is the simplest form of FIR filter, a moving average filter with constant coefficients, with an impulse response of ½, ½ and behaves as a very crude low pass filter. Equation 13*b*) is an IIR filter with an impulse response of 1,1,1…. It is a perfect integrator. FIR filters have two major advantages: they are unconditionally stable as they do not use feedback and it can be shown that provided the filter is symmetric or anti-symmetric they have linear phase response, yielding a constant group delay and zero phase distortion. Their big disadvantage is that they can often be very long with hundreds of coefficients. In complete contrast the advantage of IIR filters is that they tend to be extremely compact, however they need to be designed with care as they can become unstable. They are also not recommended if phase distortion is of importance as they are very difficult to design with constant group delay. A host of techniques are used to design such filters, however because the z transform is fundamental to the understanding of digital filters and one of the most common problems in acoustic signal processing is the removal of a sinusoid, the design of a notch filter will be discussed.

5.2. *The Z Transform and Notch Filter Design*

The z transform is defined as

$$X(z) = \sum_{n=-\infty}^{n=\infty} x[n]z^{-n} \tag{14}$$

where $x[n]$ is the signal and z is a complex number. If $z=e^{i\theta}$, then the z transform is identical to the discrete Fourier transform. The transfer function $H(z)$ for a digital filter is the z transform of the output divided by the z transform of the input. This is often represented diagrammatically on the z-plane. Every point on the z-plane will have its associated complex gain, normally however only the poles and zeros of the transfer function are plotted.

The value of $H(z)$ around the unit circle gives the steady state response to sinusoidal signals with 0 corresponding to dc and π corresponding to $f_s/2$. Points inside the unit circle correspond to decaying sinusoids, whereas those outside correspond to growing sinusoids. For a filter to be stable the poles need to be inside the unit circle.

A notch filter therefore will have zeros on the unit circle with angles of $\pm 2\pi f_c/f_s$. Poles are placed inside the unit circle with the same angle as the zeros. The width of the notch is determined by the proximity of the poles to the unit circle. If for example f_s=1kHz and f_c=100Hz a suitable transfer function is given in 15*a* and the 3dB points will be at $f_c \pm 0.05 f_s/2\pi$. The equivalent time domain representation is given in 15b). The frequency response of this filter is shown in Figure 6 left and the response to a signal buried in noise right.

$$H(z) = \frac{(z-e^{i\frac{\pi}{5}})(z-e^{-i\frac{\pi}{5}})}{(z-0.95e^{i\frac{\pi}{5}})(z-0.95e^{-i\frac{\pi}{5}})} \quad (15a) \qquad \begin{aligned} y[n] &= x[n] - 1.62x[n-1] + x[n-2] \\ &+ 1.537y[n-1] - 0.902y[n-2] \end{aligned} \quad (15b)$$

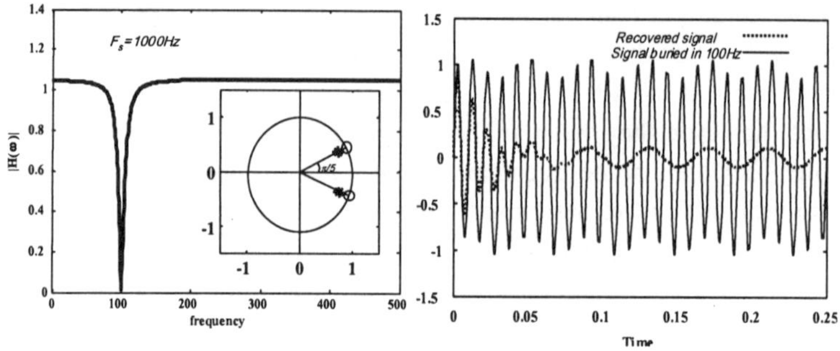

Figure 6. Left the frequency response of a notch filter. (Inset the pole and zero positions). Right Time domain response to a signal buried in noise.

6. Conclusions

In this paper a number of signal processing techniques in the area of acoustic neutrino detection have been discussed. It has been shown that these techniques offer significant advantages whether in terms of ease of use, numerical accuracy, and computational efficiency. It is predicted that the use of these techniques in this field will become more widespread over the next few years.

References

1. J. Alvarez-Muniz, E. Zas, Phys. Lett. **434** 396-406 (1998).
2. N. G. Lehtinen et al., *Astropart. Phys.* **17** 279-292 (astro-ph/0104033) (2002).
3. J. Perkin, presented at ARENA workshop 7-9th May (2005).
4. S. Danaher, D.J. Fegan, J. Hagan, Astropart. Phys. **1,** 357 (1993).
5. J. G. Learned Phys. Rev. D, v19, No. 11 (1979).
6. J D. Martin "*Signals & Processes (A foundation course)*". Pitman Publishing (1991).
7. W.S. Levine, "*Control Systems Fundamentals*", CRC Press (2000) ISBN 0273 30325 69.
8. J. P. Bentley "*Principles of Measurement systems*". 3rd Edition. Longman ISBN 0 470 23445 8.(1995).
9. V. Neiss PhD thesis, Universite de la Mediterranee, Aix-Marseille II, U.F.R de Physique (2005).
10. L.P. Huelsman *"Active and Passive Analog Filter Design"*. McGraw-Hill. ISBN 0 07 030860 (1993).
11. R.J.M Urick "Principles of Underwater Sound" McGraw-Hill (1975) ISBN 0-070-66087 5.
12. E. G. Ifeachor, B. W. Jervis, "*Digital Signal Processing (A Practical Approach)*". Addison -Wesley. 1993. ISBN 0-201-54413 X.

AN ANALYSIS APPROACH TO ACOUSTIC DETECTION OF EXTENSIVE ATMOSPHERIC SHOWERS

DMITRY ZABOROV

Institute for Theoretical and Experimental Physics (ITEP)
Bolshaya Cheremushkinskaya, 25, 117218 Moscow, Russia
E-mail: Dmitry.Zaborov@itep.ru

An experiment related to the search of acoustic signal from air showers has been made at Baikal lake. This report briefly describes the experiment setup, data analysis procedure and obtained results.

1. Introduction

It is established that a small fraction of the energy of interacting particles converts to sound waves[1]. In particular acoustic radiation is produced by extensive atmospheric showers (EAS). Below we describe a search for an acoustic signal produced by EAS in lake Baikal covered by ice.

2. Experiment

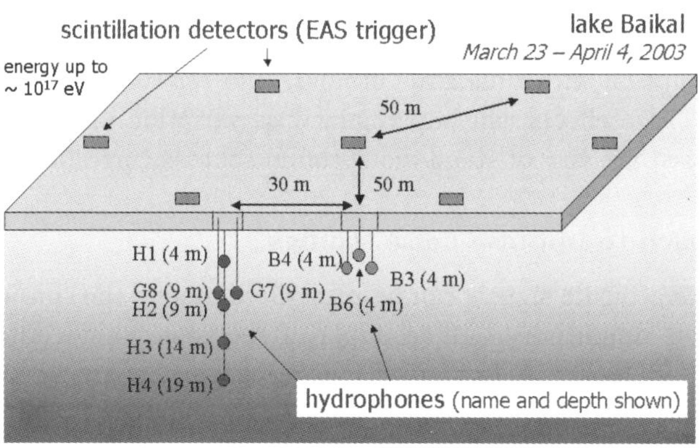

Figure 1. Schematic view of the experimental setup. Depth is measured from bottom surface of the ice. The ice was 70 cm thick.

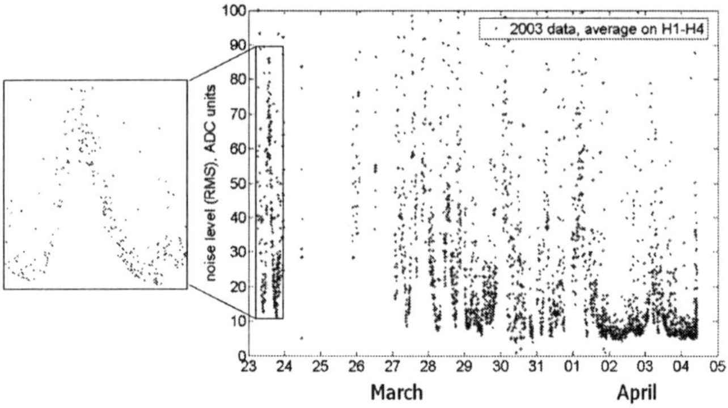

Figure 2. Acoustic noise level. 1 ADC unit roughly corresponds to 1.3 mPa.

The experiment is devoted to underice noise studies, search for acoustic signal from EAS, noise rejection techniques, etc. The experimental setup (see Fig. 1) consists of 2 parts. An array of scintillation detectors provides a trigger and enables EAS reconstruction (position, direction, energy and age). Acoustic part of the setup includes 9 hydrophones[2] deployed under ice. Each acoustic channel is read out individually. Collected statistics corresponds to 1 week live time. 1260 complete events (reconstructed shower + low noise acoustic data) have been recorded. A few showers exceeded $10^{17} eV$ in energy and fell within 20 metres from the hydrophones.

3. Continuous noise

Most of the time acoustic oscillograms represent apparently white noise. Evolution of noise level with time (see Fig. 2) demonstrates two features: smooth evolution with time scale of hours and relatively rare sudden rises of noise. These effects can be explained as variation of continuous noise and increased activity of stand-alone sound sources correspondingly.

4. Approach to impulse noise studies

Stand-alone acoustic signals can be recognized among the continuous noise due to their amplitude. An example of recorded sound data is shown in Fig. 3 (left plot). A formalized pulse recognition procedure has been adopted for this study. It consists of the following steps (see Fig. 3).
- Data exceeding in amplitude "low threshold" are chosen on oscillogram.
- Excess data points separated by less than Δt_{group} form a group (= pulse).
- If amplitude of the pulse is below "high threshold" the pulse is discarded.

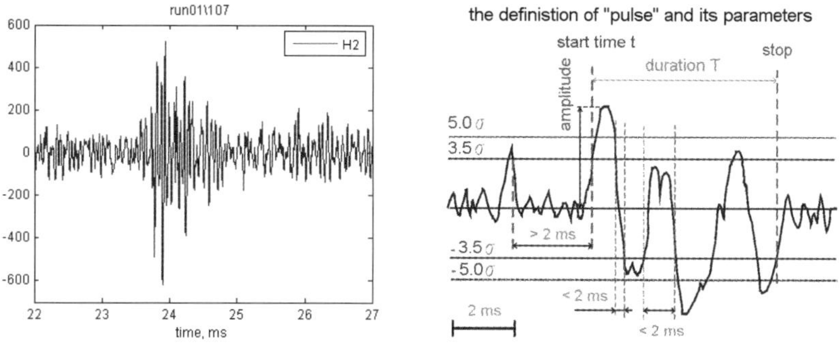

Figure 3. An example of recorded sound pulse and scheme of the pulse recognition procedure.

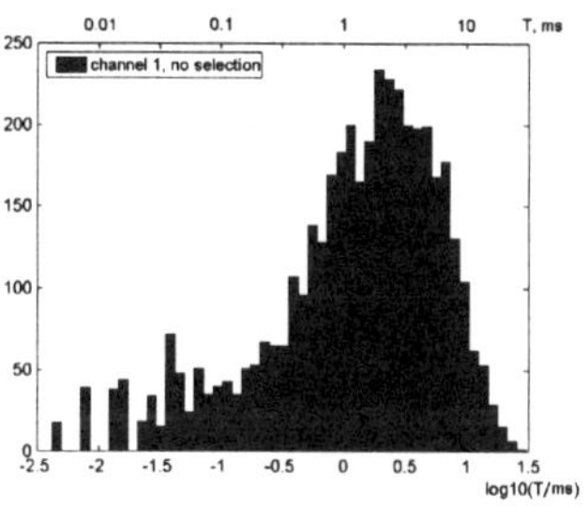

Figure 4. Duration of sound pulses.

Each acoustic channel is treated separately. Both thresholds scale with noise level which is calculated for each acoustic record. Polarity is irrelevant. Δt_{group} is chosen as a compromise between two requirements: a. long pulses should not be split; b. different pulses should not be merged. The compromise value was found $\Delta t_{group} = 2$ ms. Pulse duration is defined as time difference between first and last points of the pulse (see Fig. 4).

5. Direction reconstruction

A set of pulses recorded by several hydrophones is considered as an acoustic signal if it satisfies causality conditions. The reconstruction is performed using the timing information from the hydrophones. It was found that most of the recorded signals are well fitted as plane waves. Figure 5 shows observed delays between the hydrophones G7, H2 and G8 which are situated at equal depth (9 m) at the edges of a triangle with sides of 1.68 (G7-H2), 1.67 (H2-G8) and 1.48 (G8-G7) metres. The "ellipse" is formed by horizontally propagating sounds. Similar distribution is observed with another triangle of hydrophones (B3-B4-B6). Position on the ellipse gives

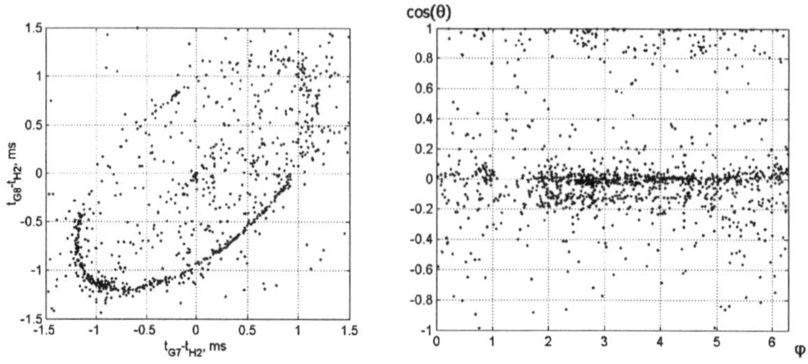

Figure 5. Delays between hydrophones H2, G7 and G8 (left) and reconstructed direction of the sound (right).

azimuth. Vertical antenna gives reliable θ information (see right plot on Fig. 5). Three clusters of events are distinguished: A. sounds coming from top ($cos(\theta) \approx 1$), mainly from the ice hole; B. sounds propagating horizontally ($cos(\theta) \approx 0$) probably originate in ice far from the setup; C. sounds, propagating from bottom to top at about 8 degree from horizon, Most likely these sounds are generated in ice and then refracted by temperature gradient, "trapped" in underice acoustic channel. It is important to note that great majority of recorded sounds are somehow related to the lake surface.

6. Position reconstruction

A spherical wave fit has been adopted for position reconstruction in this study. The position of sound source is found by maximizing a functional. Normally 6 hydrophones (H1-H4,G7,G8) were used for the fit. The method suits best for small distances compared to the size of the setup. (That's why sound deflection effects can be safely neglected.) Reconstruction of the position of an artificial sound source (pinger), situated at 75 metres from the hydrophone antenna, demonstrates good performance of the fit procedure. Direction to the pinger is reconstructed well within 1 degree. While error in distance reconstruction was found to be a few metres due to relatively large distance.

7. Search for EAS generated sound

Shower position is reconstructed from scintillator data with a few metres accuracy that allows to calculate the time and direction of the expected sound and compare them to the measurements. Figure 6 illustrates the

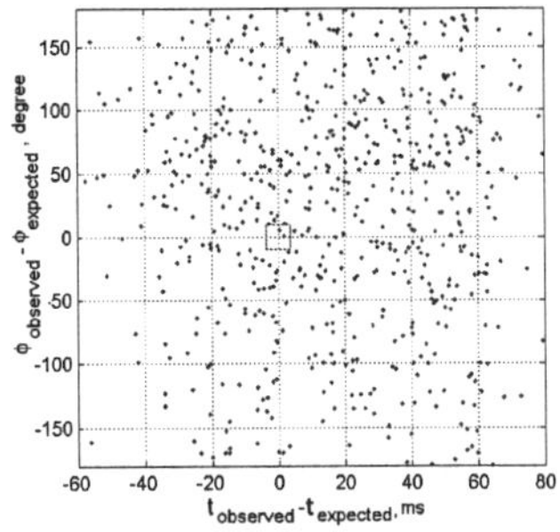

Figure 6. Search for shower induced sound pulses using time and zenith angle.

comparison. An excess around (0,0) on that plot would indicate the presence of a EAS generated signal. No excess above a randomly distributed background was observed. The most energetic events have been examined with special care. No clear signal have been identified.

8. Conclusion

A method has been developed for a search for an acoustic signal and tested with real data. Underice acoustic noise in lake Baikal has been studied. The present straightforward analysis of the experimental data does not reveal any acoustic signal produced by air showers. Due to the low signal level and generally irreducible acoustic noise more sophisticated methods of signal processing[3] must be applied in the future.

Acknowledgments

The author wishes to thank the Baikal collaboration for organizing the experiment and a cooperation. Special thanks go to Prof. N.Budnev for discussions on the underice acoustic effects. The author would like to acknowledge the important contribution of Vladimir Lyashuk who carried out the acoustic experiment.

References

1. G. Askaryan, Hydrodynamical emission of tracks of ionising particles in stable liquids. *Atomic Energy 3*, 152 (1957).
2. V. Lyashuk, ITEP investigations of acoustic phenomena from high energy particles, *these proceedings*.
3. S. Danaher, Acoustic signal processing (a tutorial), *these proceedings*.

DEVELOPMENT OF ACOUSTIC SENSORS FOR THE ANTARES EXPERIMENT

C. NAUMANN, G. ANTON, K. GRAF, J. HÖSSL, A. KAPPES, T. KARG, U. KATZ, R. LAHMANN AND K. SALOMON

Physikalisches Institut,
Friedrich-Alexander Universität Erlangen-Nürnberg,
Erwin-Rommel-Straße 1,
91058 Erlangen, Germany
E-mail: christopher.naumann@physik.uni-erlangen.de

In order to study the possibility of acoustic detection of ultra-high energy neutrinos in water, our group is planning to deploy and operate an array of acoustic sensors using the ANTARES Neutrino telescope in the Mediterranean Sea. Therefore, acoustic sensor hardware has to be developed which is both capable of operation under the hostile conditions of the deep sea and at the same time provides the high sensitivity necessary to detect the weak pressure signals resulting from the neutrino's interaction in water. In this paper, two different approaches to building such sensors, as well as performance studies in the laboratory and *in situ*, are presented.

1. Introduction

At ultra-high energies, acoustic detection of cosmic neutrinos is a very promising approach, complementary to optical detection in ice and sea water. Due to the much larger effective volumes, acoustic detectors in water, ice or salt could reach far beyond the energies accessible to current detectors[3].

Our group plans to integrate a number of customised detector units (storeys) for acoustic detection into the ANTARES detector, in which the optical sensor elements are replaced by ultrasound sensors. The aim is to perform long term studies of the acoustic background in the deep sea and to investigate the technical feasibility of acoustic particle detection in water.

2. Sensor Development and Integration

The sensors for a deep sea acoustic neutrino detector must be able to withstand the pressure of several hundred bars and at the same time be sensitive to pressure variations of the order of few mPa. For the integration into the ANTARES detector, it is also necessary to keep power consumption and changes to the overall detector design as small as possible. To meet these requirements, two different concepts have been proposed: An array of individual hydrophones, replacing the optical modules (OMs) of a "standard" storey, and modified glass spheres used as acoustic modules (Fig. 1). In both cases the data acquisition electronics is housed inside the ANTARES electronics cylinders.

Figure 1. Schematic design of an acoustic storey using hydrophones (left) and acoustic modules (right.)

The first concept uses an array of individual hydrophones, externally mounted on the support structure of an ANTARES storey with the OMs removed. As sensors, a mix of commercial and self-made hydrophones is forseen. Commercial hydrophones, both used for civilian and military naval purposes, are not specifically designed for the purpose of detecting neutrino signals with respect to their frequency response, and their sensitivity is usually insufficient for neutrino detection. On the other hand, they have the advantage of already providing the pressure- and water-resistance needed for deep sea operation. Therefore, a number of commercial hydrophones, as well as customised hydrophones bought from the Institute for Theoretical and Experimental Physics, Moscow (ITEP), will be used.

In addition to the commercial hydrophones , a range of different self-made hydrophones with piezoelectric tubes and/or discs as active elements and custom-designed internal pre-amplifiers will be employed. As a protection against the sea water, both the piezo element and the pre-amplifier are cast in polyurethane, chosen to match the surrounding water acoustically. Both the sensitivity and the frequency characteristics depend on the size and shape of the piezo elements, as well as the designs of the pre-amplifier, making it possible to build either a broad-band sensor for background studies or one with its sensitivity optimised in a narrow frequency band around the expected signal (about 5-50 kHz). The choice of geometry of the piezo element also allows for different directional characteristics, varying from narrowly beamed to nearly 4π solid angle.

As a complementary approach, the spheres of the optical modules themselves are not removed but instead of the usual photomultiplier tubes fitted with acoustic sensor elements. This makes the development of pressure- and water-resistant devices unnecessary as all equipment is situated either inside the spheres or the electronics cylinders of the respective storeys.

As active sensor devices piezoelectric discs are used, several of which are attached to the inside of each 17" sphere, using polyurethane or epoxy resin as glue to ensure good acoustic coupling to the glass. As in the case of the hydrophones, each piezo sensor is fitted with an appropriate pre-amplifier directly mounted on the sensor to minimise noise pick-up. The influence of the glass sphere on the signal shape and the overall sensitivity has yet to be studied in detail.

3. Sensitivity Studies and Sensor Calibration

To quantitatively understand data from any acoustic sensor, its frequency dependence and directionality must be known. Most of these studies can be done using a water tank in a laboratory.

As a sound source, a calibrated transducer is used, which is driven by short Gaussian voltage pulses from a signal generator. With a digital oscilloscope, the signal response of the sensor is measured and compared to the signal sent. From the Fourier transform of this transfer function, a relative calibration is calculated, which only depends on the transducer characteristics. Using the transducer calibration values provided by the manufacturer, an absolute sensitivity can be obtained.

Examples of a sensitivity spectrum for a commercial hydrophone and an acoustic module prototype are given in Fig. 2. Below the first piezo resonance, the sensitivity is nearly flat, giving a "plateau" at $-120 dB\, re\ 1V/\mu Pa$ for the sphere. This corresponds to a signal of 1 mV per mPa, which should be sufficient for the expected pressure signals for neutrino energies $\geq 10^{18} eV$ at a distance of 400 m. Both at low and high frequencies, the sensitivity is cut off by the pre-amplifier to reduce the noise influence outside the signal region and to avoid aliasing problems during digitisation. The poorer sensitivity of the commercial hydrophone can be explained by the lower sensitivity of the piezo element as well as the lower gain of its pre-amplifier.

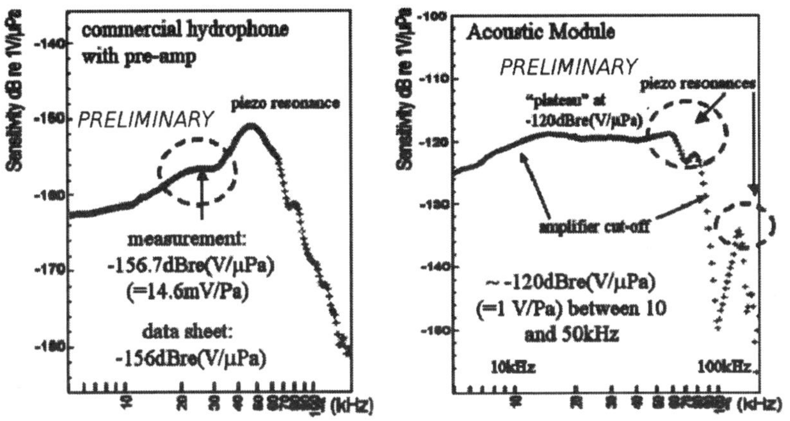

Figure 2. Frequency response of a commercial hydrophone (High Tech, Inc., left) and an acoustic module prototype (right), both equipped with a pre-amplifier.

4. First *in situ* Measurements

To test our sensor prototypes *in situ* and at the same time collect first environmental data from the ANTARES site, an autonomous detector system (AMADEUS) was built from an ANTARES electronics cylinder fitted with piezo sensors and data acquisition hardware. Five piezo discs with pre-amplifiers were glued to the inside of the titanium cylinder and read out via an autonomous PC board operated with batteries. The whole setup was placed at the base of the ANTARES test line "Line Zero" and operated for several weeks at a depth of 2400 m, yielding about 15 GByte of data. A full analysis of the data is still in progress but a first extraction of the environmental noise spectrum (shown in Fig. 3) is in good agreement with the expectations (derived form [2]).

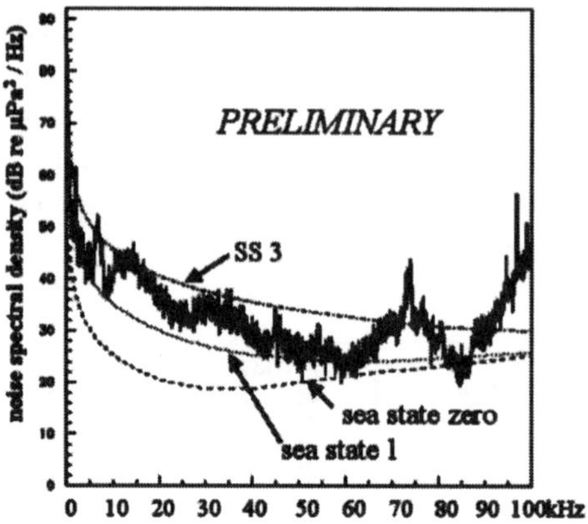

Figure 3. Preliminary noise spectrum at the ANTARES site, measured with the AMADEUS detector, compared with predictions for sea state zero, one and three. The high-frequency excess can be explained by non-acoustic sources inside the container.

5. Conclusions

For the installation of acoustic sensors into the ANTARES detector, two different types of acoustic detectors are planned, for both of which working prototypes have been built and operated successfully. Data taken in the laboratory and by an autonomous system in the deep sea show that the sensitivity needed for the acoustic detection of UHE neutrinos can be achieved using custom-built piezoelectric sensors.

Acknowledgements

This work is supported by the German BMBF Grant No. 05 CN2WE1/2.

References

1. G.A. Askariyan, *Atomnaya Energiya* **3**, 152 (1957).
2. R.J. Urick, *Ambient Noise in the Sea*, Peninsula Publishing (1981).
3. T. Karg et al., *Design considerations and sensitivity estimates for an acoustic neutrino detector* in these proceedings.

MEASUREMENTS AND SIMULATION STUDIES OF PIEZOCERAMICS FOR ACOUSTIC PARTICLE DETECTION*

K. SALOMON, G. ANTON, K. GRAF, J. HÖSSL, A. KAPPES, T. KARG, U. KATZ, R. LAHMANN AND C. NAUMANN

University of Erlangen
Erwin-Rommel-Str. 1,
91058 Erlangen, Germany
E-mail: karsten_salomon@web.de

Calibration sources are an indispensable tool for all detectors. In acoustic particle detection the goal of a calibration source is to mimic neutrino signatures as expected from hadronic cascades. A simple and promising method for the emulation of neutrino signals are piezo ceramics. We will present results of measruements and simulations on these piezo ceramics.

1. The Piezoelectric Effect and Signal Propagation in Water

The active element of a transducer is a piezoelectric material. When an electric field is applied to the material, the polarized molecules are distorted in the electric field. This distortion causes the material to change its dimensions. This phenomenon is known as electrostriction or inverse piezoelectric effect. In addition, a permanently-polarized material such as lead zirconate titanate (PZT) produces an electric field when the material changes dimensions as a result of an imposed mechanical force. This phenomenon is known as the piezoelectric effect. A piezoelectric material can be modeled by connecting Hooks law for anisotropic materials to the Gauss law for electrical displacement. This is described by tensor equations:

$$\partial_k(c_{iklm}S_{lm} + e_{lik}E_l) = \rho\ddot{u}_i \; ; \quad \partial_i(e_{ilm}S_{lm} + \epsilon_{il}E_l) = 0 \qquad (1)$$

with c_{iklm}, S_{lm}, e_{ikl}, E_l, ϵ_{il}, ρ and u_i representing the elasticity tensor, the strain tensor, the piezoelectric tensor, the electric field, the permitivity

*This work is supported by the German BMBF Grant No. 05 CN2WE1/2.

tensor, the density and the displacement[1].

With the displacement response of a piezo due to an applied voltage (described in section 2) one can study how the motion of a piezo transforms into a pressure signal. For wavelengths larger than the dimension of the transmitter the resulting pressure is given by[2]

$$p(\mathbf{r},t) = -\rho_0 \frac{\partial_t^2 V(|\mathbf{r}| - t/c)}{4\pi r},$$

where $\rho_0, r, t, V \approx Adx$ and c are the density of the medium, the location, the time, the volume of the piezo, the speed of sound, A is the area of the face of the piezo and dx the displacement at the centre of the face of the piezo, respectively. Thus the pressure pulse sent by a transducer depends on the second derivative in time of the displacement.

2. Mechanical Properties

We have studied the mechanical behaviour of the sensors via direct measurement of the displacement of the surface of a piezo when a voltage is applied with a fibre coupled interferometer[3,4] as shown in Fig. 1. Laser light

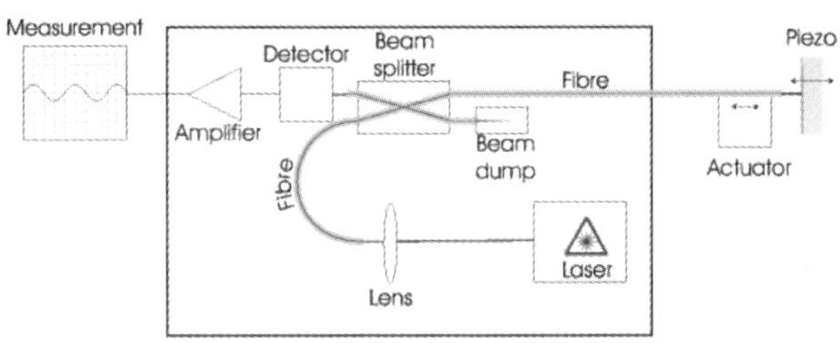

Figure 1. Schematic view of the interferometer setup

(λ=780 nm) is coupled into an optical fibre. One light path reaches the end of the fibre after a 2x2 beam splitter. The fibre exit is prepared to behave like a mirror. This results in both reflection of the light back into the fibre (about 4% for glass to air) and transmission. The face of the piezo is then positioned some micrometers away from the fibre exit. Multiple reflections between the fibre ending and the piezo[a] lead to an interference. At the beam splitter light is guided to a photo diode, where the intensity of this light is measured after amplification. To calibrate the interferometer, an

[a] A glass plate vapour deposited with gold is glued to the piezo to increase reflections

actuator is attached to the fibre ending. A resolution of 10^{-10}m is achieved with this setup.

Using the finite element program CAPA[5] one is able to solve the piezo-electric Eq. (1) numerically. In addition CAPA uses a Rayleigh model to describe damping. The equation is solved in the frequency domain such that the differentials in time are transformed to algebraic equations. Results of the comparison between simulation and measurement are shown in Fig. 2. The positions of the resonances are in excellent agreement with the

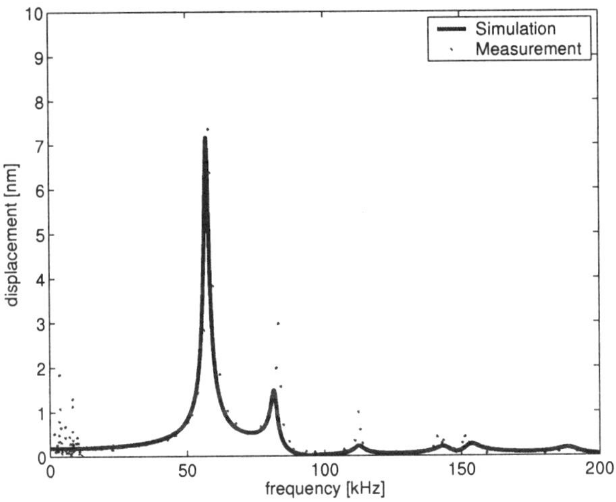

Figure 2. Displacement of a piezo disc r=12mm h=19.5mm

measurement. The amplitude of the displacement at the resonances differs from the measurement due to unknown damping parameters.

The propagation of sound in water is included in CAPA. From the resulting pressure field as a function of time in water one can extract the direction characteristics of the piezo by taking the local mean pressure field in the far field. (See Fig. 3)

3. Electrical Properties

The first resonance and antiresonance of a piezo can be emulated with the equivalent circuit diagram shown in Fig. 4 where the inductivity L, capacity C and resistance R correspond to the mass, stiffness and damping respectively. Further resonances can be emulated by additional parallel

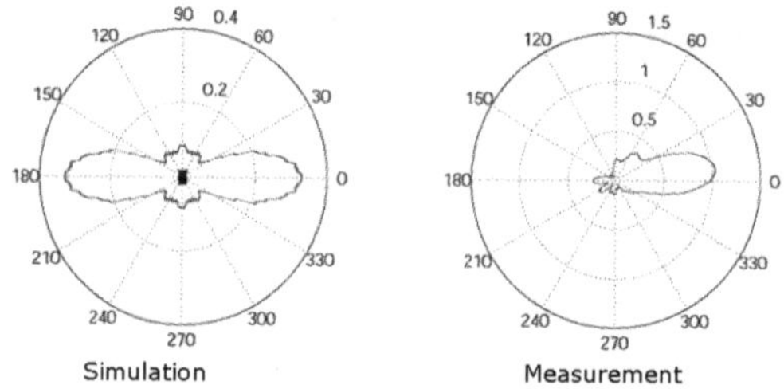

Figure 3. Direction Characteristics of a piezo disc in a.u.; Left: Simulation when sending (red: Simulation, blue: from simulation due to symmetry); Right: Measurement when receiving. Asymmetry in measurement due to shadowing of the cast integral amplifier.

circuits. The total capacity in Fig. 4 equals the capacity of a piezo. With this equivalent circuit diagram a fit to the measured impedance can be performed to extract L, R, C and C_0.

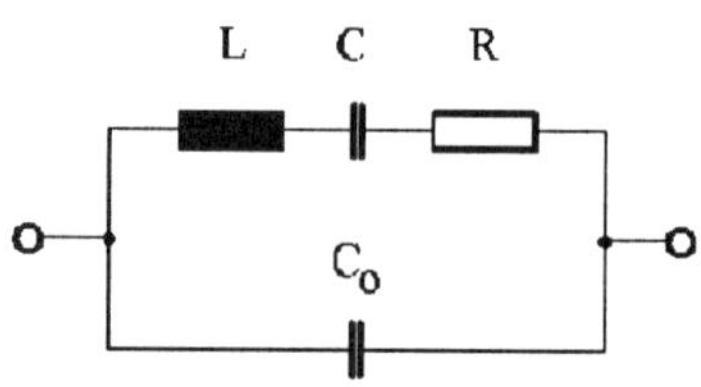

Figure 4. Equivalent circuit diagram for the first resonance of a piezo.

For the simulation of the impedance using CAPA a charge pulse $Q(t)$ has been applied to the piezo and the voltage response to this charge was calculated. In the fourierspace the impedance is then given as: $Z(\omega) = \frac{\tilde{U}(\omega)}{\tilde{I}(\omega)} = \frac{\tilde{U}(\omega)}{i\omega\tilde{Q}(\omega)}$, where U, I and Q are the calculated voltage, the current and the applied charge, respectively. A comparison of the measurement with the simulation and the results from the fit of the equivalent circuit diagram to the measurement are shown in Fig. 5.

The fit using the equivalent circuit diagram is in excellent agreement with the first resonance and antiresonance. The FE simulations fail to describe exactly the height of the resonances because of unknown damping parameters (even worse than for the displacement - see section 2).

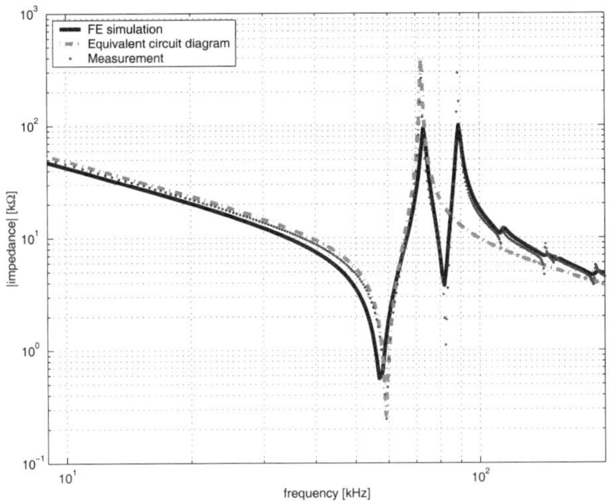

Figure 5. Impedance of a piezo disc r=12mm h=19.5mm

4. Summary and Conclusion

We have shown that Finite Element Methods are a valuable tool to study piezoceramic sensors. Comparing results of a piezo based on these methods with measurements obtained from an optic fibre interferometer yields very good agreement. Also the direction characteristics of the piezo are in good agreement with the simulation. The measured impedance can be very well reproduced using an equivalent circuit diagram on one side and using FE methods on the other. This detailed understanding of the piezo characteristics will enable us to both design the optimal sensors for acoustic particle detection and the transducers to calibrate these sensors.

References

1. H. Allik and T. J. R. Hughes, *Int. J. Num. Meth. Eng.* **2**, 151-157 (1970).
2. L. D. Landau, *Hydrodynamik*, Akademie-Verlag Berlin.
3. D. Rugar, H.J. Mamin and P. Guethner, *Appl. Phys. Lett.* **55**, 25 (1989).
4. M. Dienwiebel, et al. *Rev. Sci. Instrum.* **76**, 043704 (2005).
5. R. Lerch *IEEE Transaction on UFFC* **37 (3)**, 233-247 (1990), ISSN 0885-3010.

FIBER LASER HYDROPHONES AS PRESSURE SENSORS

P. E. BAGNOLI[1], N. BEVERINI[2,3], E. CASTORINA[2,3], E. FALCHINI[2,3], R. FALCIAI[4], V. FLAMINIO[2,3], E. MACCIONI[2,3],
M. MORGANTI[2,3], F. SORRENTINO[5], F. STEFANI[1,2], C. TRONO[2,4]

1. Dipartimento di Ingegneria dell'Informazione, Università di Pisa, Via Caruso 1, 56100 Pisa.

2. Dipartimento di Fisica "E. Fermi", Università di Pisa, Largo Pontecorvo, 3 - 56127 Pisa.

3. Istituto Nazionale di Fisica Nucleare, sezione di Pisa, Largo Pontecorvo, 3 - 56127 Pisa.

4. Istituto di Fisica Applicata " Nello Carrara", IFAC-CNR, Via Panciatichi, 64 - 50127 Firenze

5. Dipartimento di Fisica, Università di Firenze, Via Sansone 1 - 50019 Sesto F.no (FI)

The development of hydro-phonic sensors for deep see acoustic detection is described. The sensitive element is an erbium-doped single mode fiber laser, with the cavity delimited by two Bragg grating reflectors. The variations of temperature and pressure perturb the cavity, inducing a wavelength shift. The very narrow emission band of the laser, together with the interferometric detection technique, allows a dynamic pressure sensitivity in the μPa range. The devices have been characterized both optically and acoustically in a closed tub. A resin coating of the fiber laser has been experimented: this technique improves the sensitivity by more than one order of magnitude. The high sensitivity makes these sensors suitable for the detection of the acoustic waves induced in water by Ultra High Energy Neutrinos.

1. Introduction

Sensors based on optical fibers have well known advantages over conventional electro-mechanical sensors. They offer electrically passive operation, and immunity from electromagnetic fields, since the fiber is realized entirely with dielectric materials (glass and plastic). They have very small dimensions (outer diameter ~ 125 μm for a standard fiber), and have multiplexing capabilities for a quasi-distributed measurement configuration, by using a single opto-electronic control unit. Remote measurement is also possible. Indeed, the very low signal attenuation (~ 0.3 dB/km) of the fibers in the region around 1.55 μm makes it possible to place the opto-electronic control unit several km far from the measurement point. Moreover, high sensitivity and wide dynamic measurement range can usually be achieved. All these properties make fiber laser hydrophones suitable for acoustic detection of ultra-high energy neutrinos in the deep-sea large detectors. In the following, we will present the more recent progress we made in the task of developing a prototype of high-sensitivity deep-sea fiber laser hydrophone. Previous results and more details on the state of the art of fiber Bragg sensors and fiber laser sensors were reported in Ref. [1].

2. Fiber Bragg Gratings (FBG) and Fiber Laser (FL)

In a Distributed Bragg Reflector Fiber Laser (DBR-FL) the active medium is an erbium-doped optical fiber, delimited by two Bragg grating mirrors with identical pitch [2], directly inscribed on the fiber, which constitute a Fabry-Perot cavity. When pumped with 980 nm radiation, laser oscillation can be obtained with an emission peak around 1530 nm. FBGs are fabricated by using the phase-mask technique [3], usually on special photosensitive fibers, imprinting a refractive index modulation on the core of the fiber by the near-field fringe pattern of an UV writing beam.

We engraved the Bragg grating mirrors, with $\lambda_{Bragg} \cong 1532$ nm and 1 cm of length, on a single-mode erbium doped fiber by using an excimer (KrF) UV laser. The reflectivities of the mirrors are regulated for the best laser operation (typically > 99% for the total reflecting mirror, > 90% for the output coupler). The distance between the two FBGs (0.2-2 cm) is chosen in order to obtain the maximum power emission in stable single-mode regime. The emission power ranges from 100 μW up to 1 mW, with 200 mW of 980 nm pumping power.

Variations of the FBG pitch, of the cavity length, and of the fiber effective refractive index n_{eff}, induced by strain, temperature and pressure, produce a shift of the FL emission wavelength. The information is encoded on the wavelength. Thus, different lasers, with different emission wavelengths, can be inscribed on the same fiber and interrogated by the same opto-electronic unit, enabling a quasi-distributed measurement. The pressure sensitivity of a FL can be evaluated as [4]:

$$\frac{\Delta\lambda_{Bragg}}{\lambda_{Bragg}} = \left[-\frac{(1-2\nu)}{E} + \frac{n^2}{2E}(1-2\nu)(2p_{12}+p_{11}) \right] \Delta P \tag{1}$$

where n is the fiber effective refractive index, ν the Poisson's ratio, p_{ij} the Pockel's coefficients of the stress-optic tensor, and E the Young's modulus. For silica fibers at 1550 nm $\Delta\lambda_{Bragg}$ is ~ −3.6 pm/MPa [4].

For deep-sea applications, hydrophones sensitivity goal is the level of the acoustic background noise of the quiet ocean, conventionally represented by the so-called Deep Sea State Zero (DSS0) [5]. At 1 kHz, the DSS0 level is 100 $\mu Pa/Hz^{1/2}$, which corresponds (Eq. (1)) to a wavelength shift of 10^{-12} nm. An interferometric detection technique can provide an improvement of the sensitivity of a factor 10^3 - 10^4 below the linewidth of a single–mode DBR fiber laser ($\approx 4 \cdot 10^{-8}$ nm), thus approaching the DSS0 level.

3. Experimental setup

The experimental apparatus is sketched in Figure 1 (a).

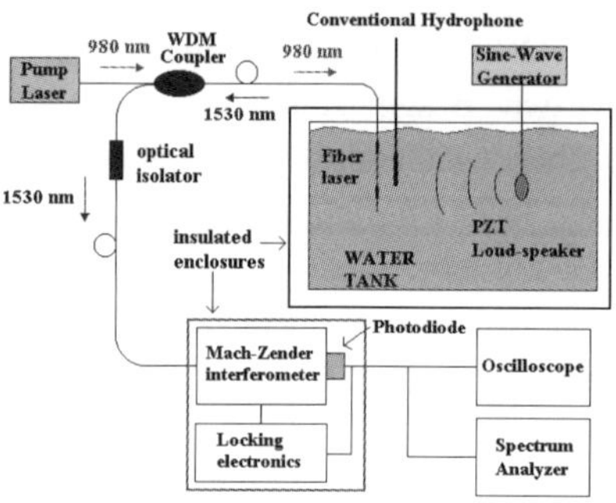

Figure 1a. Experimental setup.

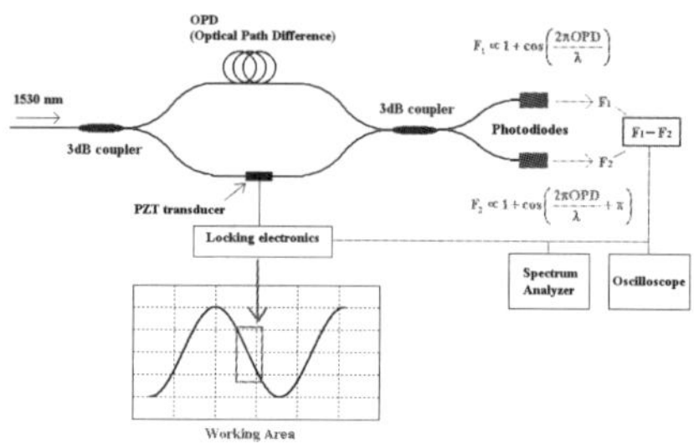

Figure 1b. Locked detection loop.

The wavelength laser line modulation, induced by the acoustic signal, was converted into phase modulation by an unbalanced Mach-Zender interferometer (MZI), and analyzed by means of a spectrum analyzer. The MZI (Fig. 1 b) converts the pressure-induced wavelength shift of the radiation, emitted by the DBR fiber laser, into a phase delay $\Delta\varphi_{MZ}$, which is a function of the FL output wavelength shift $\Delta\lambda$ and of the Optical Path Difference (OPD = nL, where L is the length unbalance of the two interferometer arms, and n is the fiber core refractive index). The following relationship holds:

$$\Delta\varphi_{MZ} = \frac{2\pi \cdot OPD}{\lambda^2}\Delta\lambda \qquad (2)$$

According to Eq. (2), the MZI sensitivity should be increased by increasing the unbalance, up to the limit of the laser coherence length, which for these lasers is of the order of some tens of km. The monotonic operative range and the dynamic range would be, however, proportionally reduced. Indeed, the phase detector signal is not univocally related to the wavelength shift, when the phase difference grows over 2π radians, and the low frequency noise of the laser exceeds this limit, yet with a few meter of unbalance. Effectively, we were then obliged to lock the MZI phase to the laser phase at low Fourier frequencies, with a cut-off of the order of several kHz, by using a servo loop that acts on the length of one arm of the interferometer by stretching the fiber through a piezoelectric actuator.

The interferometer was used in the condition usually defined as "*quadrature*" detection [6], with the output signal locked at one side of a fringe in the middle point, where the responsivity has its maximum value. Requiring $\Delta\lambda \approx 10^{-12}$ nm (DSS0 conditions), an OPD of 300 m gives $\Delta\varphi_{MZ} \approx 1$ μrad, which is hard to gain, but realistic with the present technology.
A further improvement of the FL sensor consists of the coating of the fiber with a material, which presents different mechanical characteristics. This could have several advantages: the FL is reinforced, the pressure sensitivity may increase, and the frequency response may be more uniform.

4. Results

We measured the response of our devices at several fixed frequencies in the region between about 15 kHz and 100 kHz, for different intensities of loudspeaker excitation. The FL output signal demonstrates a highly linear trend versus pressure as shown in a typical calibration curve [1].

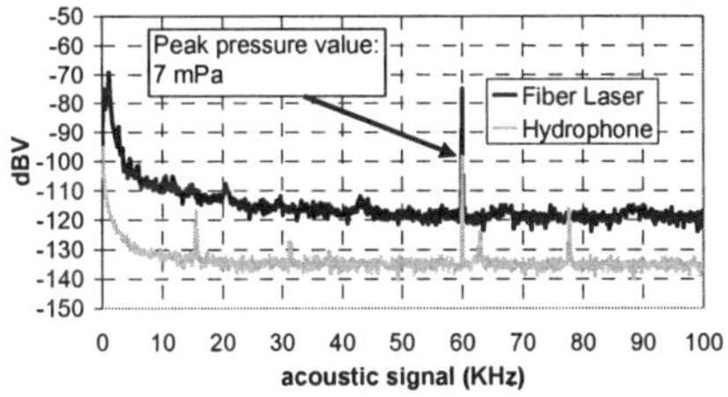

Figure 2a. FL and PZT hydrophone responses. (test note of 60 kHz and 7 mPa).

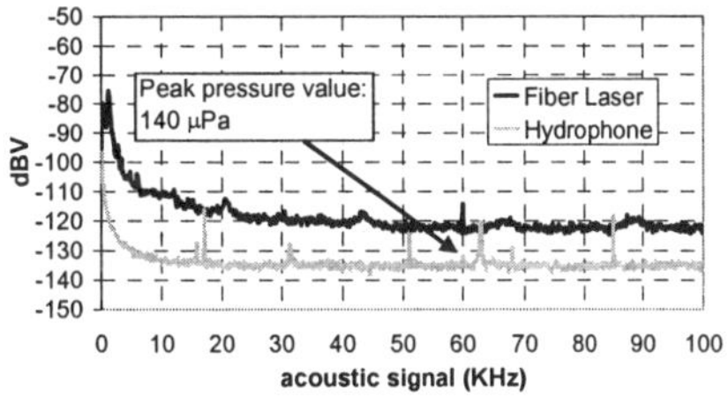

Figure 2b. FL and PZT hydrophone (test note of 60 kHz and 140 μPa).

In Figure 2a and Figure 2b the laser and the PZT hydrophone output spectra (acquired with a spectrum analyzer with a band-width of 125 Hz) are compared. In these figures, the excitation acoustic test note of 60 kHz has two different pressure value levels (7 mPa and 140 μPa, as measured by the calibrated PZT hydrophone). In our conditions the FL demonstrates an improvement of the signal-to-noise ratio at 60 kHz of about 10 dB over the PZT hydrophone.

We experimented also the performances of a fiber laser coated with an epoxy resin. In Fig. 3a a photo of a coated FL is reported, and in Fig. 3b we compare the responses to static pressure (wavelength shift) of a bare laser and of a resin-coated one. A sensitivity improvement of a factor 15 was observed. In the next future we will attempt to make this comparison in the acoustic (dynamic) range.

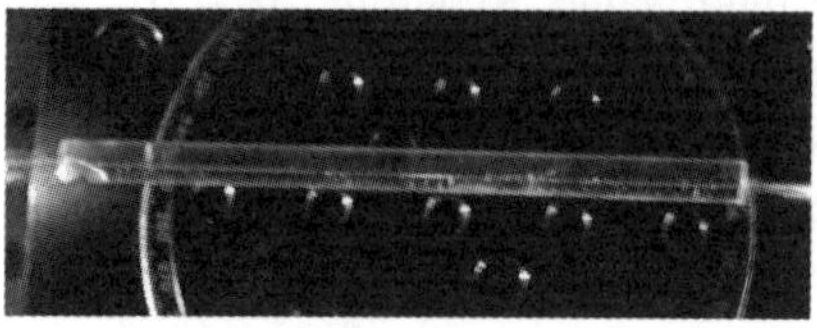

Figure 3a. Coated fiber laser.

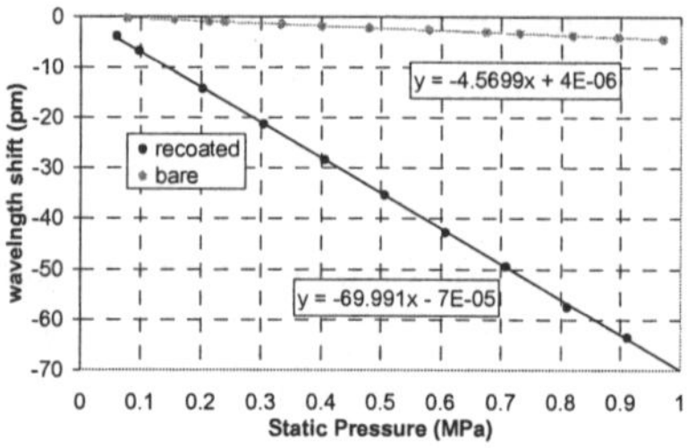

Figure 3b. Responses to static pressure of a bare laser and of a resin-coated one.

5. Conclusions

The results of the study, development and experimental validation of a DBR fiber laser for acoustic sensing in marine environment has been reported. Single mode lasers were fabricated by writing two Bragg gratings on an erbium-doped fiber core. The very narrow line-width (< 5 kHz), combined with interferometric detection, could make possible a wavelength resolution of $\sim 10^{-12}$ nm. The comparison with a calibrated PZT hydrophone (15-100 kHz frequency range) show a highly linear trend of the FL output signal versus pressure and higher sensitivity. FL sensitivity is strongly frequency dependent. This fact is probably connected to the particular experimental set-up, such as the acoustic reflections between the walls of the water tank and the mechanical resonance of the FL holder and of the fiber itself.

FL hydrophones offer a wide range of applications, spanning from the marine environmental acoustic monitoring, to the deep-sea study and the survey of dolphins and whales. In general, a network of multiplexed FL hydrophones could provide a lot of information of what happens in the "under-the-sea world", in the acoustic and ultrasound regimes, whichever could be the origin of the propagating pressure wave. Application to the acoustic detection of particle showers produced in water by very high-energy neutrinos, as an alternative method to the more conventional photo-detection, seems particularly attractive.

1. N. Beverini, R. Falciai. E. Maccioni, M. Morganti, F. Sorrentino, and C. Trono, Ann. Geophys. (to be published).
2. G.A. Ball and W.H. Glenn, J. Lightwave Techn., **10**, 1338 (1992).
3. K. O. Hill, B. Malo, F. Bilodeau, D. C. Johnson, and J. Albert, Appl. Phys. Lett., **62**, 1035 (1993)
4. M.G. Xu, L. Reekie, Y.T. Chow, J.P. Dakin, Electron. Lett., **29**, 398 (1993).
5. G.M. Wenz, J. Acoust. Soc. Am. **34**, 1936 (1962).
6. A.D. Kersey, T.A. Berkoff, W. W. Morey, Electron. Lett. 28, 236 (1992).

DEVELOPEMENT OF GLACIOPHONES AND ACOUSTIC TRANSMITTERS FOR ICE

SEBASTIAN BÖSER
DESY, Zeuthen, sboeser@ifh.de

A. HALLGREN[*], R. HELLER[†], R. NAHNHAUER[†],
M. POHL[†], K.-H. SULANKE[†]

For the detection of the feeble fluxes of neutrinos in the EeV range, recently acoustic detection is again discussed as a promising method. With the signal amplitude determined by the Greeneisen parameter, ice seems to be preferable as compared to water. However, neither suitable sensors nor transmitters existed two years ago. We present results on acoustic sensors and transmitters specially developed for that purpose. A new calibration method has been used to measure sensitivites in water, already showing some improvement with respect to commercial hydrophones.

1. Motivation

The idea of acoustic detection of highest energy neutrinos, which was first discussed more then 30 years ago [1], has had a strong revival during the last few years [2]. Due to the weak attenuation of acoustic waves in many materials very large detector volumes become possible using a rather large sensor spacing. Applications in ice and salt are favorable in comparison to water due to the much higher expected signal strength [4]. However only for water commercial sensors - hydrophones - are available. That is the reason for the development of the "glaciophones" described below.

2. Detector and Transmitter concept

The acoustic sensors are based on a piezo–ceramic as sensitive elements. Standard lead–zirconium–titanate elements with a diameter of 10 mm and a height of 5 mm are mounted in a housing which also contains a three stage low noise amplifier chain. The piezo–ceramic properties ($d_{33} = 500\,\mathrm{pC/N}$)

[*]Uppsala University, Sweden
[†]DESY, Zeuthen, Germany

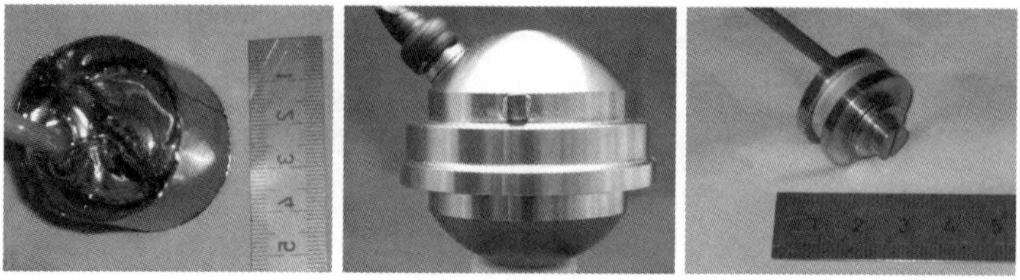

Figure 1. Designs for deep-ice application: a) amplifier and ceramic cast in epoxy, b) two sensor channels in steel housing, c) ring-shaped transmitter

together with an amplification of $\sim 80\,\mathrm{dB}$ allow to record pulses of $\sim 10\,\mathrm{mPa}$ amplitude. Several housings developed for the application in deep ice have been studied, e.g. moulding in epoxy (c.f Fig. 1a) or balls of stainless steel (c.f Fig. 1b) or glass. Apart from the impedance matching between the target material (e.g. ice) and the piezo–ceramic, the sensor response is governed by the resonance behaviour not only of the piezo–ceramic but even more of the housing. leading to a complicated frequency response of those detectors.

For the genration of short acoustic pulses, there are several options. Although thermoacoustic signal generation by dumping a laser in water has been tested successfully [3], it is much more challenging for ice, since absorption lengths for most common laser frequencies are too large. Therefore again piezo–ceramics are used. In order to achieve the required pressure amplitudes, they are driven by kV pulses. Despite the shortness of these pulses, the large capacitance of the ceramics with a typical relative premitivity of $\varepsilon_r \approx 2 - 3 \cdot 10^3$ result in peak currents of several amperes. One way to achieve this is by abruptly discharging a large inductivity ($\sim$ mH) via a power transistor into the piezo–ceramic. 1 kV pulses of $10\,\mu\mathrm{s}$ length at a peak current of ~ 8 A have been reached.
Care must also be taken in chosing the geometry of the transmitter ceramic. If, as in most deep-ice applications, the transmitter can not be oriented in the deployment process, by choosing a ring–shaped ceramic (c.f. Fig. 1c), the azimuthal uniformity can be greatly improved.

3. Calibration

Due to variations in the manufacturing process, first of all the piezoelectric ceramics themselves have to be calibrated. The ceramic is compressed using a heavy wheight m of several kg. When suddenly released (by pulling up the wheight), a charge pulse is induced that can be measured using an ADC with sufficiently high input impedance. Integration then yields the total charge Q_{tot}, which can be related to the static force from the wheight by the piezo–ceramic charge coefficient d_{33}.

$$Q_{tot} = d_{33}mg$$

The values obtained with this method are consistent with the vendor specifications up to a few percent.
Much more challenging is the calibration of a complete sensor system. All measurements of acoustic signals in laboratory sized media are strongly influenced by reflections at the media boundaries, prohibiting a calibration. In order to avoid this problem, sufficiently large and homogeneous media have to be used, which in the case of ice are not easily accessible.
In order to still get an estimate of the sensitivity of the developed sensors, a calibration in a large water tank has been performed first. At a distance of 10 cm one pair of sensors and transmitters was mounted in the center of a tank with dimensions of $12\,\mathrm{m} \times 10\,\mathrm{m} \times 5\,\mathrm{m}$. At a water temperature of $-0.1\,°\mathrm{C}$ and a salt content of 7 ppm the speed of sound [5] is $1413\,\frac{\mathrm{m}}{\mathrm{s}}$ which is well confirmed by the measured value of $1410 \pm 3\,\frac{\mathrm{m}}{\mathrm{s}}$.
A hydrophone (Sensortech SQ03) of a well known sensitivity of $163 \pm 0.3\,\mathrm{dB}$ rel. $1\mathrm{V}/\mu\mathrm{Pa}$ in the frequency range of $5 - 65\,\mathrm{kHz}$ was used as a reference and compared to the *iron ball* sensors (c.f. Fig. 1b) and *glass ball* sensor. A cylindrical piezo–ceramic of dimensions of $10\,\mathrm{mm} \times 5\,\mathrm{mm}$ cast in epoxy for electrical insulation has been used as a transmitter. For the excitation of the ceramic an arbitrary waveform generator was used. To compensate for the limited dynamic range of the sensor amplifiers and the large variation in the sensitivity, the input signal amplitude was varied in a range of $0.10\,\mathrm{V_{pp}} - 20.00\,\mathrm{V_{pp}}$ and was corrected for in the recorded signal by linear rescaling. For reduction of background noise, for each configuration 100 recordings with a length of 10 ms were taken at a sampling rate of 1.25 MHz.

Two different waveforms were used for a relative comparison of the signals, as shown in Fig. 2a. First, a gated burst of a sinusoidal wave at 6 different frequencies from $5 - 100\,\mathrm{kHz}$ was sent. With a burst length shorter than the signal traveling time of the reflection at the container walls the direct signal can be separated by application of a time window. The time for the system to overcome initial resonant excitation from the abrupt onset of the sinusoidal wave and go in a steady forced mode was estimated from send-

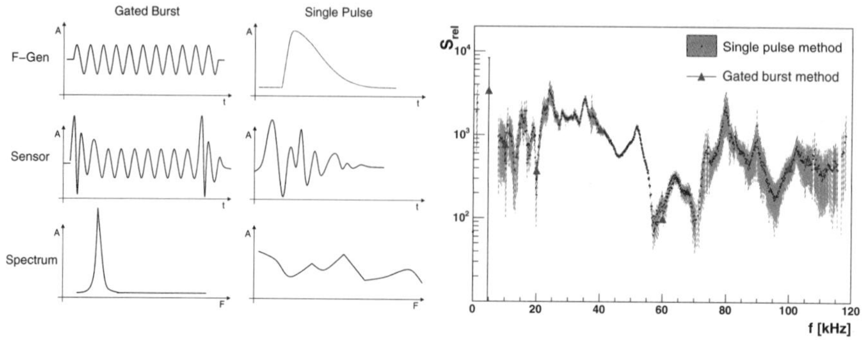

Figure 2. a) Schematic of shapes and spectra of calibration pulses and b) comparison of the two calibration methods

ing short single pulses and was subtracted from the signal time window. In the remaining time window the signal amplitude was estimated by fitting with a sine function with a fixed frequency allowing for a linear offset term to accommodate low-frequency background fluctuations. By comparison of the sensors to the reference hydrophone, a very precise measurement of the relative sensitivity with an error of $\lesssim 10\%$ could be achieved for single frequencies.

As a second waveform a fast step function with a slow return to the baseline was sent to the piezo–ceramic. This results in a short bipolar pressure pulse with a width of $\sim 10\,\mu$s containing a broad spectrum with peak frequencies up to $\sim 100\,$kHz. As in contrast to the signal the noise is not coherent, from the variation of frequency components in the single recordings the noise spectrum could be estimated and showed very good agreement with an independent direct noise spectrum measurement. Fourier spectra of measurements with the self-built sensors were then compared to the reference hydrophone, excluding frequency regions were either of the spectra is dominated by background noise. In contrast to the previous measurement, this method yields the relative sensitivity on a large frequency spectrum in one configuration, though with less precision.

Fig. 2b shows a comparison of the results from both methods for the *iron ball*. For all sensors good agreement of the methods has been observed. Please note that the errors given are only statistical. Systematic errors are probably smaller and will be accessed in further measurements. Although strongly frequency dependant, an enhanced sensitivity of both the iron ball and the glass ball in the order of $10^2 - 10^3$ can be seen, being probably mainly achieved by the larger gain of the electronic amplifiers in the self-

Table 1. Equivalent self noise level (5 – 65 kHz)

	Hydrophone	Glass Ball	Iron Ball
σ_{noise} [mPa]	40.3 ± 8.3	15.9 ± 1.7	4.7 ± 0.7

build sensors. However, also the signal to noise ratio is increased up to a factor of 50 for some frequencies. The equivalent self noise level that can be derived for the frequency range where the hydrophone is calibrated is given in table 1. In comparison to the reference hydrophone, a clear improvement for both tested sensor can be observed.

4. Outlook

Powerfull transmitters and sensitive detectors for ultrasonic signals in ice have been developed. With a new calibration method, the sensitivity of the glaciophones in water has been measured, showing a clear improvement with respect to commercial hydrophones. However, due to temperature effects and the unmatched impedance, this results can not be directly interpreted for ice. Though the same method is applicable for large ice volumes, it does not seems to be very pratical. Ice volumes of the necessary size of $10^3\,\text{m}^3$ are not easily accessible, and installation of the sensors might be problematic. One alternative to overcome the reflection problem is the option of changing boundary conditions. From a similar setup with both sensors and transmitters frozen in a small ice block, measurements can be taken with the ice block is first surrounded by air and later by water. As different amounts of the signal will be reflected on its surfaces, comparison of the measurements will yield a handle on the reflection and allow to substract it from the background.

Although still quite a lot of research needs to be done, given the short time, suitable acoustic sensors and transmitters for ice can possibly be built, opening the window for a new neutrino detection method.

References

1. G. A. Askaryan et al., NIM 164 (1979) 267;
2. R. Nahnhauer, Proc. Neutrino-2004, Paris, 2004, p.387 R. Nahnhauer, to appear in Proc. Moriond-2005, UHE penomena, La Thuille, 2005
3. K.Graf et. al., this proceedings
4. B. Price, to be published in J. Geophys. Res., (astro-ph/0506648)
5. W. Wilson, Jour. Acoust. Soc. Amer. (1960), 32(10) : 1357

PRELIMINARY RESULTS ON HYDROPHONES CALIBRATION WITH PROTON BEAM

ANTONIO CAPONE [1,2], GIULIA DE BONIS[1]

[1] *Physics Department, University "La Sapienza", Piazzale Aldo Moro 2*
Roma, 00185, Italy

[2] *INFN Sezione di Roma, Piazzale Aldo Moro 2*
Roma, 00185, Italy

Huge sensitive volumes will be required to identify interactions of U.H.E. cosmic neutrinos with energy above 10^{16}eV. New detectors and new techniques have to be developed to open this future field of H.E. astrophysics. We report on the study of acoustic signals induced by the interaction in water of a high intensity low energy proton beam (ITEP, Moscow). Acoustic signals have been collected by means of high sensitivity deep-sea hydrophones, for proton energy deposition close to 0.1 Joule. Hydrophones sensitivity has been previously characterized at the acoustic test facilities of IDAC-CNR (Rome). Results show a good linearity between the proton beam intensity and the hydrophones signal amplitude. Results are encouraging for the future detection of high-energy neutrino-induced showers developing in deep-sea water.

1. Introduction

Suggested originally by Askaryan in 1957 [1], the thermo-acoustic mechanism of energy dissipation in water causes an acoustic signal as a consequence of high-energy particles interaction. An instantaneous and localized energy deposition (following the particle interaction) produces a local heating of the medium and a local density variation [2] [3]. The perturbation propagates as a pressure wave and it can be detected with highly sensitive hydrophones. According to this picture, an acoustic signal is expected in case of neutrino interaction in water. Neutrino interactions generate electro-magnetic and/or hadronic showers; their energy is deposited in a volume that can be approximately described as cylindrical. Particles in the shower travel at about the speed of light, while the energy is radiated at the speed of sound. These two conditions (instantaneous and localized energy deposition) satisfy the hypothesis of thermo-acoustic model to generate an acoustic signal. A low energy intense proton beam interacting in water can be considered as a good calibration source to test hydrophones performances, as already reported [4].

2. Study of Hydrophones Characteristics

We used commercial piezo-electric hydrophones, produced by BENTHOS [5] and RESON [6]. Both hydrophones were pre-amplified and built to work at 2000 m depth. The BENTHOS hydrophone is a prototype (no data sheet available); the RESON hydrophone is the commercial model 4042, modified to work at 2000m depth. This hydrophone was already used for 6 months at 2000m depth at the NEMO [7] test site. The nominal sensitivity response for these hydrophones is about (-170)÷(-180) dB re 1V/1μPa. In order to have a good knowledge of the hydrophones response as a function of frequency, we performed a dedicated calibration using the facilities of the Underwater Acoustic Laboratory at the Institute of Acoustic "O.M. Corbino" [8] (UAL-IDAC, CNR) in Rome.

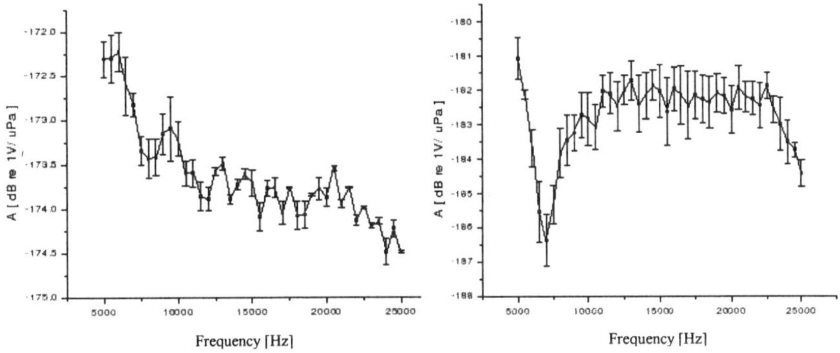

Figure 1. Hydrophones sensitivity results (BENTHOS on the left, RESON on the right).

The UAL-IDAC acoustic facilities include a water pool with dimensions 6.0m (length) • 4.0m (width) • 5.5m (depth). An Acoustic Calibration System (model RESON ACS9060) provided the calibration signal: a sinusoidal wave with frequency in the range (5-25) KHz. The ACS was coupled with a piezo-electric transmitter (model ITC 1007 [9]) in order to produce a sinusoidal pressure wave propagating in water. The pool is filled with fresh water and it is equipped with two independent remotely operated positioning systems, for the transducer and for the receiver, capable of handling weights up to 100kg. The distance between the transmitter and the receiver was set to 1m. Hydrophones response were acquired and analyzed; results are plotted in Fig. 1: on the horizontal axis is the transmitter frequency, on the vertical axis are the hydrophone sensitivities, in dB re 1V/1μPa.

3. Hydrophones Energy Calibration

Hydrophones can be used in a deep sea apparatus to identify neutrino interactions, to evaluate the interaction vertex and the primary energy. This means that the hydrophones response has to be calibrated with a well known energy source. The shape and the energy density of the high energy shower affect the signal formation as well as the environmental properties (water temperature, density and salinity). An energy calibration should be done taking into account all these parameters. After the described characterization in the frequency domain, we performed a preliminary energy calibration of the hydrophones response using a high intensity and low energy proton beam at the accelerator facilities of Institute of Theoretical and Experimental Physics (ITEP, Moscow).

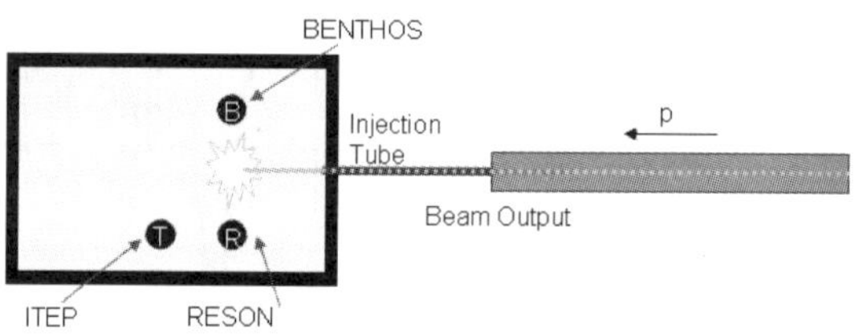

Figure 2. The ITEP test experimental set-up.

The ITEP proton beam is actually used for cancer therapy since the energy deposition of low energy protons in water can be easily tuned: a proton with energy in the range 100-200 MeV looses the most of its energy at the end of its path, at the so-called "Bragg peak" [10]. Therefore, by means of intense proton beams (about 10^{10} protons per spill), one can obtain a good approximation of a localized high-density energy deposition in water and can simulate a case where the thermo-acoustic description of the acoustic pulse generation can be applied. The "acoustic source" can be controlled varying the beam parameters: for a fixed beam momentum the total energy deposited in water is related to the beam intensity (i.e. the number of protons interacting in water) and the position of the Bragg peak is a function of the primary particle's energy. Penetration depth is about 7cm and 25cm for 100 MeV and 200 MeV protons respectively [11].

3.1. *Experimental Set-up*

The test was performed sending the proton beam into a tank filled with fresh water at room temperature. Water temperature and salinity values have not been varied during data taking. The basin dimensions were 50.8cm•52.3cm•94.5cm.

Three hydrophones have been placed inside the tank arranged as shown in Figure 2. Two hydrophones were the ones previously indicated as BENTHOS and RESON; a third hydrophone was provided by the ITEP group. The hydrophones have been positioned close to the Bragg peak region, therefore the hydrophones coordinates was accordingly changed with the proton energy. We have been able to select different beam conditions ranging from 100 to 200 MeV in energy and from 10^9 to $2.5 \cdot 10^{10}$ in proton intensity per spill. Such settings lead to proton energy deposition in water close to 0.1 Joule. Proton bunches were following a $\cos^2(t)$ distribution in time, with a duration of approximately 70 ns. The beam profile was gaussian in space with r.m.s. of about 2cm (100 MeV) and 1.5cm (200 MeV).

3.2. *Data Acquisition and Results*

Data acquisition was performed recording the pulses induced in the various hydrophones by means of a multi-channel oscilloscope. Bipolar shaped pulses

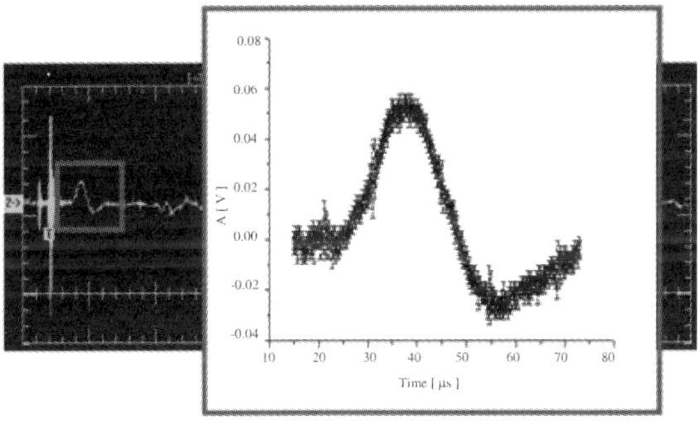

Figure 3. Bipolar pulse (hydrophone is RESON, E=200 MeV, $N_{protons} = 2.8 \cdot 10^{10}$).

have been observed (Figure 3) related to the ejection of proton spill into the water tank. Pulses time occurrence allows us to explain them as acoustic signals due to thermo-acoustic pressure wave emission at the Bragg peak.

Signals, in Volt, have been analyzed using fitting tools in order to get information on pulses amplitude and width as a function of the number of proton interacting in water (Figure 4).

4. Discussion

A deep understanding of the linear behaviour shown in Figure 4 will be completed after evaluating the distribution of the energy density deposited in water by the proton beam; this will be achieved by means of a Montecarlo simulation based on Geant4 [12]. This work is in progress. Knowing the energy

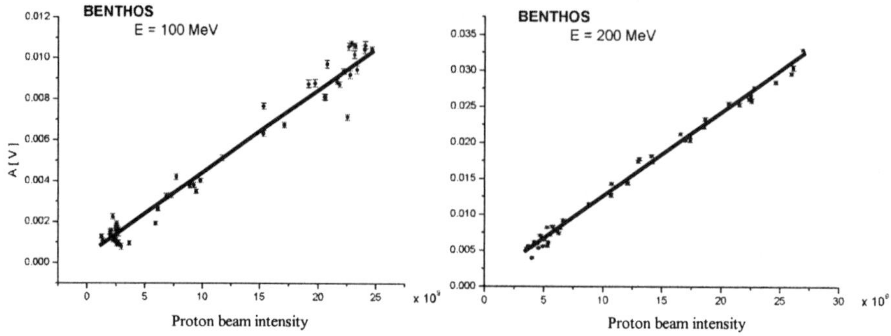

Figure 4. BENTHOS hydrophone signal (in Volt) as a function of proton beam intensity.

density distribution and using an analytical approach, i.e. solving the d'Alembert equation of wave propagation, to evaluate the pressure wave amplitude at the hydrophone position, we will be able to compare the signals acquired in ITEP with what is foreseen from the thermo-acoustic mechanism. This will allow us also to evaluate acoustic signals expected from high-energy neutrino-induced showers in water and finally to define the hydrophone sensitivity and the best layout required for a deep-sea acoustic neutrino telescope.

Acknowledgments

We thank Rocco Masullo (University "La Sapienza") and Giorgio Riccobene (INFN-LNS) for their precious help; the UAL-IDAC and Silvano Buogo who helped us for hydrophones characterization in the frequency domain. We finally thank Victor Lyashuk and the Andrei Rostovstev's ITEP group for their precious collaboration during the test beam.

References

1. G. A. Askaryan, *Atomic Energy* **3**, 152 (1957).
2. G. A. Askaryan *et al.*, *Nucl. Instr. Meth.* **164**, 267 (1979).
3. J. G. Learned, *Phys. Rev.* **D19**, 3293 (1979).
4. L. Sulak *et al.*, *Nucl. Instr. Meth.* **161**, 203 (1979).
5. http://www.benthos.com/
6. http://www.reson.com/
7. http://nemoweb.lns.infn.it/
8. http://www.idac.rm.cnr.it/
9. http://www.itc-transducers.com/default.html
10. K. Prelec, *FIZIKA* **B6**, 177 (1997).
11. T. Satogata *et al.*, C-A/AP/#120 (2003).
12. http://wwwasd.web.cern.ch/wwwasd/geant4/geant4.html

HIGH FREQUENCY NOISE IN LAKE BAIKAL AS A BACKGROUND FOR THE ACOUSTIC DETECTION OF HIGH ENERGY NEUTRINOS *

V.M.AYNUTDINOV[A], V.A.BALKANOV[A], I.A.BELOLAPTIKOV[D], L.B.BEZRUKOV[A], D.A.BORSCHEV[A], N.M.BUDNEV[B], K.V.BURMISTROV[A], A.G.CHENSKY[B], I.A.DANILCHENKO[A], YA.I.DAVIDOV[A], A.A.DOROSHENKO[A], ZH.-A.M.DJILKIBAEV[A], G.V.DOMOGATSKY[A], A.N.DYACHOK[B], O.N.GAPONENKO[A], K.V.GOLUBKOV[A], O.A.GRESS[B], T.I.GRESS[B], O.G.GRISHIN[B], S.V.FIALKOVSKI[F], A.M.KLABUKOV[A], A.I.KLIMOV[L], A.A.KOCHANOV[B], K.V.KONISCHEV[D], A.P.KOSHECHKIN[A], VY.E.KUZNETZOV[A], V.F.KULEPOV[F], L.A.KUZMICHEV[C], B.K.LUBSANDORZHIEV[A], S.P.MIKHEYEV[A], T.MIKOLAJSKI[E], M.B.MILENIN[F], R.R.MIRGAZOV[B], E.A.OSIPOVA[C], A.I.PANFILOV[A], G.L.PAN'KOV[B], L.V.PAN'KOV[B], YU.V.PARFENOV[B], A.A.PAVLOV[B], D.P.PETUHOV[A], E.N.PLISKOVSKY[D], P.G.POKHIL[A], V.A.POLESCHUK[A], E.G.POPOVA[C], V.V.PROSIN[C], M.I.ROZANOV[G], V.YU.RUBTZOV[B], B.A.SHAIBONOV[A], A.SHIROKOV[C], CH.SPIERING[E], B.A.TARASHANSKY[B], R.V.VASILJEV[D], R.WISCHNEWSKI[E], V.A.ZHUKOV[A], I.V.YASHIN[C]

(A) Institute for Nuclear Research, Moscow, Russia
(B) Irkutsk State University, Irkutsk, Russia
(C) Skobeltsyn Institute of Nuclear Physics MSU, Moscow, Russia
(D) Joint Institute for Nuclear Research, Dubna, Russia
(E) DESY, Zeuthen, Germany
(F) Nizhni Novgorod State Technical University, Nizhni Novgorod, Russia
(G) St.Petersburg State Marine University, St.Petersburg, Russia
(L) Kurchatov Institute, Moscow, Russia

*This work is supported by the Russian Ministry of Education and Science, the German Ministry of Education and Research, the Russian Fund of Fundamental Research (grants 02-02-14427, 05-02-16593), INTAS grant 2001-2309

We review the ongoing work of the Baikal Collaboration on acoustic detection of super high energy neutrinos. Some results of the study of high frequency acoustic noise in Lake Baikal are presented. A lot of short impulses with different amplitudes and shapes were detected, which should be considered as a background for acoustic neutrino detection. However, most of the short impulses appear to be due to noise sound waves interference and can be eliminated by a correlation analysis. Only a few detected bipolar impulses probably were produced by quasi local sources.

1. Introduction

Since several years, feasibility studies towards acoustic detection of particle cascades [1,2] are performed in Lake Baikal [3,4]. They follow two lines: a) the detection of acoustic signals coinciding with Extensive Air Showers (EAS); b) study the feasibility of acoustic detection of cascades generated in neutrino interactions in deep water. In 2000, we detected characteristic bipolar acoustic signals with about 150 μs duration, which probably have been generated in the region where the EAS core hit the water [3]. In 2001-2003, special experiments have been performed to search for acoustic signals produced by EAS in water. In order to measure noise characteristics an autonomous hydro-acoustic recorder with two input channels has been developed. We have performed a series of hydro-acoustic measurements in Lake Baikal in order to investigate the spectrum of acoustic noise, its dependence on depth and angle of incidence, and its correlation with external factors like wind or processes in the ice cover of the Lake [4]. In 2005, an acoustic system with four input channels was successfully tested. It can be regarded as a prototype of an elementary unit of a future underwater acoustic detector for super-high energy neutrinos.

2. The search for correlation between EAS and acoustic signals

To measure showers parameters a scintillator EAS array was deployed on the ice cover of Lake Baikal. The array consisted of 7 scintillation detectors with area 1 m^2 each. They were arranged uniformly along a circle with radius 80 m, one detector was placed in the centre of the circle. The EAS array was used as a trigger system for two acoustic data acquisition systems. The first one was deployed by ITEP group [5], the second one by Baikal-collaboration. In 2001, four our hydrophones were placed at the corners of a square with 40 meters side length, at a depth of 4 m. Fig.1a shows, for a typical data run, the distribution of the duration of detected bipolar acoustic impulses with amplitudes larger than $n\sigma$ (σ is the standard

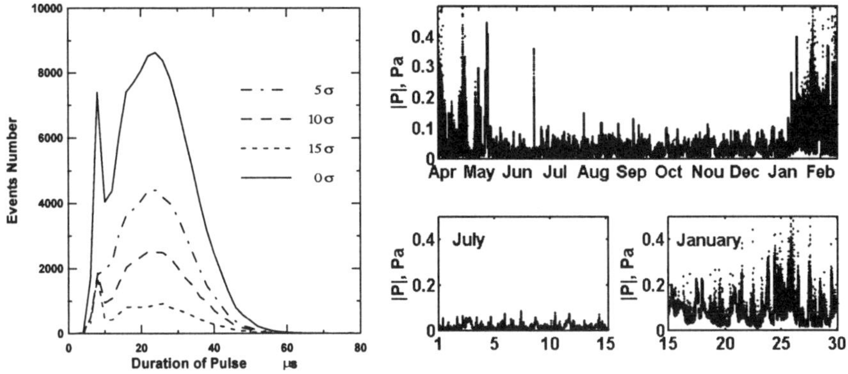

Figure 1. Left (a): Distribution of the duration of bipolar acoustic pulses with amplitude larger than $n \cdot \sigma$. Right (b): Time variation of the integral noise over one year at a depth of 1000 m.

deviation of an integral recorder's noise). Most of the impulses appear as a result of the interference of acoustic waves generated by numerous sources. It was very difficult to extract and to identify weak signals in data collected by an antenna with widely spaced hydrophones. So, in what followed, we have been using rather compact acoustic antennas with a distance between hydrophones of about only 1 m. It allows to decrease the time window for the search of acoustic signals and to have a suitable accuracy of the reconstructed direction to the quasi-local source. To check the quality of operation of the acoustic system and of the reconstruction procedure, a special source of short acoustic signals was deployed at shallow depths below the EAS array. The typical difference between true and reconstructed position of the pinger then turned out to be less than 1 m. Only in a few cases the pinger signals was modified or time-shifted due to interferences with the acoustic background, resulting in larger errors. In total about 5000 EAS and a large number of acoustic pulses with widely varying form, duration and amplitude have been detected. A few short bipolar impulses probably were generated close to EAS core at the moment when it entered the water. However, taking into account the high noise level at shallow depths, it was not possible to decide whether at least one of them was really produced by an EAS. The similar conclusion was done by ITEP group [6].

3. The hydro-acoustic recorder

The autonomous hydro-acoustic recorder has two spherical piezo-ceramic hydrophones. Their signals are processed by preamplifiers with 80 dB amplification and frequency correction. The further processing is performed by a micro-controller which includes a 12-bit Flash-ADC with a maximum conversion rate of 0.2 Msamples/sec and a multi-channel analog multiplexer. The cut frequency of the low pass filter was set to 50 kHz, in accordance with the number of channels and the maximum conversion rate of the Flash ADC. Data are written to a 1 Gbyte flash card. The integrated instrument noise between 1 and 50 kHz is about 12.5 mPa.

4. The Results

Fig.1b shows the integral noise measured over one year, starting from April 1, 2003. One observes significant seasonal and daily variations of noise depending on meteorological conditions. Contrary to what one would expect intuitively, the average noise is larger when the Lake is covered by ice. This is since cracking ice provides a more powerful source of high frequency noise than wind and waves. In most cases, at winter time one or two maxima per day are observed due to changing of the ice temperature. Examples of the

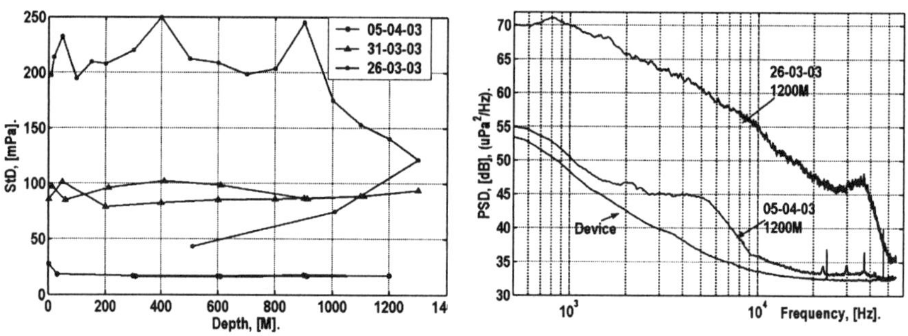

Figure 2. Left (a): Dependence of the integral noise on depth and time. Right (b): Frequency spectra of acoustic signals.

depth dependence of noise integrated over the bandwidth of the recorder are shown in Fig.2a. The curves give the effective fluctuation of the acoustic noise field, expressed in milli-Pascal (mPa), for different meteorological conditions. It turns out to be difficult to draw generalized conclusions on the depth dependence of the noise since the depth effects are combined with meteorological effects. Data obtained at the same depth but at different time can differ considerably. Comparing data taken at windless, sunny weather, starting from (00:45 pm) at (26.03.2003) with the data which have been taken practically at the same time (05.04.2003) but at a cloudy day, we may say that: in the first case sun shine changed the ice temperature and as a result the power of acoustic noise; in the second case the ice temperature was stable and the noise level at all depths was practically the same and very low. Fig.2b shows examples of the power spectral density (PSD) of the recorded noise signals. The spectrum obtained at(26.03.2003) is typical for stable meteorological conditions when any specific sources of noise like rain, gas seeps, ships and so on are absent. The bump in the spectrum (05.04.2003) probably appears due to the release of methane bubbles from the bottom of the Lake .

5. Summary and Outlook

The acoustic detection of high energy neutrinos and EAS in Lake Baikal as well in other natural basins is far from being trivial due to large level of background [7]. The results of first measurements of acoustic noise in Lake Baikal show its complicated structure and strong dependence from different factors. We observed daily and seasonal variations which are stronger than the dependence on the depth. Occasional effects like rain or gas seep do also strongly change the acoustic background - the integral noise as well as the spectral characteristics. At stationary and homogenous meteorological conditions the integral noise power in the frequency range 1-50 kHz are depth-independent and as a rule its amplitude is around 10-200 mPa and rarely higher. The noise power spectrum decreases typically by 4-6 dB per octave. Many short impulses with different amplitudes and shapes are observed and should be considered as a background for acoustic neutrino detection. However, most of the short pulses appear to be due to interferences of noise sound waves and can be eliminated by a correlation analysis. Only a small part of bipolar pulses probably was produced by quasi-local sources. Rather compact antennas are preferable. A new acoustic recorder with four hydrophones for more detailed long-term studies of noise was designed and tested. The results of the test are presented in a second talk.

References

1. G.A.Askaryan, Atomnaya Energiya 3 (1957) 152.
2. J.G.Learned, Phys. Rev-. D19 (1979) 3293.
3. G.V. Domogatsky, Proc. Neutrino 2000.
4. N.Budnev et al, http://saund.stanford.edu/saund1/workshop.
5. R.Nahnhauer, A.Rostovtsev http://saund.stanford.edu/saund1/workshop.
6. D.Zaborov, http://www-zeuthen.desy.de/arena.
7. J.Vandenbroucke, G.Gratta, N.Lehtinen astro-ph/0406105.

ITEP INVESTIGATION OF ACOUSTIC PHENOMENA FROM HIGH ENERGY PARTICLES*

V.S. DEMIDOV, E.V. DEMIDOVA, K.E. GUSEV, V.E. LUKYASHIN, V.I. LYASHUK, E.G. NOVIKOV, A.A. ROSTOVTSEV, A.YU. SOKOLOV

Russian State Research Center, Institute for Theoretical and Experimental Physics, Bol'shaya Cheremushkinskaya st. 25, Moscow, 117259, Russia

V.I. ALBUL, V.B. BYCHKOV, N.K. KRASNOV, A.F. KURCHANOV

Mendeleev All-Russia Research Institute for Physics-Technical and Radiotechnical Measurements, Mendeleevo, Moscow region, 141570, Russia

The mechanisms of sound wave generation from high energy protons and features of detector response are investigated in ITEP. The attention is attracted to the different possible mechanisms than the dominant thermoradiation excitation of sound. The search of sonic signals from cosmic particles, acoustic accelerator experiments and specific hardware - are the three vectors of ITEP acoustic activity.

1. Guidelines

1.1. *Introduction*

Along with the straight way for search of acoustic waves from high energy cosmic particles the accelerator experiments push forward understanding of acoustic physical phenomena arise from traverse of beam particle through condense media. The features of acoustic investigation at accelerators and today's large scale experiments (as BAIKAL, ANTARES, AMANDA, NEMO, NESTOR) dictate some special demands to experimental devises and first of all it refers to the sensors. This article is devoted to experiments at the ITEP synchrotrone and used hardware (hydrophones and devises). Activity of ITEP and BAIKAL Collaboration for search of hydroacoustic effects from extensive atmospheric showers (EAS) is presented in reports [1],[2] and [3]. The results which indicate on the evidence of the acoustic effects from EAS are discussed in the work [3].

* This work is supported by grant 03-02-16755 of the Russian Foundation for Basic Research.

1.2. *Hydroacoustic experiments at the ITEP synchrotrone*

ITEP accelerator experiments are directed on investigation of sound wave generation from proton at different beam parameters, searching of effects from different mechanisms in these processes, study of different hydrophones application for these task.

The proton beam had the following parameters: a particle energy of either 200 or 125 MeV, repetition period of the beam pulses of 4.2 s, intensity of 5×10^7 to 6×10^{10} protons per pulse, duration of the beam spill-out of 70 ns, and beam diameter of 10, 20, or 45 mm. The energy of the protons extracted from the accelerator is equal to 200 MeV and remains unchanged. The energy of protons falling onto the target can be lowered to 125 MeV with help of a moderating plexiglas filter set in front of the target.

The acoustic pressure, as function of the energy release in the target for energies ranging from 10^{16} to 1.4×10^{18} eV, is shown in Figure 1 [4].

Experimental results for energy deposition in water $E = 3.2 \times 10^{17}$ eV (at distance 0.32 m from the beam axe) was simulated according to the G.A.Askarian model [5] (Figure 2). The model underestimates the experimental data by factor of two.

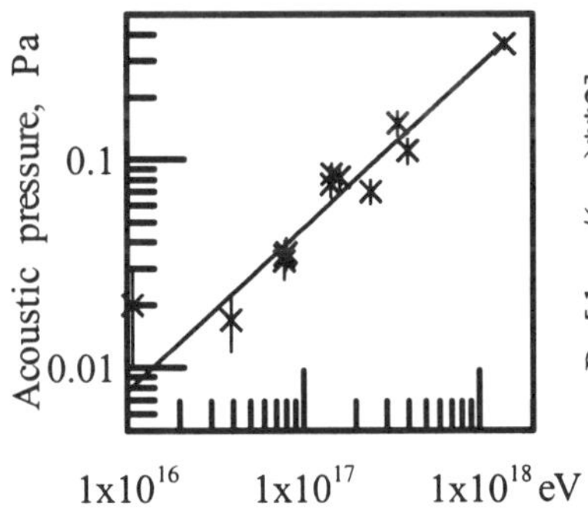

Figure 1. The acoustic pressure as function of the energy released in the water target.

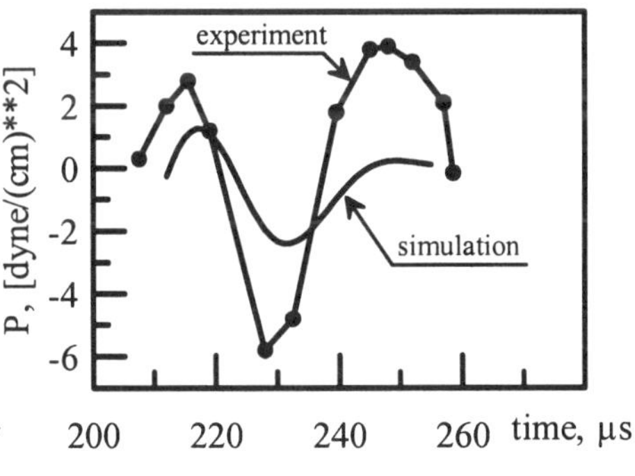

Figure 2. Simulation compare to the experimental data.

The results of our measurements are similar to the data presented earlier in [6]. From the results of this work, for an energy release of 10^{16} eV and a beam diameter of 1 cm, the amplitude of the acoustic beam recorded at a distance of 8 cm from the hydrophone is 0.071 ± 0.01 Pa. In our experiment, for the same values of the beam diameter and the energy release, the amplitude of the signal recorded at a distance four times larger (at $R = 32$ cm) was 0.02 ± 0.01 Pa.

Two next accelerator experiments were devoted to investigation of signal evolution at temperature change from about 1° to 10°C [7]. The purpose of these experiments is investigation of signal forming at the maximal water density close to 4°C. The registered change of the amplitude sign at a saddle-point through temperature 4 °C is also the "instrument" of checkout for the thermoradiation mechanism and its contribution to the sound pulse generation (Figure 3). In the last "temperature" experiment the water temperature was "on-line" controlled by ten electronic detectors along the beam trajectory and written to the hard disk. This investigation is directed towards detecting of another process yield (including the microstriction one) to acoustic signal forming from high energy particles in the water media.

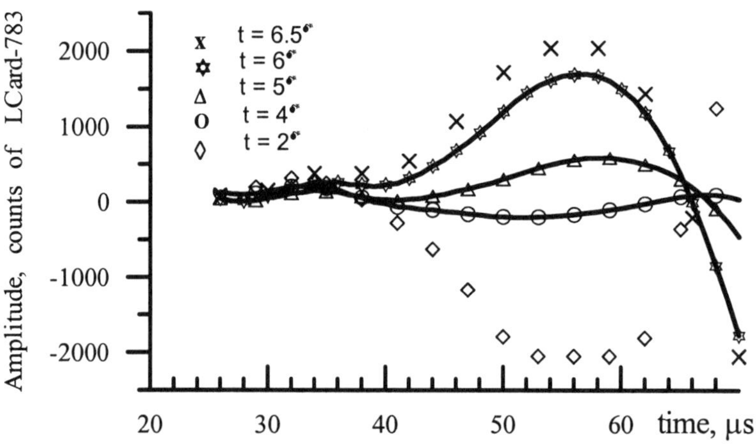

Figure 3. The evolution of the first-half wave at increase of the water temperature from 2 ° to 6.5 °C. Change of the amplitude sign when the temperature goes through the 4 °C saddle-point. Time of signal arrival from the beam is ~ 40 μs.

1.3. *Hardware*

Hydrophones, used in accelerator experiments and for seach of acoustic effects from extensive atmospheric showers in the Baikal lake have very sensitive (about 1000 μV/Pa) piezoceramic, that allows to apply the rather small built in preamplifier with gain coefficient about 200 (in single hydrophones). Signal amplification in hydroacoustic antenne (four hydrophones at the 5 m distance from each other) is provided in two steps: in built in preamplifiers (coefficient is about 26-27) and in the matching unit of the antenne (amplifier with gain about

8-10). Such "two-step" solution is need for use a rather long cable (210 m length at present time). All the antenne metal parts are produced from titanium allow to ensure the chemical stability and to prevent the noise that originates from differences in galvanic potentials. The antenne and cables were tested in the pressure camera for work at the 3 km water depth. It was provided the suppression of the temperature dependence in the electronic scheme to ensure the operation at negative temperatures in Baikal condition.

Working frequency diapasone of the antenne (Figure 4) and single hydrophones are about 16÷50 kHz. The gain flatness of antenne hydrophones is in the ±2 dB. For the single hydrophones the gain flatness is not worse than ±3dB. The flatness of directional pattern for antenne hydrophones at frequency 12.5, 20, 40 kHz is the next: not more then 2.0, 2.5, 3 dB in horizontal plane; not more then 7.0, 4.5 и 5 dB in vertical plane for the opening angle ±30°. For the single hydrophones (Figure 5) at frequences 16, 25 and 40 kHz: the directivity patterns in the horizontal plane lies within 6.5, 1.5, 3.5 dB; the directivity patterns in the vertical plane for the opening angle (0÷+30)° lies within 3.0, 5.0 and 4.0 dB; for the angle (0÷-30)° the flatness is not worst than 4.5, 5.0 and 4.0 dB.

Acoustic signal path include 12-th channel differential amplifiers on the base of 24 standard operational ones. The amplifier block is included to the acoustic path by means the long screened differential cable to take away the amplifier block beyond the bounds of intensive electromagnetic interferences. The given solution of acoustic signal path allows to decrease the signal-detection threshold up to natural noises.

The analog signals from hydrophones are digitized by means the LCard-783 board and written to vinchester. The LCard-783 board allows to

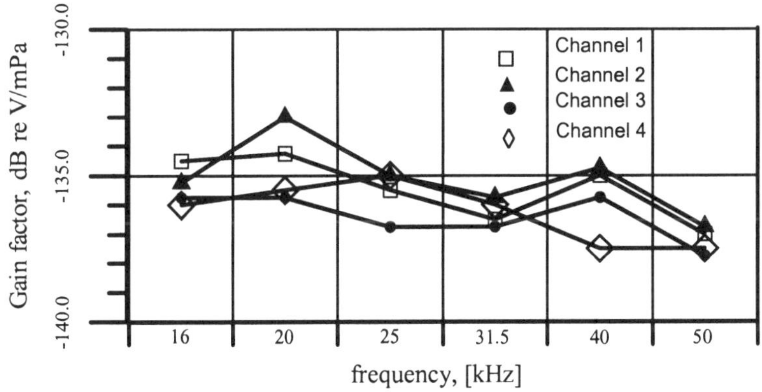

Figure 4. A Gain factor for the four antenne channels in case of 210 m main cable length.

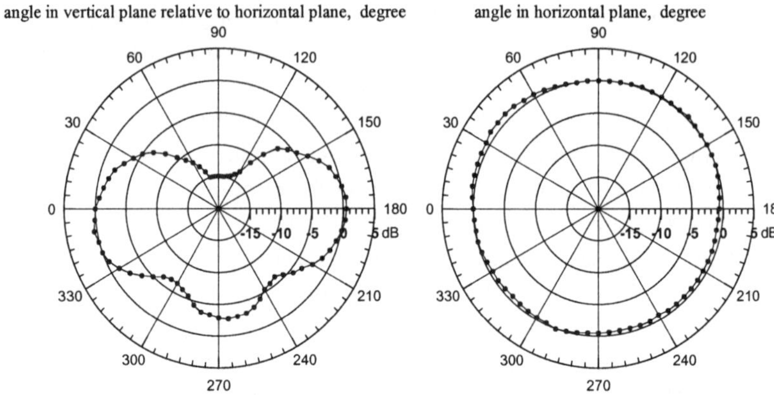

Figure 5. Directional pattern for the single hydrophone № 8 at the frequency 40 kHz.

connect up tp 16 channels with total frequency of digitization up to 3.2 MHz.

Data taking starts from synchronizing pulse. To prevent the possible overflow of signal queue and digitization failures it was developed the trigger with function of storage and time filter. The trigger is reprogrammable and based on the FPGA ALTERA logic of the BASELINE MAX PLUS 2 package.

References

1. V.Balkanov et al. The Baikal neutrino project: status report // Nucl. Phys. B (proc. Suppl.) 118 (2003) 363-370.
2. D. Zaborov, "Analysis methods of ITEP Baikal data". Report at International ARENA Workshop, DEZY, Zeuthen, Germany, May 17-19, 2005; http://www-zeuthen.desy.de/arena/
3. V. Lyashuk, "Search of acoustic effects from extensive atmospheric showers in Baikal". Report at V International Conference on Non-Accelerator New Physics (NANP-2005), Dubna, Russia, June 20-25, 2005; http://nanp.dubna.ru
4. V.I. Albul, V.B. Bychkov, K.E. Gusev, et. al., *Instrum.Exp.Tech.* **44**, 327 (2001); *Prib.Tekh.Eksp.* **N3**, 50 (2001).
5. G. A. Askariyan, B. A. Dolgoshein, A. N. Kalinovsky and N. V. Mokhov, Nucl. Instrum. Methods. **164**, 267, (1979).
6. L. Sulak, T. Armstrong, H. Baranger, et. al., *Nucl. Instrum. Methods.* **161**, 203, (1979).
7. V.I. Albul, V.B. Bychkov, K.E. Gusev, et. al., *Instrum.Exp.Tech.* **47**, 507 (2004); *Prib.Tekh.Eksp.* **N4**, 94 (2004).

TESTING THERMO-ACOUSTIC SOUND GENERATION IN WATER WITH PROTON AND LASER BEAMS*

K. GRAF, G. ANTON, J. HÖSSL, A. KAPPES, T. KARG, U. KATZ, R. LAHMANN, C. NAUMANN, K. SALOMON AND C. STEGMANN

Physikalisches Institut,
Friedrich-Alexander-Universität Erlangen-Nürnberg,
Erwin-Rommel-Straße 1,
91058 Erlangen, Germany
E-mail: kay.graf@physik.uni-erlangen.de

Experiments were performed at a proton accelerator and an infrared laser facility to investigate the sound generation caused by the energy deposition of pulsed particle and laser beams in water. The beams with an energy range of 1 PeV to 400 PeV per proton beam spill and up to 10 EeV for the laser pulse were dumped into a water volume and the resulting acoustic signals were recorded with pressure sensitive sensors. Measurements were performed at varying pulse energies, sensor positions, beam diameters and temperatures. The data is well described by simulations based on the thermo-acoustic model. This implies that the primary mechanism for sound generation by the energy deposition of particles propagating in water is the local heating of the media giving rise to an expansion or contraction of the medium resulting in a pressure pulse with bipolar shape.

1. Introduction

The production of hydrodynamic radiation (ultrasonic pressure waves) by fast particles passing through liquids was first predicted already in 1957 leading to the development of the so-called thermo-acoustic model in 1979[1,2]. The model allowed to describe the primary production mechanism of the bipolar shaped acoustic signals measured in an experiment with proton pulses in fluid media[3]. According to the model, the energy deposition of particles traversing liquids leads to a local heating of the medium which can be regarded as instantaneous with respect to the hydrodynamic time scale. Due to the temperature change the medium expands or contracts according to its volume expansion coefficient α. The accelerated motion

*This work is supported by the German BMBF Grant No. 05 CN2WE1/2.

of the heated medium forms an ultrasonic pulse which propagates in the volume. The wave equation describing the pulse is[2]

$$\Delta p(\vec{r},t) - \frac{1}{c_s^2} \cdot \frac{\partial^2 p(\vec{r},t)}{\partial t^2} = -\frac{\alpha}{C_p} \cdot \frac{\partial^2 \epsilon(\vec{r},t)}{\partial t^2} \tag{1}$$

and can be solved using the Kirchhoff integral:

$$p(\vec{r},t) = \frac{1}{4\pi} \frac{\alpha}{C_p} \int_V \frac{\mathrm{d}V'}{|\vec{r}-\vec{r}'|} \frac{\partial^2}{\partial t^2} \epsilon\left(\vec{r}', t - \frac{|\vec{r}-\vec{r}'|}{c_s}\right). \tag{2}$$

Here $p(\vec{r},t)$ denotes the hydrodynamic pressure at a given place and time, c_s the speed of sound in the medium, C_p its specific heat capacity and $\epsilon(\vec{r},t)$ the energy deposition density of the particles. The resulting pressure field is determined by the spatial and temporal distribution of ϵ and by c_s, C_p and α, the latter three depending on the temperature. A controlled variation of these parameters in laboratory experiments and a study of the resulting pressure signals allows therefore a precise test of the thermo-acoustic model. One decisive test is the disappearance of the signal at 4°C in water, the medium considered in the following, due to the vanishing α at this temperature.

However, in previously conducted experiments investigating this effect in different liquids, the observed pulses could not be unambiguously verified as thermo-acoustical[3,4,5,6]. There the variation of the pulse amplitude with the temperature in water showed not the predicted dependency and particularly not the disappearance of the signal at 4°C.

2. Conducted Experiments

The experiments presented in this paper were performed with a pulsed 1064 nm Nd:YAG laser facility located at our institute, and the 177 MeV proton beam of the "Gustaf Werner Cyclotron" at the "Theodor Svedberg Laboratory" in Uppsala, Sweden. The beams were dumped into a $150 \times 60 \times 60\,\mathrm{cm}^3$ water tank, where the acoustic field was measured with several position-adjustable hydrophones (pressure sensitive sensors based on the piezo-electric effect). The temperature of the water was varied between 1°C and 20°C with a precision of 0.1°C by cooling and gradual controlled homogeneous reheating of the whole water volume.

The spill energy of the proton beam was varied from 10 PeV to 400 PeV, the beam diameter was approx. 1 cm and the spill time 30 μs. For 177 MeV protons, the energy deposition in the water along the beam axis (z-axis, beam entry into the water at $z = 0$ cm) is relatively uniform up to $z = 20$ cm ending in the prominent Bragg-peak at $z \approx 22\,\mathrm{cm}$. The laser pulse energy was adjusted between 0.1 EeV and 10 EeV at a beam diameter of a few mm, the pulse length was fixed at 9 ns. The laser energy density deposited along the beam axis had an exponential decrease with a absorption length of $(6.0 \pm 0.1)\,\mathrm{cm}$. The two experiments enabled us to use different spatial and temporal distributions of the energy deposition as well as two different kinds of energy transfer into the medium, i.e. by excitation by both beams and additionally by ionisation by the proton beam.

The sensors used were characterised and found to be linear in amplitude response, the frequency response was flat up to the main resonance at 50 kHz with a sensitivity of approx. $-150\,\mathrm{dB\,re\,1V/\mu Pa}$. The sensitivity dependence on temperature was measured and the relative decrease was found to be less than 1.5% per 1°C.

For every set of experimental parameters the signals of 1000 beam pulses were recorded, with a sampling rate of 10 MHz, sufficient for the typical frequency range of the signals of 5 kHz to 100 kHz.

3. Results

The measured bipolar signals were found to be in good agreement with simulations based on the thermo-acoustic signal generation mechanism. The hydrodynamic nature of the signal was proven by determining the propagation times of the signals at different positions in the x-direction perpendicular to the beam axis. They were consistent with the expected propagation times for sound in water. Also the investigated signal dependencies on beam energy, beam width and sensor distance from the beam show very good agreement with the simulation based on the thermo-acoustic model.

Figures 1 and 2 show the temperature dependence of the peak-to-peak amplitude of the bipolar signals for the two experiments, where a positive (negative) sign denotes a leading positive (negative) peak of the signal. The two data sets shown in each figure were recorded by two sensors positioned at $x = 10\,\mathrm{cm}$ perpendicular to the beam axis and at $z = 12\,\mathrm{cm}$ and $z = 22\,\mathrm{cm}$ along the beam axis, respectively. In the case of the proton beam setup the hydrophone positions correspond roughly to the z-position of the Bragg-peak and a z-position half way between the Bragg-peak and the beam entry into the water, respectively. For comparability, the same positions and the same sensors were chosen for the laser experiment.

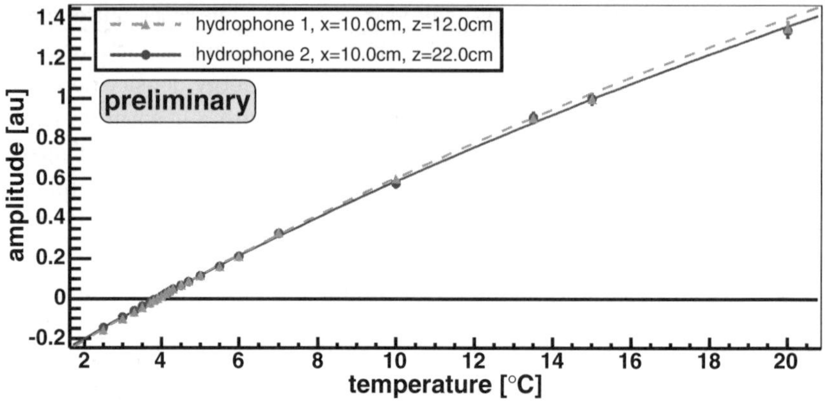

Figure 1. Measured signal amplitude of the bipolar acoustic signal produced by laser pulses at different temperatures fitted with the model expectation as described in the text. All amplitudes were normalised to 1 at 15.0°C.

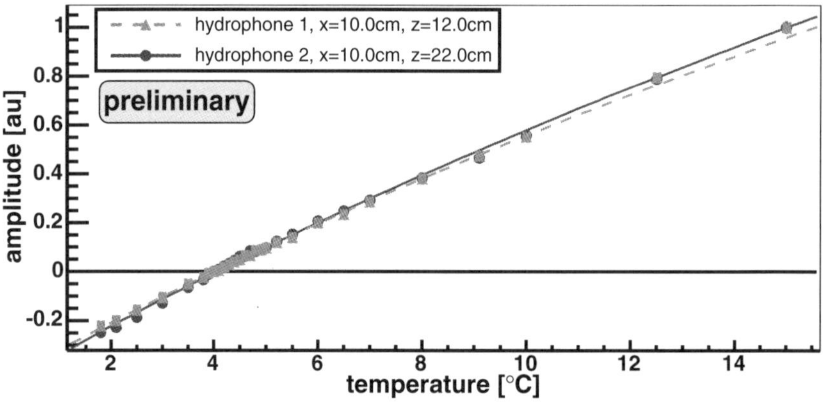

Figure 2. Measured signal amplitude of the bipolar acoustic signal produced by proton pulses at different temperatures fitted with the model expectation as described in the text. The non-thermo-acoustic signal at 4.0°C was subtracted at every temperature. The amplitudes were afterwards normalised to 1 at 15.0°C.

The laser beam signal shown in Fig. 1 changes its polarity around 4°C, as expected from the thermo-acoustic model. The model expectation for the signal amplitude, which is proportional to α/C_p and vanishes at 4°C, is fitted to the experimental data. In the fit an overall scaling factor and a constant temperature shift were left free as fit parameters. The fit yielded a zero-crossing of the amplitude at (3.9 ± 0.1)°C, where the error is dominated by the systematic uncertainty in the temperature setting. The zero-crossing is in good agreement with the expectation of 4.0°C. Analysing the proton data in the same way yielded a shape slightly deviating from the model expectation, and a zero-crossing significantly different from 4.0°C at (4.5 ± 0.1)°C. In view of the results from the laser beam measurements, we subtracted the residual signal at 4.0°C, which has an amplitude of approx. 5% of the 15.0°C signal, from all signals, assuming a non-temperature dependent effect on top of the thermo-acoustic signal. The resulting amplitudes shown in Fig. 2 are well described by the model prediction.

The source of the small non-thermo-acoustic signal which was only seen in the proton experiment could not be unambiguously verified with these experiments. The obvious difference to the laser experiment are the charges involved both from the protons themselves and the ionisation of the water. For clarification further experiments are needed either with ionising neutral particles (e.g. synchrotron radiation) or with charged particles (e.g. protons, α-particles) with more sensors positioned around the Bragg-peak. With such experiments it might be possible to distinguish between the effect of ionisation in the water and of net charge introduced by charged particles.

4. Conclusions

We have demonstrated that the sound generation mechanism of intense pulsed beams is well described by the thermo-acoustic model. In almost all aspects investigated, the signal properties are consistent with the model. Relying on the model allows to calculate the characteristics of sound pulses generated in the interaction of high energy particles in water with the input of the energy deposition of the resulting cascade. A possible application of this technique would be the detection of neutrinos with energies $\gtrsim 1\,\mathrm{EeV}$[7].

References

1. G.A. Askariyan, *Atomnaya Energiya* **3**, 152 (1957).
2. G.A. Askariyan et al., *Nucl. Inst. Meth.* **164**, 267 (1979).
3. L. Sulak et al., *Nucl. Inst. Meth.* **161**, 203 (1979).
4. S.D. Hunter et al., *J. Acoust. Soc. Am.* **69**, 1557 (1981).
5. S.D. Hunter et al., *J. Acoust. Soc. Am.* **69**, 1563 (1981).
6. V.I. Albul et al., *Instr. Exp. Tech.* **44**, 327 (2001).
7. T. Karg et al., *Design Considerations and Sensitivity Estimates for an Acoustic Neutrino Detector* in these proceedings.

THE NEMO ACOUSTIC TEST FACILITY

G. RICCOBENE *
Laboratori Nazionali del Sud - INFN,
Via S. Sofia 62,
Catania 95123, Italy
E-mail: riccobene@lns.infn.it

The NEMO (NEutrino Mediterranean Observatory) Collaboration is constructing, 25 km E from Catania (Sicily) at 2000 m depth, an underwater test site to perform long-term tests of prototypes and new technologies for astrophysical HE neutrino telescopes. In this framework the collaboration deployed an electro-optical cable equipped with several e.o. terminations. An experimental apparatus for the measurement of underwater acoustic background was also installed and connected to shore on 22 Jan 2005, allowing continuous on-line monitoring of deep-sea noise in the range 30 Hz - 40 kHz. Underwater noise spectra were produced and classification of transient signals is under way.

1. Introduction

The scientific interest in neutrino astronomy is leading to the installation of km^3-scale Čerenkov neutrino telescopes: ICECUBE[1] is under construction in the South Pole, another one is planned to be located in the Mediterranean Sea. Since 1998 the NEutrino Mediterranean Observatory Collaboration (NEMO[2]) is conducting an intense activity to select and monitor abyssal sites for the installation of the detector. Data strongly indicate that a large marine region $\sim$ 80 km SE off the Sicilian Coast of *Capo Passero*, is optimal[3,4]. In the same time the Collaboration is installing and operating a test laboratory at 2000 m depth, 25 km E offshore the port of Catania (Italy). Using the test site infrastructures, the Collaboration deployed and is operating the OνDE (Ocean noise Detection Experiment) station, a real-time experiment to monitor acoustic signals at 2000 m depth[5].

*For the NEMO collaboration

2. The NEMO test site

The NEMO test site consists of a shore infrastructure, a 28 km long electro-optical cable connecting the shore to the abyssal test site, the underwater laboratory. The shore building hosts the land termination of the cable, mechanics and electronics workshops, the counting room and power supplies for underwater instrumentation. The 28 km long electro-optical cable that connects the counting room to deep sea is an underwater telecommunication cable, with external steel armor, containing 10 optical fibers and 6 electrical conductors (3 mm^2). At about 20 km from the shore, the cable is divided into two branches, roughly 5 km long. The Test Site North branch has 2 conductors and 4 fibres directly connected to shore, the Test Site South (TSS) branch has 6 fibers and 4 conductors. A sea campaign was conducted on January 2005 to equip the cable submarine terminations with electro-optical connectors mounted on titanium frames. The frames are deployed on the seabed at $\sim$ 2050 m depth. During the sea campaign two experimental devices were deployed and connected to the cable: a seismic monitoring station called Submarine Network 1, is operational on Test Site North; the OνDE station, described in the following paragraph, is connected to TSS.

3. The OνDE acoustic station

The measurement of ambient acoustic background is fundamental in order to carry out a feasibility study for an underwater acoustic neutrino detector. Noise in the sea may arise from different sources[6]. Biological and human noises could reach very high pressure level (up to 250 db re 1μPa at 1 m), but they are, generally, produced by local and impulsive sources. In the frequency range of interest, average noise is due to surface agitation and to navigation that, close to ports or commercial ship routes, generates a diffuse acoustic background. At present only few data of acoustic noise at large depth are available in literature. For this reason the NEMO Collaboration deployed and is operating the OνDE (Ocean Noise Detection Experiment) station in the NEMO underwater test site. The station is devoted to perform real-time monitoring of the level of acoustic noise at depth $\geq$ 2000 m. It is equipped with four hydrophones, whose signals are digitized and translated into optical by underwater electronics. Data are sent to shore through optical fibers and acquired on a PC. The hydrophones come from a special series, TC-4042C, manufactured by *RESON* for NEMO to operate at 250 bar pressure. The TC-4042C are piezoelectric omnidirectional sensors with a nominal receiving sensitivity of -193±3 dB re 1V/μPa, linear over a wide range of frequencies: from few tens Hz to 50 kHz. Each detector (hereafter H1,H2 H3 and H4) is mounted on an Al-alloy vessel, pressure resistant, which also contains the hydrophone preamplifier (20 db gain). Two preamps (for hydrophones H2 and H4) where modified applying a hi-pass filter ($>$ 1 kHz, 6 db per octave) to reduce most of the ambient noise, which has a $1/f$ spectrum. The detectors are fixed on the TSS frame to form a tetragon of $\sim$ 1 m side. In figure 1 we show the station set-up: hydrophone H3, mounted in the uppermost position ($\sim$ 3.2 m above seabed), is placed close to the frame top; H1,H2 and H4 are attached at the same height ($\sim$ 2.6 m above seabed) on the top frame edges. Hydrophones signals are fed into two stereo A/D Boards *Crystal* CS5396. H1 and H3 (large bandwidth) reach the same A/D stereo board, H2 and H4 (hipass filtered) the other one. The two A/Ds are driven by the same clock: the four signals are, then, synchronized. The CS5396 samples the analog data at a rate of 96 kHz with 24 bit resolution. The digital outputs of the two A/Ds are sent to two fiber optic data transmitters. Optical data are reconverted, on shore, into electrical by two fiber optic receivers and, then, acquired using two PCI audio boards *RME DIGI96-8 PAD* (96 kHz, 24 bits) mounted on a PC, Pentium4 3GHz. On January 22nd 2005 OνDE was deployed together with the TSS frame and then connected and switched on. The orientation of the frame front with respect to North is $\sim$ 110°, measured both by the ROV and the on-board compass.

4. Preliminary results and discussion

OνDE continuously transmits to shore some MB of data per sec, the present data acquisition strategy is to record the acoustic data stream for 5 minutes every hour. This permits to limit the data storage on disk and to record enough statistics to measure the noise level as a function of time. In figure 2 we plot, as an example, the average value of power spectral density of noise measured from April 10 to July 17 at 4:00 am using channel H1. For this channel (hydrophone with electronic chain) a preliminary calibration curve was measured to be -175 ± 2db re V/μPa, a value consistent with data provided from RESON in the interval 4÷22 kHz. This value of response sensitivity is used for the whole frequency range plotted (1 ÷ 42 kHz). Further data analysis is anyhow required for frequencies lower than 4 kHz and higher than 22 kHz. Data recorded by the other hydrophones are consistent with H1. The periodogram shown in figure is obtained cal-

Figure 1. The titanium frame installed on TSS. The ROV operable electro-optical connectors are visible on the front panel. The hydrophones and electronics housing of the OνDE station are also shown (see text).

culating the 2048 points Fast Fourier Transform of data sample (2 minutes per each day), windowed using a Hanning window, shifting by 1024 points. Plot shows that average ambient noise at 2000 m depth is below sea state 2 curve[6]. Fluctuation (1 $\sigma \simeq 6$ db) are present at frequency below 20 kHz, depending on sea state and biological and human activity. Fluctuations are smaller at higher frequencies. The highest noise levels are recorded in coincidence with the passage of ships and/or with the presence of marine mammals. Correlations with sea state and ship traffic data are in progress.

5. Conclusions

The OνDE station is fully operational at NEMO test site of Catania since Jan 22nd 2005, transmitting data continuously. Data analysis is presently addressed to characterize noise variations as a function of time. Several impulsive signals were also found in the data set, most of them being connected to human or biological activities. Localization of noise sources using the 4 hydrophones data together is under development. Recorded data set is also extensively used for marine mammals research.

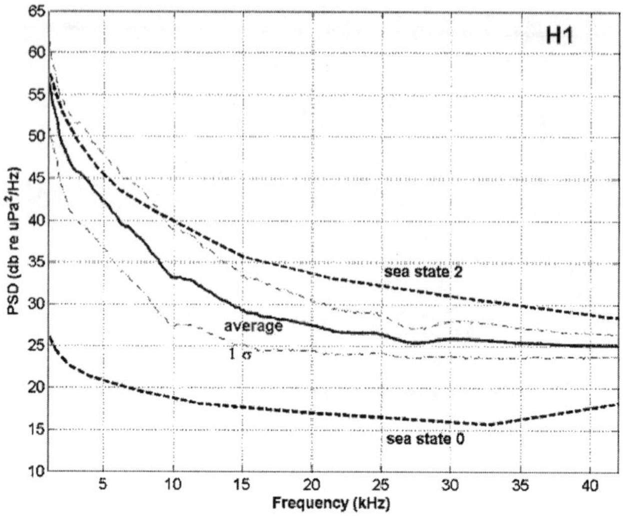

Figure 2. Average power spectral density (PSD) of underwater noise measured by Hydrophone H1 every day, from April 10 to July 17 2005, at 4:00 a.m. for 5 minutes (solid thick line). The value of average PSD $\pm 1\sigma$ is also shown (dashed thin line). Expected noise due to sea surface agitation is plotted (thick dashed line) for comparison.

References

1. `http://icecube.wisc.edu`.
2. M. Circella for the NEMO Collaboration, *Proceedings of 29th ICRC* ita-circella-M-abs1-og25-oral, Pune, India (2005).
3. G. Riccobene for the NEMO Collaboration, *Proceedings of VLVnT Workshop*, Amsterdam, The Netherlands (2003).
4. T. Chiarusi for the NEMO Collaboration, *Proceedings of 29th ICRC* ita-chiarusi-T-abs1-og26-oral, Pune, India (2005).
5. G. Riccobene for the NEMO Collaboration, *Proceedings of 29th ICRC* ita-riccobene-G-abs1-he24-oral, Pune, India (2005).
6. R.J. Urick, *Principles od underwater sound*, McGraw-Hill (1983).

FIRST ACTIVITIES IN ACOUSTIC DETECTION OF PARTICLES IN UPV

M. ARDID†, J. RAMIS, V. ESPINOSA, J.A. MARTÍNEZ-MORA, F. CAMARENA, J. ALBA, V. SANCHEZ-MORCILLO

Departament de Física Aplicada, EPS Gandia, Universitat Politècnica de València, Carretera Nazaret-Oliva s/n, E-46730 Gandia, Spain

The first activities related to acoustic detection of particles by DISAO research group in the Univesitat Politècnica de València are described. We are applying some techniques from physic, engineering and oceanographic acoustics to face the high energy neutrino underwater acoustic detection challenge. The work is focused mainly in two topics: design, characterization and calibration of hydrophones, and simulation of the propagation of the signal in the sea. We present also some examples for these two topics: piezoelectric modelling and transducer simulation, calibration of hydrophones using MLS signals, and evaluation of the contribution of the sea surface noise to the deep water noise in the Mediterranean Sea by means of simulations of propagation of sound.

1. Introduction

Recently, there has been an increase of interest in high energy neutrino astronomy, and several detection techniques and experiments in this field have been developed, see for example [1]. Acoustic detection of neutrinos is one of the promising techniques that could be used for larger neutrino detector in a future. However, more research and development in this technique are required before it could be used in a neutrino telescope, and there is a lively activity for this purpose [2]. In this context, the paper describes the first activities related to acoustic detection of particles by DISAO research group in the Universitat Politècnica de València. This group is focused in different fields of acoustic research: applications of ultrasounds, nonlinear acoustics, room acoustics, etc. On the other hand, the group has also an important background in experimental particle physics. The combination of these skills could be very convenient to complement the work in acoustic detection of neutrinos carried out by the astroparticle groups, and a wider collaboration is foreseen.

† Corresponding author. Tel..: +34 962849314; fax: +34 962849309; E-mail: mardid@fis.upv.es

2. Activities Related to Acoustic Neutrino Detection

The activities carried out by DISAO group related to acoustic neutrino detection are focused in two topics: design, characterization and calibration of hydrophones, and simulation of the propagation of the signal in the sea. We show some examples of activities in these fields.

2.1. *Design of Piezoelectric Transducers*

We are developing a software package in MATLAB for the design of piezoelectric transducers based on a modified KLM model [3], which uses the localized constants method. A scheme of the model is shown in Figure 1. A detailed description of the model and the software package can be found in [4]. In the model, there are three gates: the electric gate and the backward and forward acoustic gates. The model gives the response of the whole piezoelectric transducer as functions of the materials and geometry of the transducer by means of translating these components into the corresponding electric and acoustic parameters. The influence of the backing material, the medium of irradiation, and the electric feeding are also considered in the model software package.

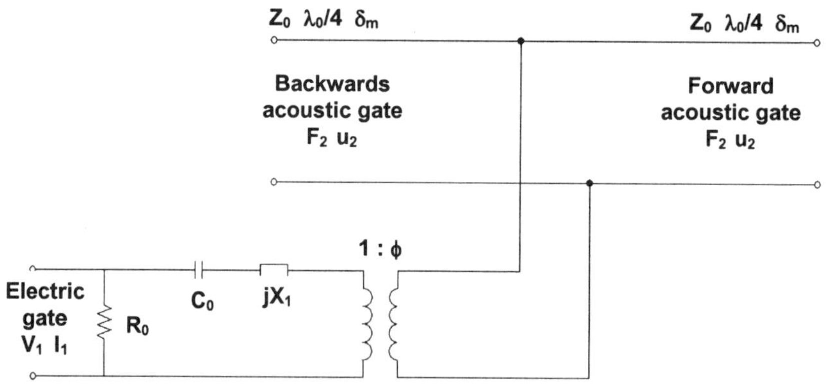

Figure 1. Scheme of the model used for the piezoelectric design with the different parameters involved.

The model has been implemented in a friendly and easy to use interface which involves several steps: the piezoelectric can be selected from a database or introduced manually; the properties of the backing and forward materials have to be introduced; it is also possible to study the effect of the feeding cable by including its properties; finally, several options for the calculations can be selected. Once all the parameters have been introduced, the simulation of the response of the transducer is done in a few seconds, and the results appear both numerically and graphically. For example, the input electric or acoustic impedances as a function of the frequency are obtained. Other possible interesting results are the emitting and receiving transfer functions, and the response to a delta excitation (in time and frequency).

From above, we could conclude that the status of the software is good, but there are still different tasks in the software that are in progress. It needs more work to validate them: a comparison between simulation and experimental results is needed, and comparisons between these kinds of simulations with those using finite elements methods would also be convenient. An upgrade of the model with secondary circuits in the gates is needed if more effects would like to be included, or more different piezoelectric geometries have to be used (not only discs). Although the limitations, we consider the present software very useful in the study of a large variety of piezoelectric transducers that could be used for acoustic detection of neutrinos and, therefore, an important tool for the optimization of this kind of detectors. The aim is that, at the end, the upgraded software could be part of the simulation package for acoustic detection of neutrinos.

2.2. *Characterization and Calibration of Hydrophones*

The characterization and calibration of hydrophones in the lab is not an easy task because there are reflections and diffraction, which could affect well-known methods of calibration like the reciprocity method. We are working in designing a better method for hydrophone calibration in two directions: improvement in the tank and improvement in the signal used. With respect to the tank, the walls, the floor and the water surface have been covered in order to make it anechoic. On the other hand, the Maximum Length Sequence (MLS) signal is used for calibration [5]. This signal is extensively used in different acoustic fields like in room acoustics. It is a pseudorandom signal, analog version of a digital sequence of values 1 and -1. It is periodic with period 2^N-1, where N is the order of the sequence. The most interesting aspects of this signal are that it has a flat frequency distribution and the circular autocorrelation gives a delta function, allowing simulating the response to an impulse easily. These aspects result in that the calibration with this signal in is not affected by noise. In figure 2 the time and frequency response of the system (two hydrophones plus the tank) using the MLS technique and the reciprocity method are shown. Notice that the response depends on the hydrophones to be studied plus the medium. Moreover, it is possible to design an easy system for calibration of hydrophones in the neutrino detection sites using the same detectors as receivers and transmitters of MLS signals. The above mentioned characteristics of flat response and no influence of noise makes this method of calibration very convenient in neutrino detection sites not only for calibration of hydrophones but also for the online calibration of the acoustic properties of the sea.

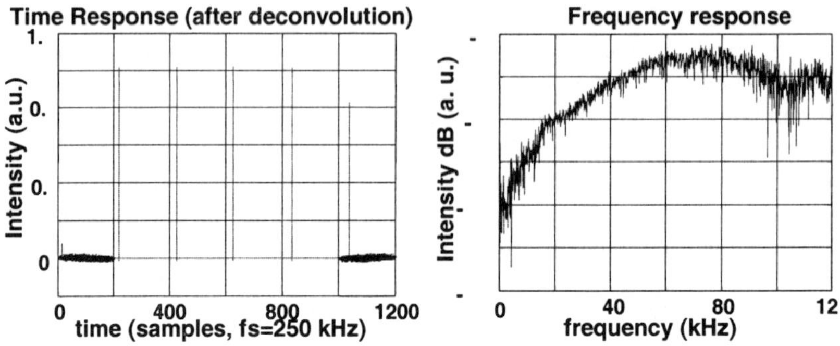

Figure 2. Time and frequency response using the MLS signal for a system composed of an emitter and a receiver hydrophone in a tank.

2.3. *Simulation of the Propagation of the Signal in the Sea*

The last topic related to underwater acoustic neutrino detection we are working on is the simulation of propagation of acoustic signals in the sea. The understanding of this aspect is essential to determine the number of acoustic sensors needed in a neutrino detection site and how separated they could be. On the other hand, development in this topic is also crucial for the neutrino source location. We are using the acoustic toolbox written by M. Porter in our simulations, which includes four acoustic models: a beam/ray trace code (BELLHOP), a normal mode code (KRAKEN), a finite element FFP code (SCOOTER), and a time domain FFP code (SPARC). Next, we show the application of this code to learn about and evaluate the contribution of the sea surface noise to the deep-water noise in the Mediterranean Sea, which could be an important aspect for the feasibility of underwater acoustic neutrino detection.

The first aspect to consider in this kind of simulations is the sound speed profile since the results are very dependent of them. The sound speed profile depends very much on the salinity and temperature; therefore it could vary considerable for different seasons and for different seas. Our simulations have been done for a typical speed profile of the Mediterranean Sea. BELLHOP code shows easily that rays are bended and rays emitted in small angles do not reach deep water locations. This could result in a small contribution of the surface noise to deep water noise, especially in case of directive sources.

The transmission loss as a function of the range has been studied with the KRAKEN normal code [6] considering a source in the surface and measuring in the sea floor for two different depths: 2400, and 4100 m. Results for a 1 kHz and a 15 kHz sources respectively are shown in figure 3. Naturally, the transmission loss is higher at high frequencies. With respect to the depth, the transmission loss is in average a little higher for the deeper site, however there could be significant variations depending on the range, that is, the distance considered, which could have some importance in the election of the neutrino detection site.

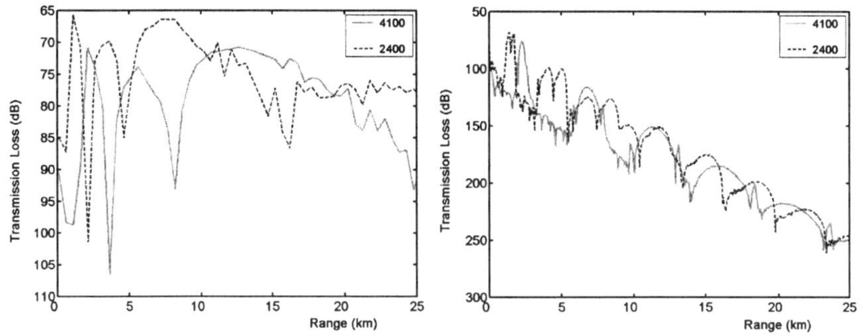

Figure 3. Transmission loss for a source of 1 kHz (left) and 15 kHz (right) in the surface and measuring in the sea floor at 2400 and 4100 m.

Acknowledgments

We would like to acknowledge the Spanish Ministerio de Educación y Ciencia for supporting this work, project number FPA2002-12566-E.

References

1. C. Spiering, *Nucl. Phys. B-Proc. Sup.* **125**, 1 (2003).
2. This issue. Proceedings of the Acoustic and Radio EeV-Neutrino Activities Workshop, DESY-Zeuthen 2005, *Int. J. Mod. Phys.* **A** (2006).
3. R. Krimholtz, D. Leedom and G. Matthaei, *Electron. Lett.* **6**, 398 (1970).
4. J. Ramis, J. Alba and D. Peiró, *Ultrans 1_0: Software para el diseño de transductores ultrasónicos*, Actas de Tecniacústica 2005, Terrassa, Spain (2005) (in Spanish).
5. J. Borish and J.B. Angell, *J. Audio Eng. Soc.* **31**, 478 (1983).
6. M.B. Porter and E.L. Reiss, *J. Acoust. Soc. Am.* **76**, 244 (1984).

THE UPPER LIMIT TO THE EHE NEUTRINO FLUX FROM OBSERVATIONS OF THE MOON WITH KALYAZIN RADIO TELESCOPE

R.D.DAGKESAMANSKII, A.R.BERESNYAK[†], A.V.KOVALENKO

Pushchino Radio Astronomy Observatory, Lebedev Physical Institute, Russian Academy of Sciences, Pushchino, Moscow region, 142290, Russia

I.M.ZHELEZNYKH

Institute of Nuclear Research, Russian Academy of Sciences, 60-letiya Oktyabrya prosp.,7a, Moscow, 117312, Russia

Very brief history of the RAMHAND-type experiments is presented. Some distinctive features of the Kalyazin experiment is described, and the first results obtained in it are discussed.

1. Some milestones on the way to high-energy neutrino detection

In his papers published in 1961 and 1965 Gurgen Askaryan [1,2] had showed that a cascade that develops in a dense media should has a negative charge excess, so the corresponding pulse of coherent Cherenkov radio emission could be expected from such cascade. In these papers Askaryan estimated roughly a spectrum and an intensity of the pulses.

Basing on Askaryan's idea, G.A.Gusev and I.M.Zheleznykh proposed in 1983 [3] the Radio Antarctic Muon And Neutrino Detector (RAMAND) with the radio antennas "listening" to the Antarctic ice massive. Realization of the RAMAND project started at Soviet Antarctic Station "Vostok" in second half of 1980th, but unfortunately almost stopped in the beginning of 1990th.

The next step on the way to detection of super-high energy neutrinos by radio method was made by Zheleznykh in 1988 [4], when he suggested the new RAdio Moon Hadron And Neutrino Detector (RAMHAND). In 1989, Dagkesamanskii & Zheleznykh [5] had obtained the first estimates of the Moon target volume for the neutrinos of extremely high energies and a sensitivity of the large ground-based radio telescopes as a detector of the coherent Cherenkov radio emission pulses.

[†] Present address is University of Wisconsin-Madison, Dept. of Astronomy

Beginning from the mid of 1990s the RICE (Radio Ice Cherenkov Experiment), ANITA-project and some other neutrino radio detection experiments were suggested (most of them propose to use the Antarctic ice as a target). However, in this paper we will concentrate our attention on the monitoring of the nanosecond pulses from the Moon, *i.e.* on the RAMHAND-type experiments for high-energy neutrino detection.

2. Parks and GLUE experiments

In 1996 T.Hankins, R.Ekers and J.D.O'Sullivan [6] had made a first attempt to register the nanosecond pulses from the Moon, using the 64-meter radio telescope of the Parks Radio Observatory in Australia. The authors had not registered any nanosecond radio pulses from the Moon at decimeters wavelengths and had put the upper limit to a rate of the events observed from central part of the lunar disk.

The second RAMHAND-type experiment was made by P.Gorham, D.Saltzberg and their colleagues. For observations, they used 70-meter and 34-metr radio dishes of the Goldstone DSN station. In this experiment, a point of the lunar disk that closes to the limb, was observed, where the maximum events should be expected, as the authors had showed. There was not registered any event that could be considered unambiguously as a high-energy neutrino detection in the experiment, and the authors of the GLUE (Goldstone LUnar Experiment) put the first strong upper limit to the flux of cosmic neutrinos with energies above 10^{20} eV [7].

The upper limit to the neutrino flux, found in GLUE-experiment, closed only some exotic and not very popular models of the Universe. On the other hand, the excellent SLAC experiment, made by D.Saltzberg, P.Gorham *et al.* [8], entirely confirms the results of numerous theoretical predictions of the Cherenkov pulses from SHE neutrino cascades and their parameters (see, for example, [1,2,4,9-11]). For this reason, our team from Pushchino Radio Astronomy Observatory decided to continue the preparations to the RAMHAND-type experiment and started corresponding observations of the Moon in 2002.

3. Description of the Kalyazin experiment

Some of the first observations we made with 22-meter dish of the Pushchino Radio Astronomy Observatory, but the most fruitful results have been obtained with Kalyazin 64-meter radio telescope. Though the idea of Kalyazin experiment is the same as in the Parks and Goldstone observations, there are several distinctive features in it. First and the main, Kalyazin 64-meter radio telescope has a multi-frequency feed system, so the simultaneous observations of the same point of the Moon at several different radio bands are available with it. This is a great advantage, because in this case the rather large delay of the signal at lower frequency due to dispersion in the Earth ionosphere could be used to separate any cosmic signal from the local interferences. We use it and make simultaneous observations of the same region of the Moon at two (sometimes at three) frequencies. Our main frequencies are 1.4 and 2.3 GHz, and it seems, they are not so far from the optimal frequency for such observations. The antenna beamwidths are ~ 11' at 1.4 GHz and ~ 7' at 2.3 GHz. The circular polarization feeds is used at all frequencies. As it has mentioned above, the expected events should be near the limb of the Moon, so we used the pointing that was at 14' apart from the center of the Moon. Trigger system with a time resolution of 2 ns based on 4-channels digital oscilloscope (Tektronix TD3034) was used to record the all suspicious events. During the monitoring, we used several a little bit different options in different observational sessions. Flux density threshold, which corresponded to the most sensitive option, was 3500 Jy.

4. Discussion of our first results

The total "live-time on the Moon" for Kalyazin experiment is slightly more than 60 hrs up to now. With this "exposition" we had not found any reliable radio pulse that could be considered as Cherenkov emission pulse from the Moon. Using the method described by Beresnyak [12], we estimated effective volume (aperture) of the lunar target. Taking into account the estimate of the effective volume, the differences in sensitivity, realized in different observational sessions, and some other specific conditions of the Kalyazin experiment we derived an upper limit to the extremely-high energy (> 10^{20} eV) neutrino flux. Figure 1, that corresponds to figure 2 from our paper [13], show some theoretical predictions of high-energy neutrino flux together with the experimental estimates of the flux by different groups. It could be easily seen that our estimate is more conservative than the results of GLUE experiment. As discussed in papers [12] and [13], the main cause of the difference are the different estimates of the effective target volume. Another possible cause could be the difference in the slopes of the neutrino spectrum above 10^{20} eV, as well as the differences in some other model parameters, suggested by two groups.

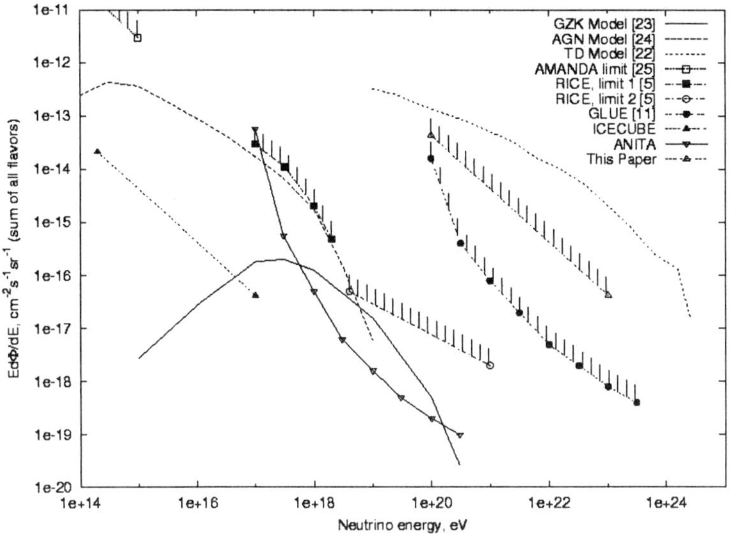

Figure 1. Theoretical predictions and experimental limits of extremely high energy neutrino flux.

As it was mentioned above, the estimates of the extremely high-energy neutrino flux obtained up to now exclude only most exotic models of the Universe. To reduce an upper limit to the neutrino flux or to measure it we need in more "live-time on the Moon", as well as in higher sensitivity and reliability of the observations. We consider the following ways to achieve these: to make more and more observations, to use wider receivers bandwidths (now the bandwidths of the receivers used in Kalyazin observations are only about 120 MHz), to increase the numbers of frequency channels with coincidence scheme between them. The last but not the least, much higher reliability of the results could be achieved in international cooperative observations, such as suggested by I.Zheleznykh in 1988 [4]. Indeed, using simultaneous observations of the Moon with several high sensitive radio telescopes, we can get the results that will be very important not only for astrophysics and cosmology, but for neutrino physics, too.

Acknowledgments

This work supported by Civilian Research Development Foundation (CRDF grant No. 2624), and by Russian Academy of Sciences (Progran "No stationary processes in the Universe"). R.D.D., A.V.K. and I.M.Zh. are grateful the

ARENA-2005 Organizing Committee for financial support during the conference.

References

1. I.G. Askaryan, 1961, Zh. Eksp. Teor. Fiz., V.41, p.616 [Sov. Phys. JETP V.14, p.441 (1961)].
2. G.Askaryan, 1965, Zh. Eksp. Teor. Fiz., V.48, p.988 [Sov. Phys. JETP, V.21, p.707, (1965)].
3. G.A. Gusev and I.M. Zheleznykh, 1983, Pis'ma Zh. Eksp. Teor. Fiz., V.38, p.505 [JETP Lett., V.38, p.611, (1983)].
4. I.M.Zheleznykh, 1988, Proc. 13th Intl. Conf. Neutrino Physics and Astrophysics, p.528.
5. R.D.Dagkesamanskii and I.M.Zheleznykh, 1989, Pis'ma Zh. Eksp. Teor. Fiz., V.50, p.233 [JETP Lett., V.50, p.259 (1989)].
6. T.H. Hankins, R.D. Ekers and J.D. O'Sullivan, 1996, Mon. Not. R. Astron. Soc., V.283, p.1027.
7. P.W. Gorham, K.M. Liewer, C.J.Naudet, D.P.Saltzberg and D.Williams, 2001, Proc. RADHEP-2000, eds. D.Saltzberg & P.Gorham, AIP Conf. Proc., V.579, p.177.
8. D. Saltzberg, P.W. Gorham, D. Walz, C. Field, R. Iverson, A. Odian, G. Resch, P.Schoessow, and D. Williams, 2001, Proc. RADHEP-2000, eds. D.Saltzberg & P.Gorham, AIP Conf. Proc., V.579, p.225.
9. M.A. Markov and I.M. Zheleznykh, 1986, Nucl. Instrum. Methods Phys. Res. A, V.248, p.242.
10. E. Zas, F. Halzen, and T.Stanev, 1992, Phys. Rev. D 45, p.362.
11. J. Alvarez-Muniz and E. Zas, 2001, Proc. RADHEP-2000, eds. D.Saltzberg & P.Gorham, AIP Conf. Proc., V.579, p.128.
12. A.R.Beresnyak, 2003, astro-ph/0310295.
13. A.R.Beresnyak, R.D.Dagkesamanskii, I.M.Zheleznykh A.V.Kovalenko, and V.V.Oreshko, 2005, Astronomy Reports, V.49, p.127 [Astronomicheskii Zhurnal, V.82, p.149 (2005)].

USING THE WESTERBORK RADIO OBSERVATORY TO DETECT UHE COSMIC PARTICLES INTERACTING ON THE MOON

J. BACELAR, O. SCHOLTEN

KVI, University of Groningen, The Netherlands

A.G. DE BRUYN, H. FALCKE

ASTRON, Dwingeloo, The Netherlands

Ultra-High-Energy (UHE) particles of cosmological origin (cosmic-rays and neutrinos), carry information on the most spectacular events known. These extremely energetic (energies larger than 1 ZeV = 10^{21} eV) cosmic-rays or neutrinos initiate in the lunar regolith a cascade of charged particles which acts as a radio pulse emitter. The instantaneous power produced can be detected here at the Earth, with a radio telescope operating at the optimal frequency window around 150 MHz. Using 12 telescopes of the Westerbork Synthesis Radio Telescope, WSRT, with a field of view covering the whole lunar surface, our calculations show that one should identify 10 UHE events within an observation time of 500 hours, assuming an extrapolated power law dependence of the highest ever measured cosmic-ray events, around an energy of 10^{20} eV. A null result will determine unambiguously the GKZ effect for the cosmic-ray flux and improve the present world upper limit on the neutrino flux above 1 ZeV, by three orders of magnitude, allowing for the first time to test the Waxman-Bahcall neutrino flux limit.

1. Introduction

Cosmic-rays arriving at the Earth show a continuous flux spectrum which decreases exponentially with the energy of the particle (Flux $\propto E^{-2.7}$) [1]. When an ultra-high-energy particle (cosmic-ray or neutrino) penetrates the lunar surface there is a substantial chance that in the lunar regolith it will interact producing a shower of charged particles. In matter (the lunar rock) the shower develops a leading cloud of negatively charged particles moving at velocities close to the speed of light in vacuum. Since this velocity is much higher than the speed of light in matter, Čerenkov radiation is emitted. Since the charged particles in the shower move simultaneously the emitted radiation is coherently enhanced for wavelengths which are of the same order of magnitude as the dimension of the shower: radio waves in our case (the so-called Askaryan effect). As pointed out in [2], at frequencies around 150 MHz, this coherent emission is isotropic, allowing essentially all particles impinging on the surface of the moon to be observed from the Earth. Furthermore, the lunar sub-surface attenuation is sufficiently low at these frequencies, that a significant interaction depth (>100m) is still visible. The very large interaction volume makes this the most competitive method of detecting UHE events in the range 10^{21}-10^{24} eV.

Several different kinds of astrophysical phenomena can produce particles (protons) of UHE energies predominantly through shock-wave acceleration [3]. At energies exceeding 0.05 ZeV (the so-called Greisen-Zatsepin-Kuzmin (GZK)-limit) these protons produce pions when they interact with a low-energy photon from the omni-present Cosmic Microwave Background. Such an interaction has a mean free path of only 10 Mpc. The occurrence of this process has two important consequences: *i)* it implies that UHE protons with energies exceeding the GZK limit which are observed on earth can only come from relatively nearby sources [4]; *ii)* pions are produced with energies of the order of the original proton. The decay of these pions gives rise to UHE neutrinos which may traverse the universe basically un-attenuated. Thus it is expected that, for energies larger than 1 ZeV, the composition of the cosmic-ray spectrum will contain a significant fraction of neutrinos. The experimental determination of the GZK-cutoff effect is still under debate.

2. Production and transmission of the radio signal from the Moon

At energies around 1 ZeV, the interaction of cosmic rays (protons) in matter is practically instantaneous. Neutrinos on the other hand have a mean free path in rock given by $\lambda_\nu = 60\ E_\nu^{-1/3}$ km (where E_ν is the neutrino energy in ZeV) based on measured neutrino-matter interaction lengths supplemented by theoretical modeling. We treat the first interaction point accordingly and let the shower develop into the regolith in the same direction as the incoming particle, the Askaryan effect. This effect has been measured in the laboratory. At SLAC, Stanford USA, experiments were performed [6] with pulsed-beams of electrons and photons impinging on a large sand container and measuring the radio waves emitted from this container. In this way the amplitude predicted by the Askaryan effect was checked, up to an equivalent incoming cosmic-particle energy of 10^{19} eV.

The power density scales with E_s^2 (the incoming particle energy) and has a peak at around 3 to 5 GHz. In the frequency range of interest (0.1-0.2 GHz) it increases quadratically with frequency and linearly with the bandwidth of the measurement $\Delta\nu$.

The coherence requirement for the emitted radiation depends on the size of the shower and the wavelength. The lateral extension of the shower reaches sizes of order 10 cm, which determines the peak of the power emitted at the Čerenkov angle . The longitudinal extension of the shower is of order of meters, and only weakly (logarithmically) dependent on the energy. Therefore, for radiation with a wavelength of meters, the shower starts to look like a point source and coherence is obtained at all angles. One can approximate the spreading width of the Čerenkov angle by the following formula: Δ_c= 2.5 (3/ν), where the frequency ν is given in GHz. For the emitted radio waves to leave the lunar surface, they need to be transmitted across this boundary, where the internal reflection angle is the complementary angle of θ_c (for the lunar regolith $\theta_c = 56^0$). This means that at frequencies around the maximum emitted power, 3-5 GHz, most of the radiation is internally reflected. On the other hand, at 0.1 GHz, the Čerenkov cone has a spreading angle $\Delta_c = 75^0$, which ensures emitted power radiation from the lunar surface for all incoming particle angles. This radiation can then be detected at the Earth if the power density is larger than the detection threshold [2].

A calculation of the total effective volume of the moon to detect UHE cosmic-rays and/or neutrinos involves a proper numerical integration of the neutrino flux, the interaction depth, the propagation of the radio-wave through the medium and the interface at the lunar surface, and the roughness of the surface for structures with size characteristic of the radio-wavelength. Such a code was developed in order to allow count rate estimates to be calculated. The code, clearly predicts that the best frequency window of observation is 100-200 MHz (see [2] for details).

3. Detection of radio-pulses with the WSRT array

The radio-observatory of Westerbork, WSRT, consists of 14 dishes of 25 meter diameter. Ten are at fixed locations, linearly arranged with a separation of 144 meters, and four dishes can be moved collinearly with the rest. In the so-called tied array mode the field of view corresponds to a beam which is narrow in one dimension, covering half the moon for low declination observations. The backend electronics of the WSRT, PuMa II, allows 8 different beams to be measured simultaneously, each with a bandwidth of 20 MHz. The data is sampled every 25 nsec and both polarization amplitudes are measured. Another feature that makes this array unique is its LFFE, low frequency band receiver, operating at just the right frequency band of 117-175 MHz. The celestial noise of 7000 Jy per antenna at this low-frequency band is reduced to 580 Jy by phasing 12 dishes of the array. This determines the low energy threshold of our measurement (see figure 1). The phases of the different antennas are set for a specific object in space, which is continuously tracked during the observing time. At present, using discretionary observational time at WSRT, some possible

setups of the tied-array are being tested in a series of one hour observations tracking the moon. The data obtained so far, at frequencies of 117 and 147 MHz, show the predicted celestial noise, with little ground based interference.
We propose to track the Moon, with the WSRT operating in the mode described above, over a period of three years, accumulating 500 hours of observational time. The expected flux limit for 500 hour measuring time is given in figure 1 below for both cosmic-ray and neutrino interactions. The flux limit is defined as the flux needed to observe one count during the accumulated observational time.

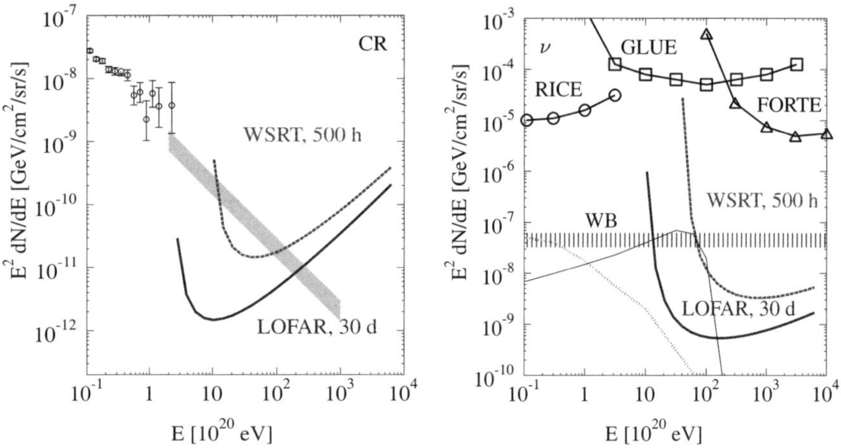

Figure 1: Flux limits for 500 hours observational time for cosmic-rays (l.h.s.) and neutrino (r.h.s) events with the Westerbork facility. Also given are the predicted results for the future LOFAR facility. Published data for cosmic rays [8] and present best limits published for neutrino fluxes [9-11] are also shown. Theoretical predictions (see text) are shown for the neutrino flux.

If one looks at the left hand side of figure 1, then the 500 hour period of observation could provide 10 events of cosmic-ray origin at the energy around 3 10^{21} eV, if the power law of the published cosmic-ray flux observed at the lower energy region is extrapolated to higher energies (gray band in the l.h.s. panel). If the GZK process indeed occurs, then the extrapolation to higher energies is model dependent [4], i.e. it depends on the cosmic-ray source distribution in the cosmos. Our data will for the first time, measure the magnitude of the drop in flux above the GZK cut-off.

For the neutrino flux, the limit from the proposed experiment, will for the first time, test the Waxman and Bahcall prediction [12] (curve labeled WB in the r.h.s. panel of figure 2). This theoretical prediction is based on the observed cosmic-ray and photon fluxes at lower energies, and is viewed as a rather sturdy upper limit. Other theoretical model predictions shown in the r.h.s. panel of figure 1 are the GZK neutrino flux (green line) and a typical topological-defect model prediction [5] (blue thin line).

In the past some experiments have been performed [9,13] attempting to measure the radio-pulses from cosmic-particle interactions on the moon. These experiments were performed at 2.2 GHz, a frequency close to the peak of the amplitude which in the case of the lunar regolith corresponds to 3-5 GHz. The best result was obtained by the GLUE collaboration [9]. Our calculations, when performed at the frequency of the GLUE experiment, and with their energy thresholds, yields a neutrino flux limit which is in very good agreement [2] with their published results. As is shown in figure 1, the efficiency at these high frequencies is orders of magnitude worse than the proposed low frequency region of 100-200 MHz.

Other large international collaborations have performed experiments to measure neutrino flux limits at UHE. Among these, the best limits were set by: i) RICE [10], using radio antennas in the Antarctica, looking for neutrino interaction in the ice-cap; ii) FORTE [11] using antennas in a satellite orbiting above Iceland, looking at the north-polar ice-cap. As seen in figure 3, our predicted results improve all attempts performed so far by three orders of magnitude. A planned experiment, ANITA, attempts to fly a balloon for 45 days around the South Pole, measuring radio signals from neutrino interactions on the ice-cap, improving the RICE results by two orders of magnitude.

The LOFAR (LOw Frequency Array) antenna is perfectly designed for this work. With its low frequency band, of 100-240 MHz and high rate sampling mode, with a time sampling of 5 nsec, it is suited to repeat this experiment and set even more stringent flux limits. Predictions for a 30 day observational period are given also in figure 1. Although the full capability of LOFAR is only expected in 2008, we are planning to use the test station of LOFAR, available from 2006, to test the LOFAR system for possible future lunar observations.

References

[1] A.A. Watson, Contemporary Physics 43, 181 (2002) ; M. Nagano and A.A. Watson, Reviews of Modern Physics 72, 689 (2000).
[2] O. Scholten, J. Bacelar, R.Braun, A.G. de Bruyn, H. Falcke, B. Stappers and R.G. Strom, Astroparticle Physics (submitted July 2005),astro-ph/0508580.
[3] P. Bierman, J. Phys. G23, 1 (1997); A.M. Hillas, Ann. Rev. Astron. Astrophys. 22, 425 (1984)
[4] A. Achterberg et al., MNRAS 328, 393 (2001).
[5] S. Yoshida et al., Astrophysics Journal 479, 547 (1997), P. Chattacharjee, C.T. Hill and D.N. Schramm, Phys. Rev. Lett.. 69, 567 (1992), T. Stanev, astro-ph/0411113.
[6] D. Saltzberg et al., Phys. Rev. Lett. 86, 2802 (2001); and P.W. Gorham et al., Phys. Rev. E62, 8590 (2000).
[7] G.R. Olhoeft and D.W. Strangway, Earth Plan. Sci. Lett. 24, 394 (1975).
[8] M. Takeda et al., Astropart. Phys. 19, 447 (2003).
[9] The GLUE experiment: P. Gorham et al., Phys. Rev. Lett. 93, 41101 (2004).
[10] The RICE experiment: I. Kravchenko et al., Astropart. Phys. 20, 195 (2003)
[11] The FORTE experiment: N.G. Lehtinen et al., Phys. Rev. D69, 013008 (2004)
[12] J. Bachall and E. Waxman, Phys. Rev. D64, 64 (2001).
[13] R.D. Dagkesamanskii and I.M. Zheleznyk, Sov. Phys. JETP 50, 233 (1989), T.H. Hankins, R.D. Ekers and J.D. OSullivan, Mon. Not. R. Astron. Soc. 283, 1027 (1996).

UPDATED LIMITS ON THE ULTRA-HIGH ENERGY (UHE) NEUTRINO FLUX FROM THE RICE EXPERIMENT

I. KRAVCHENKO

M.I.T Lab. for Nuclear Science, Cambridge, MA 02139

C. COOLEY

Whitman College Dept. of Physics, Walla Walla , WA 99362

D. SECKEL

Bartol Research Institute, U. of Delaware, Newark, DE 19716

J. ADAMS, S. CHURCHWELL, P. HARRIS, S. SEUNARINE, P. WAHRLICH

Department of Physics and Astronomy, Private Bag 4800, U. of Canterbury, Christchurch, New Zealand

A. BEAN, D. BESSON, S. GRAHAM, S. HOLT, S. HUSSAIN, D. MARFATIA, D. MCKAY, J. MEYERS, J. RALSTON, R.SCHIEL, H. SWIFT

U. Kansas Dept. of Physics and Astronomy, Lawrence KS 66045-2151

J. LEDFORD, K. RATZLAFF

U. Kansas Instrumentation Design Laboratory, Lawrence KS 66045-2151

The RICE experiment (Radio Ice Cherenkov Experiment) at South Pole consists of an array of dipole antennas designed to detect the coherent radio frequency radiation produced by neutrino-induced showers in the Antarctic ice. We report updated limits on the ultra-high energy neutrino flux, based on RICE data taken between 2000 and 2004. These limits also reflect improvements in Monte Carlo simulations and detector modeling.

1. Introduction and Overview of RICE

The RICE experiment[1,2,3] is designed to detect UHE neutrinos[4] with energies above ~10 PeV. UHECRs are observed at energies above $10^{19.5}$ eV and are guaranteed to produce so called "GZK" neutrinos[7] with EeV energies during propagation. Such neutrinos are probes of the evolution of UHECR sources and provide a floor prediction useful for baseline detector design.

Early results from RICE[2] were based on one month (333 hours live) of data analyzed for the presence of electron neutrinos. Subsequent studies included the possibility of hadronic showers and expanded the data collection time to greater than two years[3]. Here we describe the result of an improved analysis performed on data collected over a five year time frame. To date the experiment has not detected neutrinos, but places interesting limits on models of neutrino production in the energy range of 100 PeV - 1 ZeV.

During its operation RICE has consisted of an array of 16-20 radio antennas deployed within a roughly 200m$\times$200m footprint at depths of 100m-300m near South Pole. The array is designed to intercept the Cherenkov cone of coherent, radio-frequency radiation from an UHE shower produced by a cosmic ray neutrino interaction in the Antarctic icecap. After pickup by the antenna, the signal is amplified and transmitted by coax cable to the surface. In the surface DAQ the signals are filtered, amplified again, and split into two copies: one for triggering and one for digitizing and analysis of the pulse waveform.

Data acquisition is triggered by the arrival of 4 pulse hits within a 1.2 microsecond window. The pulses must exceed a common discriminator threshold V_d which coarsely tracks the background noise level. The pattern of arrival times is used to form an on-line veto against noise sources located on the surface. The off-line analysis includes tests based on wave form quality, vertex location and the ability to reconstruct a Cherenkov radiation pattern based on signal amplitudes in the receiver channels.

Effective Volume. Expectations for the RICE experiment are determined by Monte Carlo simulation. Details of the neutrino interaction determine the spectrum and radiation pattern of the shower. Showers initiated by electrons (e) are elongated by the LPM effect, whereas hadronic (h) showers initiated by quark jets are not. As a result electron initiated showers have narrow radiation patterns and exhibit reduced detection efficiency. We have refined the modeling of the showers and RF production, resulting in a modest increase in the signal strength[8]. Once produced, the radiation propagates to the antenna array, suffering both refraction and attenuation. Since the RICE antennas are located in the firn, a region of changing density and index of refraction[9], ray tracing calculations lead to reduction of the ice volume visible to the array. Recent *in situ* measurements[10] lead to a reduced attenuation length. These effects reduce $V_{\rm eff}$ as compared to previous publications[1,2,3].

The final result of the Monte Carlo is an effective volume V_{eff}, defined as the volume integral of the detection efficiency for interaction vertices near the array. This integral is averaged over the solid angle of the neutrino beam, taken to be 2π for a diffuse flux shadowed by the Earth. The new results are somewhat more sensitive at low energies due to adjustments for various instrument settings, but the improvements in modeling ice propagation reduce the sensitivity for $E_s > 10^{20}$ by a factor of ~ 3 when compared to previous studies.

Online Veto and Analysis Efficiency. Monte Carlo events which would otherwise trigger the detector are used to estimate the efficiency for neutrino events to pass the online veto and the analysis filters. Modeled waveforms for a MC event are embedded in randomly selected "unbiased" events, collected to monitor noise characteristics. These events pass a code simulating the online veto with an efficiency of 0.86. These events are also given to the analysis chain, and pass with an efficiency of 0.70. The combined efficiency is $\epsilon = 0.60$.

Operations & Results. During the five years covered in this analysis, RICE was operational approximately 80% of the time. For about half of this time, the South Pole Station satellite uplink operated at an amplitude which overwhelms the RICE DAQ. Additional deadtime arose while applying the online surface veto and readout of the oscilliscopes which comprise the digitization side of the DAQ, leaving a total of approximately 1.5 year of useful data.

The final data set includes of order 10^6 events that are analysed for neutrino interactions. After applying the analysis cuts, 43 events remained. These were examined by hand, and all were found to have defects that eliminated their consideration as neutrino events. Consequently, we can only report flux limits at this time.

2. Limits on the flux of UHE neutrinos

With no events, the RICE results may be used to place bounds on the number of events expected for a given model. Let the model have an overall normalization, A. If the expected number of events for $A = 1$ is N, then with zero events the 95% CL upper limit on the normalization is $A_{95} = \frac{3}{N}$.

For a given flux model ϕ, the number of events expected during the

RICE exposure is given by

$$N = \sum_{i,a,j} \int dt_j dE_\nu dy \; 2\pi\epsilon V_j^a(E_s(E_\nu)) \frac{d\sigma_i^a}{dy}(E_\nu) \frac{d\phi_i}{dE_\nu} \tag{1}$$

where E_ν is the neutrino energy, σ is the νN cross-section, y is the inelasticity, ϵ is the combined efficiency, and V_j is the effective volume for the experient configuration during time interval $_j$. The event rate includes a sum over flavors $_i$ and event types ^{a}. All flavors of neutrino create h-showers as recoil jets in charged (CC) and neutral (NC) current reactions. One flavor (ν_e) creates e-showers in CC events. For h-showers the shower energy is $E_s^a = yE_\nu$, whereas for the e-shower in ν_e CC events $E_s^a = (1-y)E_\nu$.

For cross-sections we use isoscalar-target SM cross sections evolved to high energy. The ingredients include the tree-level parton amplitudes and CTEQ 6.2 parton distribution functions, with Q^2 extrapolation where required. We include a 20% reduction due to the nuclear effects in oxygen. For a given flux model, neutrino mixing is assumed to distribute the total flux equally across all three flavors. Due to the competition between the LPM effect and an average inelasticity of $y \sim 0.2$, for $E_\nu < 1$ EeV detection of an isoflavor flux is dominated by e-showers, but h-showers dominate above 1 EeV.

Our 95% C.L. bounds on representative ν-flux models are shown in Fig. 1. The illustrative AGN models are ruled out at 95% C.L., but the Waxmann-Bahcall standard is below our limits. The GZK flux models differ substantially. ESS and PJ, keyed to models of the star formation rate, are below the RICE sensitivity. The KKSS flux, constructed to saturate bounds derived from EGRET observations, is just barely consistent with our 95% C.L. limit, i.e. RICE should have detected 2 events for this model but saw none. Direct UHE neutrino detection is competitive with limits derived from complementary photon and cosmic ray observations.

3. Acknowledgements

We gratefully acknowledge the NSF Office of Polar Programs for support.

References

1. RICE Collaboration, I. Kravchenko *et al.*, Astroparticle Physics **19**, 15 (2003).
2. RICE Collaboration, I. Kravchenko *et al.*, Astroparticle Physics, **20**, 195 (2003).

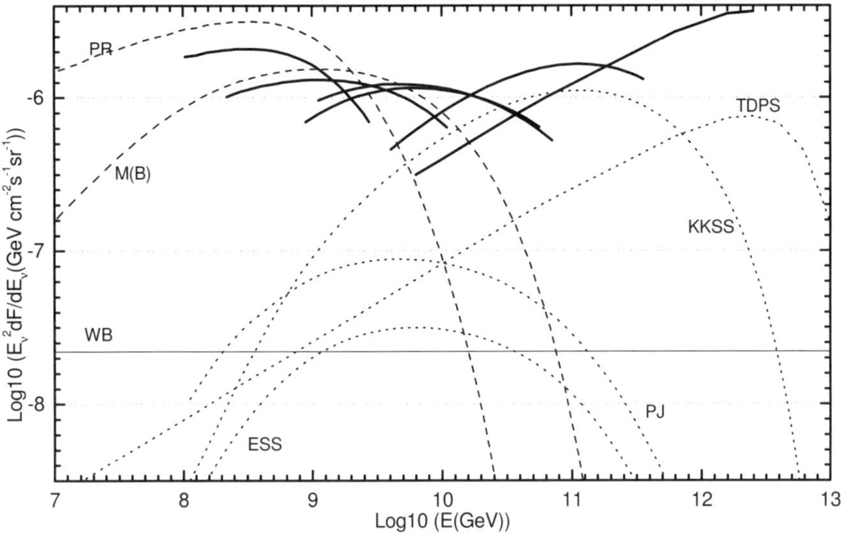

Figure 1. Upper bounds on total (all flavor) neutrino fluxes for AGN models of PR and MB GZK neutrino models of ESS, PJ, and KKSS, and the topological defect model of PS, due to all flavor NC+CC interactions, based on 2000-2004 RICE livetime of about 13200 hrs. Thin curves are for model fluxes and the thick curves are the corresponding bounds. The energy range covered by a bound represents the central 80% of the event rate.

3. RICE collaboration, I. Kravchenko *et. al*, in proceedings of 28th ICRC, Tokyo, 2003 (Universal Academy Press, Tokyo, 2003).
4. V.S. Berezinsky and G.T. Zatsepin, Phys. Lett. **28B**, 423 (1969); F.W. Stecker, Astrophys. J. **228**, 919 (1979)
5. R. J. Protheroe, astro-ph/9607165; K. Mannheim, Astropart. Phys. **3**, 295 (1995).
6. R. J. Protheroe and T. Stanev, Phys.Rev.Lett. 77 (1996) 3708-3711; Erratum-ibid. 78 (1997) 3420.
7. R. Protheroe and P. Johnson, astro-ph/9506119; R. Engel, D. Seckel and T. Stanev, Phys. ReV. D **64**, 093010 (2001); O. Kalashev *et al.*, Phys. Rev. D **66**, 063004 (2002).
8. S. Hussain and D. McKay, Phys. Rev. D **70**, 103003 (2004); S. Hussain and D. Seckel, in preparation.
9. Kravchenko, I, *et al.* "*In situ* measurements of the index of refraction of the South Polar firn with the RICE detector", *J. Glaciol.* in press.
10. S. W. Barwick et al., submitted to J. Glaciol (2004).
11. J. Vandenbroucke, Proceedings of 29th ICRC (2005).

THE ANITA COSMOGENIC NEUTRINO EXPERIMENT

PETER W. GORHAM, FOR THE ANITA COLLABORATION
Department of Physics and Astronomy, University of Hawaii at Manoa, 2505 Correa Road, Honolulu, HI, 96822; E-mail: gorham@phys.hawaii.edu

We report on new limits on cosmic neutrino fluxes from the flight of the Antarctic Impulsive Transient Antenna (ANITA) prototype, dubbed ANITA-lite, which completed an 18.4 day flight of a long-duration balloon (LDB) payload in early 2004.

Cosmic rays of energy above 3×10^{19} eV are almost certain to be of extragalactic origin. Their gyroradius far exceeds that required for magnetic confinement in our galaxy. At this energy however, pion photoproduction losses on the cosmic microwave background radiation (CMBR) via the Greisen-Zatsepin-Kuzmin (GZK [1]) process limit their propagation distances to the local supercluster, of order 40 Mpc or less. Within this volume there are no known sources yet identified. Compounding the mystery, the measured cosmic ray spectrum to date shows hints but as yet no compelling feature due to the GZK-process absorption. This leads to uncertainty both on the nature of the sources and their cosmic spatial distribution, and has fueled a variety of new experiments to elucidate this mystery [2,3].

Neutrinos are coupled to the highest energy cosmic rays both as a direct byproduct, and perhaps as a potential source of them. In the former case, simple arguments (which are relatively insenstive to the uncertainty surrounding the cosmic ray spectrum) lead to the conclusion that there is a "guaranteed" cosmogenic neutrino flux [5] with a broad peak in the energy range of 10^{17-19} eV. Neutrinos may not only be cosmogenic byproducts, but could also be closely associated with sources of the UHECR, though this possibility is far more speculative. If there are large fluxes of neutrinos at energies of order 10^{22-23} eV, they can annihilate with Big-Bang relic cosmic background ($T_\nu \sim 1.9$K) neutrinos in our own Galactic halo via the interaction $\nu\bar{\nu} \rightarrow Z^0$, the *Z-burst* process [12,11,13,14]. The neutral weak vector boson Z^0 then decays immediately into hadrons in part, yielding UHECRs in the process, and overcoming the GZK cutoff because of the nearby production. Analogous to this are Topological Defect (TD) models [10], which postulate a flux of super-heavy (10^{24} eV) relic particles decaying in our current epoch, and within the Earth's GZK sphere, to yield both neutrinos and UHECR hadrons in the process.

The NASA-sponsored ANITA mission, now completing construction for a first launch as a long duration balloon (LDB) payload in 2006, has a primary design goal of detecting EeV cosmogenic neutrinos, or providing a compelling limit on their flux at a level which would require re-evaluation of the standard assumptions above, challenging our current understanding of the physics or astrophysics involved. We report here on the first flight of a prototype instrument that was developed as a proof-of-concept for the ANITA mission. ANITA detects neutrino interactions through coherent radio Cherenkov emission from neutrino-induced electromagnetic (EM) particle cascades within the ice sheet exploiting a property of EM cascades that has become known as the Askaryan effect [7]. The prototype payload, known as ANITA-lite, flew as a piggyback instrument aboard the Trans-Iron Galactic Element Recorder (TIGER) LDB payload. The flight made 1.3 circuits around the Antarctic continent, from Dec. 18, 2003 to Jan. 6 2004, for a total 18.4 days aloft. ANITA-lite was intended both to investigate possible backgrounds to neutrino detection in Antarctica, and to verify as many of the systems used by the full ANITA mission as possible.

From balloon altitudes of 37 km, the horizon is at nearly 700 km distance, giving a synoptic view of more than 2 M km^3 of ice to a depth of order one radio attenuation length. ANITA will consist of a 2π array of dual- polarization antennas designed to monitor this entire ice target. ANITA-lite flew only two first-generation ANITA antennas, with a field-of-view covering about 12% of the $\sim$ 1.5M km^2 ice sheet area within its horizon at any time, but the $\sim$170,000 km^2 area of ice in view still represents an enormous monitored volume for the uppermost km of ice to which we were primarily sensitive. This leads to the strongest current limit on neutrino fluxes within its energy regime, as we will describe.

ANITA-lite was launched from Williams field, McMurdo station Antarctica along with its host payload on Dec. 18, 2004, and achieved its float altitude of about 40 km a few hours later. The payload stayed at relatively high latitudes during its flight, and was terminated Jan. 6, 2005, landing on the ice sheet several hundred km from Mawson Station (Australia) at an altitude of 2500 m.

For ANITA-lite, the trigger required a coincidence between the four possible statistically independent channels (two antennas and two polarizations) at a variable level of between one and four channels required to exceed a power threshold within a 25 ns window. The pulse-height spectrum of received voltages due to ideal thermal noise is nearly gaussian, and ANITA-lite was operated with an average threshold corresponding to 4.3 σ_V, where $\sigma_V = \sqrt{k\langle T_{sys}\rangle Z\Delta\nu}$ for bandpass-averaged system temperature values of $\langle T_{sys}\rangle \approx 700$ K during the flight. Here k is Boltzmann's constant, $Z = 50\ \Omega$, and $\Delta\nu = 800$ MHz is the system bandwidth.

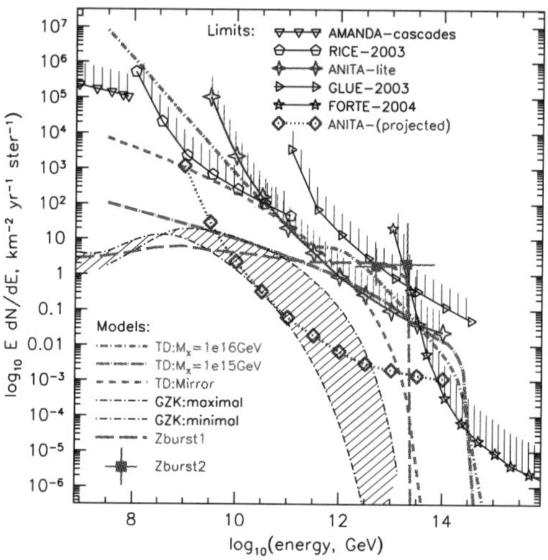

Figure 1. Limits on various models for neutrino fluxes at EeV to ZeV energies. The limits are: AMANDA cascades [24], from the Radio Ice Cherenkov Experiment (RICE) [26], the current work, The Goldstone Lunar Ultra-high energy neutrino Experiment [27], the Fast On-orbit Recorder of Transient Events (FORTE) satellite [28], and projected sensitivity for the full ANITA. Models shown are Topological Defects for two values of the X-particle mass [10], a TD model involving mirror matter [15], a range of models for GZK cosmogenic neutrinos [4,20,21], and several models for Z-bursts [12,19]. In the Zburst models plotted as points, the flux is a narrow spectral feature in energy, and the error-bars shown indicate the range possible for the central energy and peak flux values.

ANITA-lite recorded about 113,000 3-fold-coincident triggers. Other than our own ground calibration signals, we also detected no sources of impulsive noise that could be established to be external to our own payload. Several types of triggers were investigated for correlations to known Antarctic stations, and no such correlations were found. These events were analyzed in a variety of ways to establish their correlation to expected coherent Cherenkov events, which were simulated by convolving the known impulse response of the system with the theoretically expected intrinsic pulse shape of the radio Cherenkov, established at accelerator experiments. This analysis looked at both temporal- and frequency spectral-domain figures of merit. Two independent analyses were performed by two different subgroups within the ANITA collaboration, including blinding of half the data in one of the analyses. The end results of both analyses were virtually identical: none of the triggers passed the analysis, while typically 50% of the simulated signal was detected. We thus conclude that no events consistent with neutrino cascades were observed, where we had a 50% efficiency for detection.

Accounting for the net 10 days above the ice sheet, along with the 40% livetime and 50% analysis efficiency, the resulting limit on neutrino fluxes with standard-model cross sections is shown in Fig. 1. ANITA-lite approaches the highest energy cosmogenic neutrino flux model [21], and now appears to have entirely excluded the Z-burst model [9,12,14] at a level required to account for the fluxes of the highest energy cosmic rays, as represented by the three crosses in the figure, with vertical and horizontal bars indicating the range of allowed model parameters for this case. Prior limits from the GLUE and FORTE experiments had constrained most but not all of this range. Our limits rule out all of the remaining range for two of the highest standard topological defect models, shown in Fig. 1, both of which were constrained already by other experiments. We also provide the first experimental limits on the highest mirror-matter TD model [15].

Although designed primarily as an engineering test, ANITA-lite has set the best current limits above $10^{18.5}$ eV, and has improved constraints by more than an order of magnitude over the GLUE results [27]. This demonstrates the power of the radio Cherenkov technique applied to the balloon-based observations of the Antarctic continent. Simulations for ANITA, shown in Fig. 1 indicate an order of magnitude lower energy threshold and an order of magnitude greater sensitivity, sufficient to constrain or detect all current GZK neutrino models.

This work has been supported by the National Aeronautics and Space Administration. We thank the National Scientific Balloon Facility and the National Science Foundation for their excellent support of the Antarctic campaign.

References

1. K. Greisen, *Phys. Rev Lett.* **16**,748 (1966); G.T. Zatsepin and V. A. Kuz'min, *JETP Letters* **4**, 78 (1966).

2. R. U. Abbasi *et al.* [High Resolution Fly's Eye Collaboration], Phys. Rev. Lett. **92**, 151101 (2004) [arXiv:astro-ph/0208243].
3. J. W. Cronin, Nucl. Phys. Proc. Suppl. **138**, 465 (2005) [arXiv:astro-ph/0402487].
4. R. Engel, D. Seckel, and T. Stanev, Phys. Rev. D 64, 093010 (2001).
5. V. S. Berezinsky, & G. T. Zatsepin, Phys. Lett. 28B, 423 (1969); Sov. J. Nucl. Phys. 11, 111 (1970); F.W. Stecker 1973, Astrophys. Space Sci. 20, 47; F.W. Stecker 1979, Ap.J. 238, 919.
6. M. Ave, N. Busca, A. V. Olinto, A. A. Watson, T. Yamamoto, Astropart.Phys. 23 (2005) 19.
7. G. A. Askaryan, 1962, JETP 14, 441; 1965, JETP 21, 658.
8. D. Saltzberg, P. Gorham, D. Walz, *et al.* Phys. Rev. Lett., **86**, 2802 (2001); P. W. Gorham, D. P. Saltzberg, P. Schoessow, *et al.*, Phys. Rev. E. **62**, 8590 (2000).
9. B. Eberle, A. Ringwald, L. Song, T. J. Weiler, Phys.Rev. D70 (2004) 023007; hep-ph/0401203
10. S. Yoshida, H. Dai, C.C.H. Jui, & P. Sommers, Ap. J. 479 (1997) 547.
11. O. E. Kalashev, V. A. Kuzmin, D. V. Semikoz, G. Sigl, Phys. Rev. D65 (2002) 103003.
12. Z. Fodor, S. D. Katz, & A. Ringwald, Phys. Rev. Lett. 88 (2002), 171101; hep-ph/0105064.
13. T. J. Weiler, Astropart. Phys. 11 (1999) 303; hep-ph/9710431.
14. T. Weiler, Phys. Rev. Lett. (1982), 49, 234.
15. V. Berezinsky, Proc. of 11th Int. Workshop Neutrino Telescopes (ed. Milla Baldo Ceolin) p. 339, 2005; astro-ph/0509675.
16. S. Barwick, D. Besson, P. Gorham, and D. Saltzberg, J. Glaciol. (2005), in press.
17. R. Gandhi, Nucl. Phys. Proc. Suppl. 91 (2000) 453.
18. E. Zas, F. Halzen, & T. Stanev, 1992, Phys Rev D 45, 362.
19. O. E. Kalashev, V. A. Kuzmin, D. V. Semikoz and G. Sigl, Phys. Rev. D **66**, 063004 (2002).
20. R. J. Protheroe & P. A. Johnson, Astropart. Phys. 4, 253 (1996).
21. C. Aramo, A. Insolia, A. Leonardi, G. Miele, L. Perrone, O. Pisanti, D.V. Semikoz, Astropart.Phys. 23 (2005) 65.
22. J. Alvarez–Muñiz, & E. Zas, 1996, Proc. 25th ICRC, ed. M.S. Potgeiter et al.,7,309.
23. J. Alvarez–Muñiz, & E. Zas, 1997, Phys. Lett. B, 411, 218.
24. M. Ackermann *et. al*, Astropart. Phys. 22 (2004) 127.
25. L. A. Anchordoqui et al., Phys. Rev. D 66, 103002 (2002).
26. I. Kravchenko et al., Astropart.Phys. 20 195-213 (2003).
27. P. W. Gorham, C. L. Hebert, K. M. Liewer, C. J. Naudet, D. Saltzberg, D. Williams, Phys. Rev. Lett. 93 (2004) 041101.
28. N. Lehtinen, P. Gorham, A. Jacobson, & R. Roussel-Dupŕe, Phys.Rev.D 69 (2004) 013008; astro-ph/030965.

MEASURING THE NEUTRINO-NUCLEON CROSS SECTION WITH SALSA

A. CONNOLLY

University of California at Los Angeles
Department of Physics and Astronomy,
Box 951547 475 Portola Plaza, Los Angeles, CA 90095-1547 USA
E-mail: connolly@physics.ucla.edu

These proceedings describe a study of the expected sensitivity of the SalSA experiment to the neutrino-nucleon cross section. We expect the measurement to be statistics limited for the events rates expected from SalSA. With 100 measured events, we expect to measure a standard model cross section with a 38% uncertainty that is dominantly statistical.

1. Introduction

Cosmic rays above $10^{19.5}$ eV should produce neutrinos when they interact with the cosmic microwave background through through the Greisen-Zatsepin-Kuzmin [7] (GZK) process. Those neutrinos may provide important clues about the origin of the highest energy cosmic rays and hence a few experiments seek to discover them; ANITA expects to see approximately 5-15 neutrinos from the GZK process, while IceCube may detect of order one such neutrino event [1,2]. The Salt dome Shower Array (SalSA) aims to move beyond the discovery of these neutrinos, and to measure a large sample of neutrinos above 10^{17} eV to study their properties [3].

The center of mass energy of an interaction between a 10^{17} eV neutrino with a nucleon at rest is 14 TeV, beyond the center of mass of typical parton interactions at the LHC. SalSA's sensitivity to neutrinos above 10^{17} eV would put it in a unique position to probe particle physics at unprecedented energy scales. For example, SalSA could measure the neutrino-nucleon cross section by measuring the rate of neutrinos as a function of zenith angle. Here we outline the feasibility of such a measurement.

2. Description of SalSA

SalSA would be an array of antennas embedded in a naturally occurring, large volume (10's of km^3) of salt with high purity. Salt domes are good candidates for SalSA's medium; a few candidates have been found in the Southeastern United States. Antennas would be deployed on strings lowered into holes drilled into the salt. The data acquisition and trigger would be wireless and solar powered, with each string digitized either within the hole or at a station at the surface. [4].

For this study we use a rectangular detector with 100 strings arranged in a square grid with 250 m spacing. Along each string is 10 equally spaced nodes with 12 antennas per node separated by 0.75 m. Within each node, the antennas alternate between dipole antennas, with their polarization aligned in the vertical direction along the length of the hole, and slot antennas with alternating horizontal polarizations. The signal and antenna response are frequency dependent, and we consider the frequency band from 100 MHz to 300 MHz. A module is triggered when five antennas out of 12 read a voltage that surpasses 3.0 times the expected noise. Five triggered modules trigger the event.

The top edge of the salt dome is 500 m below the surface and the strings begin 750 m below the surface. The salt "dome" itself goes 10 km deeper than its top edge and it is 4 km by 4 km square in the aerial view.

Simulations of the SalSA detector have shown dozens of events detected over approximately 3 years. These simulations have also shown an angular resolution of 0.5 degrees to be achievable for events where the interaction occurs within the detector volume and 1 degree for events where the interaction occurred outside the detector volume.

3. Description of the Measurement

Most of the neutrinos detected with SalSA will be down-going and so will have traversed a distance in the earth much smaller than the neutrino's expected interaction length. Therefore, across much of the acceptance, the rate of neutrino events vs. incident angle does not show an observable dependence on the cross section. For $10^{17.5}$ eV neutrinos that interact at 2 km depth, only the neutrinos originating from between 1.6° above and 6.1° below the horizon, 12% of the events, are attenuated between 10% and 90% of the time in their path through the earth and thus impact the cross section measurement. For these numbers and throughout this paper, a constant crust density of 2900 km/m^3 is assumed. Therefore, given SalSA's expected event rates and angular resolution, the cross section measurement is statistics limited.

4. Cross Section Dependence in $\cos\theta_{\text{zenith}}$ Distribution

Here, we derive the functional dependence that we expect for the number of observed events binned in the $\cos\theta_{\text{zenith}}$ variable. Here, we assume the detector response is flat in $\cos\theta_z$. Although not true across the entire range in $\cos\theta_z$, it is a good approximation in the narrow range of angles where the neutrinos provide information about the cross section. For a given interaction depth δ, the distribution will follow:

$$\frac{dN}{d(\cos\theta_z)} = \exp[-y(\cos\theta_z,\ \delta)/L(E)] \tag{1}$$

where $y(\cos\theta_z,\ \delta)$ is the length of the chord that the neutrino traverses along its path through the earth before reaching the interaction point. The neutrino's interaction length at an energy E is $L(E) = \sigma(E)/\rho$ where $\sigma(E)$ is the neutrino cross section and ρ is the density of the earth's crust. We derive the following equation for the chord length as a function of $\cos\theta_z$ for a given δ:

$$y = -(R-\delta)\cdot\cos\theta_z + R\cdot\cos\theta_z\sqrt{(R^2-2R\delta)\cdot\cos^2\theta_z + 2\delta R} \tag{2}$$

Here, R is the radius of the earth. For this study, we always set $\delta = 2$ km in Equation 1 when we perform the fits, although in the simulation, the interactions are allowed to occur anywhere within the volume of the salt dome. The results presented here are only weakly dependent on the depth chosen. It is interesting to note that, at $\theta_z = \pi/2$, this function is steepest when $L(E) = \sqrt{2\delta R}$. For $\delta = 2$ km, this corresponds to 4.95 times the standard model cross section predicted in Gandhi *et al* [5].

Any measured sample will be composed of neutrinos in a range of energies, and so Equation 1 more generally becomes:

$$\frac{dN}{d(\cos\theta_z)} = \sum_i a_i\cdot\exp[-y(\cos\theta_z,\ \delta)/L_i(E)] \tag{3}$$

where the sum is over energy bins (we take one bin per decade), and a_i is the fraction of detected events in each energy bin. We derive a_i by simulating a standard spectrum of neutrinos from the GZK process [6], hereafter called the baseline GZK spectrum. We take the cross section to be proportional to $E^{0.363}$ as in Gandhi *et al.* so that

$$L_{i+1}(E) = L_i(10\cdot E) = L_i(E)\cdot E^{0.363}. \tag{4}$$

That leaves the interaction length at a single energy, which we choose to be 10^{18} eV, and the overall normalization as the two degrees of freedom in the fit.

Figure 1 shows the functional dependence of the rate of neutrinos on $\cos\theta_z$. The line labeled "GQR&C σ" is the derived functional dependence of this distribution for the cross section given in Gandhi *et al.* The lower (higher) curves are the prediction corresponding to a model where the cross section at these energies is enhanced (reduced) by a factor of 2 compared to the GQR&C value.

5. Pseudo-Experiments

In order to predict the expected resolution on the measured cross section given an expected number of events observed, we simulate 1,000 pseudo-experiments. For each pseudo-experiment, we select events according to Equation 3 assuming the true cross section is the GQR&C cross section. We then smear the data according to a 1.0 degree angular resolution. Each distribution in zenith angle resulting from a pseudo-experiment is fit to the functional form in Equation 1 with two free variables: the interaction length at 10^{18} eV, $L(E = 10^{18}eV)$, and the overall normalization. The fit is a minimization of a Poisson likelihood chi-square [8]. We take 50 bins in the region $-1 < \cos\theta_z < 1$.

6. Results

Figure 1 shows the result of a typical pseudo-experiment where the expectation was 100 events. Figure 2 shows the distribution of fitted values of the interaction length at 10^{18} eV for all pseudo-experiments. Recall that although we assume that the injected neutrino spectrum is the baseline GZK spectrum, we only allow the interaction at a single energy to be a free variable after we assume a specific energy dependence for the cross section. That interaction length is 395 km according to the GQR&C cross section. These pseudo-experiments measure on average 430 km for the interaction length with a 38% RMS. The mean tends slightly toward higher interaction lengths when the data is smeared by the detector's angular resolution. For 75 events measured with SalSA, the RMS on the fitted interaction length increases to 44.9%.

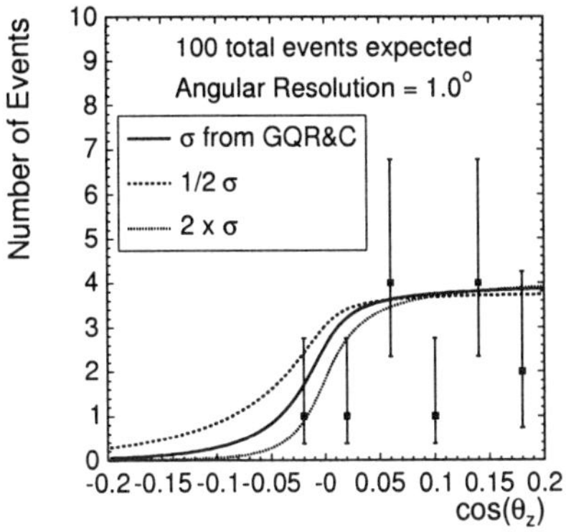

Figure 1. A typical SalSA pseudo-experiment with 100 total events expected, zooming in on the region where the cross section dependence is greatest. We compare hypothetical data to the prediction from Gandhi *et al.* and what would be expected if that cross section were reduced (enlarged) by a factor of 2. All three curves are normalized to 100 events in the range $-1 < \cos\theta_z < 1$.

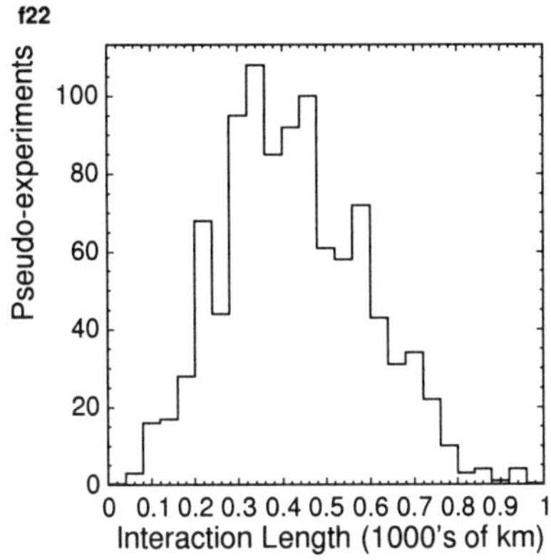

Figure 2. Distribution of interaction lengths measured from each pseudo-experiment, with 100 events expected in each experiment, from fitting the measured distribution to Equation 3.

Acknowledgments

This work was supported by the DOE and NASA. The author is grateful to David Saltzberg, Steve Barwick and Gary Varner for helpful discussions and suggestions.

References

1. P. Miocinovic *et al.* [The ANITA Collaboration], eConf **C041213**, 2516 (2004)
2. A. Silvestri *et al.* [The ANITA Collaboration], arXiv:astro-ph/0411007.
3. P. Gorham, *et al.*, *Nucl. Instrum. Meth.* **A490**, 476-491 (2002).
4. G. Varner *et al.*, *Nucl. Instrum. Meth.* **A554**, 437-443 (2005).
5. Gandhi *et al.*, *Phys. Rev.***D58**, 093009 (1998).
6. R. Engel, D. Seckel and T. Stanev, *Phys. Rev.* **D64**, 093010 (2001).
7. K. Greisen, *Phys. Rev. Lett.***16**7481966.
8. S. Baker, R. Cousins, *Nucl. Instrum. Meth.***A221**, 437-442 (1984).

RADIO DETECTION OF COSMIC RAYS WITH LOPES

A. HORNEFFER[H,E], W.D. APEL[A], F. BADEA[A], L. BÄHREN[B], K. BEKK[A], A. BERCUCI[C], M. BERTAINA[D], P.L. BIERMANN[E], J. BLÜMER[A,F], H. BOZDOG[A], I.M. BRANCUS[C], M. BRÜGGEMANN[G], P. BUCHHOLZ[G], S. BUITINK[H], H. BUTCHER[B], A. CHIAVASSA[D], K. DAUMILLER[A], A.G. DE BRUYN[B], C.M. DE VOS[B], F. DI PIERRO[D], P. DOLL[A], R. ENGEL[A], H. FALCKE[B,E,H], H. GEMMEKE[I], P.L. GHIA[J], R. GLASSTETTER[K], C. GRUPEN[G], A. HAUNGS[A], D. HECK[A], J.R. HÖRANDEL[F], T. HUEGE[A,E], K.-H. KAMPERT[K], G.W. KANT[B], U. KLEIN[L], Y. KOLOTAEV[G], Y. KOOPMAN[B], O. KRÖMER[I], J. KUIJPERS[H], S. LAFEBRE[H], G. MAIER[A], H.J. MATHES[A], H.J. MAYER[A], J. MILKE[A], B. MITRICA[C], C. MORELLO[J], G. NAVARRA[D], S. NEHLS[A], A. NIGL[H], R. OBENLAND[A], J. OEHLSCHLÄGER[A], S. OSTAPCHENKO[A], S. OVER[G], H.J. PEPPING[B], M. PETCU[C], J. PETROVIC[H], T. PIEROG[A], S. PLEWNIA[A], H. REBEL[A], A. RISSE[M], M. ROTH[F], H. SCHIELER[A], G. SCHOONDERBEEK[B], O. SIMA[C], M. STÜMPERT[F], G. TOMA[C], G.C. TRINCHERO[J], H. ULRICH[A], J. VAN BUREN[A], W. VAN CAPELLEN[B], W. WALKOWIAK[G], A. WEINDL[A], S. WIJNHOLDS[B], J. WOCHELE[A], J. ZABIEROWSKI[M], J.A. ZENSUS[E], D. ZIMMERMANN[G]

[A] *Institut für Kernphysik, Forschungszentrum Karlsruhe, Germany*
[B] *ASTRON Dwingeloo, The Netherlands*
[C] *NIPNE Bucharest, Romania*
[D] *Dpt di Fisica Generale dell'Università Torino, Italy*
[E] *Max-Planck-Institut für Radioastronomie, Bonn, Germany*
[F] *Institut für Experimentelle Kernphysik, Uni Karlsruhe, Germany,*
[G] *Fachbereich Physik, Universität Siegen, Germany*
[H] *Dpt of Astrophysics, Radboud Uni Nijmegen, The Netherlands*
[I] *IPE, Forschungszentrum Karlsruhe, Germany*
[J] *Ist di Fisica dello Spazio Interplanetario INAF, Torino, Italy*
[K] *Fachbereich Physik, Uni Wuppertal, Germany*
[L] *Radioastronomisches Institut der Uni Bonn, Germany*
[M] *Soltan Institute for Nuclear Studies, Lodz, Poland*

Measuring radio pulses from cosmic ray air showers offers various new opportunities. New digital radio receivers allow measurements of these radio pulses even in environments that have lots of radio interference. With high bandwidth ADCs and fast data processing it is possible to store the whole waveform information in digital form and analyse transient events like air showers even after they have been recorded. Digital filtering and beam forming can be used to suppress the radio interference so that it is possible to measure the radio pulses even in radio loud environments. LOPES is a prototype station for the new digital radio interferometer LOFAR and is tailored to measure air showers. For this it is located at the site of the KASCADE-Grande air shower experiment. Already with the first phase of LOPES we have been able to measure radio pulses from air showers and show correlations between the radio pulse height and air shower parameters.
The first part gives an introduction and presents the science results of LOPES, while the second part presents the hard- and software that enables LOPES to detect air short pulses.

1. Radio Detection of Cosmic Rays

1.1. *Introduction*

Radio emission from particle showers was first proposed by Askaryan[1]. Radio pulses from cosmic ray air showers were discovered for the first time by Jelley[2] at 44 MHz. The results were soon verified and in the late 1960's emission from 2 MHz up to 520 MHz was found. These experiments suffered from difficulty with radio interference, low spatial and time resolution, and uncertainty about the interpretation of the results. So in the following years the interest of radio astronomy moved to higher frequencies and the cosmic ray community to other, more successful methods.

Today a new generation of radio telescopes is being built. LOFAR is a new digital radio telescope for the frequency range of 30-240 MHz. It is being built in The Netherlands to study the high redshift universe, cosmology, the bursting universe and cosmic rays & neutrinos. The basic idea is to build a large array of separate stations with quasi-omnidirectional dipole antennas in which the received waves are digitised and sent to a central super-cluster of computers. A new feature of this design is the possibility to store the entire data stream for a certain period of time. If one detects a transient phenomenon one can then retrospectively form a beam in the desired direction. This combines the advantages of a low-gain antenna (large field of view) and a high-gain antenna (high sensitivity and background suppression). Thus LOFAR will be well suited to study radio emission from cosmic ray air showers in an energy range around 10^{18} eV.

Measuring the radio emission from air showers has several advantages compared to other methods of measuring air showers. The radio signal is only slightly attenuated in the atmosphere, so one can also measure distant and inclined showers which makes it interesting for neutrino measurements. The signal is integrated over the whole shower evolution, so the pulse height gives a bolometric measurement of the air shower. Radio measurements can have a high duty cycle, measuring 24 h/day and only stopping during thunderstorms or other bad weather conditions. Measuring with radio antennas and (even unshielded) particle detectors on the ground could give hybrid measurements that allows the determination of the cosmic ray composition. The radio measurements also have some potential problems. Most noteworthy: Radio frequency interference (RFI) that has to be filtered, the size of the footprint on the ground which is presumably smaller than the useful particle disk, the correlation with other air shower parameters is still unclear, and it is only practical for primary energies above ca. 10^{17} eV.

To address these issues and test the technology we set up LOPES, a **LO**FAR **P**rototyp**e** **S**tation at the site of the existing air shower experiment KASCADE-Grande. The data from a well tested air shower experiment allows us to calibrate the radio data with other air shower parameters, and helps us by providing starting points for the air shower reconstruction. A more detailed description of LOPES and the used hard- and software follows in chapter 2. Together with the experimental work there are also theoretical studies modelling the radio emission from air showers as coherent synchrotron radiation in the earth's magnetic field[3] which led to a new Monte Carlo code[4]. A related experiment, with a similar technology, uses part of the Nançay decametric array[5].

1.2. *Results*

LOPES detected its first air shower pulse in January 2004[6]. From January to September 2004 LOPES collected ca. 630 thousand events. For this analysis we selected the largest events in which: a) the KASCADE array processor did not fail, b) the distance of the shower core to the array centre was less that 91 m, and c) the electron number was greater than 5×10^6 **or** the truncated muon number was greater than 2×10^5. This selected 412 events, in 228 of which events we found a coherent pulse from the air shower.

The height of the radio pulses correlates with several air shower parameters: It rises with shower size (i.e. with the electron number or the muon number), it falls with increasing distance of the shower axis to the antennas, and it rises with increasing angle to the geomagnetic field (see

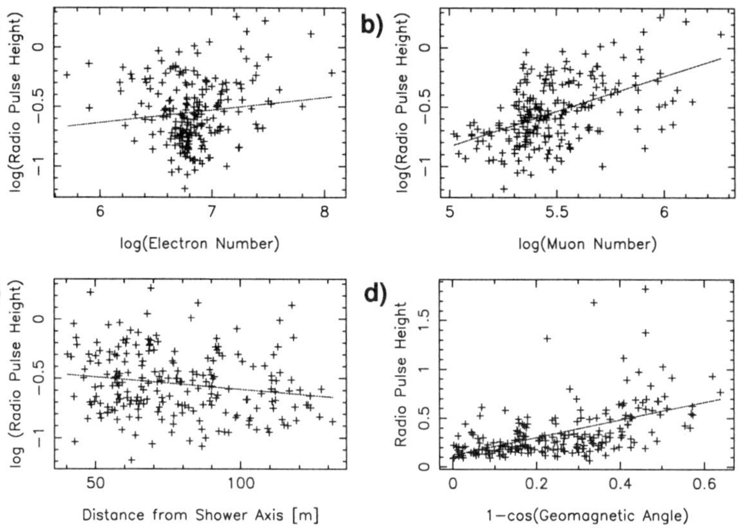

Figure 1. The raw radio pulse height, plotted against a) number of electrons, b) number of muons, c) distance to the shower axis, and d) cosine of the angle to the geomagnetic field.

figure 1).

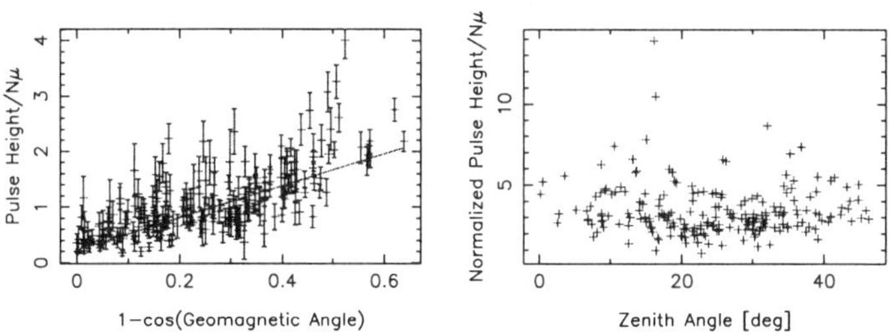

Figure 2. Left: Height of the radio pulse, divided by the muon number, against the cosine of the geomagnetic angle. Right: Radio pulse height scaled with the results of the fit in the left panel. After taking out the effect of the geomagnetic angle, no further dependence on the zenith angle can be seen.

As the dependence on the shower size is most pronounced, in a first step we normalised the pulse height by dividing by the truncated muon number. The left panel of figure 2 shows the dependence of the thus normalised pulse height on the cosine of the geomagnetic angle. In the right panel of figure 2 we additionally normalised the pulse height with the values of the fit to the geomagnetic angle, by multiplying with the fraction of the fit results at 90° to those at the angle of the air shower. After taking out the effect of the angle to the geomagnetic field, no further dependence on the zenith angle can be seen. The same is true for the azimuthal angle.

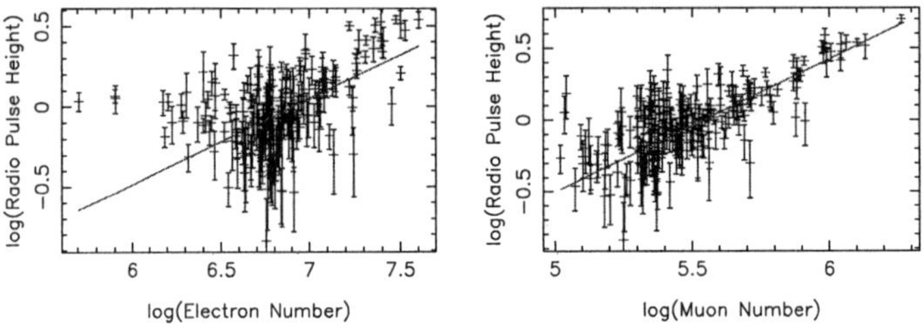

Figure 3. Normalised radio pulse height after scaling by the fit to the geomagnetic angle and distance to the shower axis. Plotted against the electron number (left) and muon number (right).

In figure 3 the pulse height was scaled with the results of the fit to the geomagnetic angle and the results of a fit to the distance of the antennas to the shower axis. The difference between the left and the right panel shows that the radio pulse height is better correlated with the muon number than with the electron number. This is expected as at the KASCADE-Grande experiment the muon number is a better tracer of the total number of particles during the shower evolution than the electron number. The slope of the linear fit to the log(pulse height) vs. log(muon number) plot is close to one. That means that the field strength indeed rises nearly linearly with primary energy and thus the received power rises quadratically.

Figure 4 shows the sum over the formed beams of the eight least ($\theta < 6°$) and the eight most inclined ($45° < \theta < 56°$) air showers. While the vertical air showers show more noise from the particle detectors in the range of -1.7 to $-1\,\mu s$, the inclined air showers show a stronger radio signal from the air showers at $-1.8\,\mu s$[7]. This shows that for inclined air showers the particle signal is already strongly attenuated, while the radio signal is not.

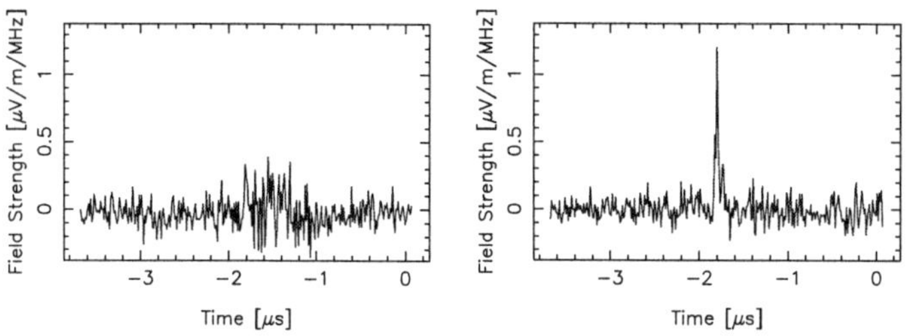

Figure 4. Left: Sum over the formed beams of the eight least inclined air showers in the selection. Right: Same sum but over the eight most inclined air showers. The average muon number in both groups is ca. 3×10^5.

1.3. *Summary and Conclusions*

LOPES is able to reliably measure radio emission from air showers, and already took enough data for a first science analysis. The radio signal is a faithful tracer of air showers and gives good energy information. It can also give precise ($\Delta\alpha < 1^\circ$) arrival directions. Inclined air showers give a strong radio signal, thus radio detection will be a good tool to search for neutrino induced air showers.

2. Detecting radio pulses from air showers with LOPES

2.1. *LOPES at KASCADE-Grande*

The layout of the LOPES antennas inside the KASCADE array can be seen in the left panel of fig. 5. In the first phase LOPES10 had 10 antennas, arranged in a cross like pattern. In the second phase LOPES30 has now 30 antennas, of which 26 are inside the KASCADE array and 4 are outside on a meadow next to the KASCADE array. The position of the antennas in relation to each other has been measured with high precision (better than 10 cm) with a differential GPS system.

LOPES is triggered by a large event trigger from KASCADE. This trigger is formed, when 10 out of the 16 KASCADE array clusters had an internal trigger. This gives ca. two triggers per minute or 2500–3000 events per day. To get good frequency resolution 2^{16} samples (or ~ 0.82 ms of data) are saved per antenna for each event, with half the data from before and the other half after the trigger. This gives a frequency resolution of ~ 1.2 kHz and places the radio pulse in the middle of the data block.

The correlation of KASCADE-Grande and LOPES events is done offline.

Results of the KASCADE air shower reconstruction are used as starting points for the LOPES event analysis, primarily the direction of the air shower, the shower size (electron and muon numbers), and the core position of the air shower.

2.2. *Hardware of LOPES*

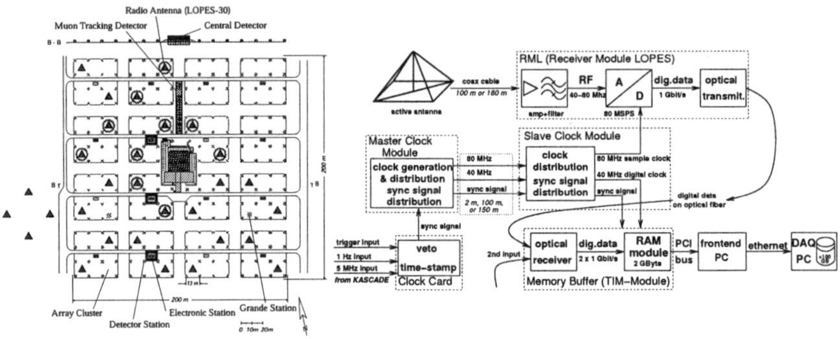

Figure 5. Left: Layout of LOPES inside the KASCADE array. The triangles show the positions of the 30 LOPES antennas, the red circles highlight the 10 antennas of LOPES10. The blue squares mark the electronic stations that house the LOPES electronics. Right: Outline of the LOPES hardware. The signal is picked up by the antenna, sent via a coaxial cable to the receiver module, digitised and sent to the memory module. The clock signals are generated by the master clock module and, together with the sync-signal, sent to the slave modules for further distribution.

LOPES operates in the frequency range of 40–80 MHz. This is a band where there are few strong man made radio transmitters, as it lies between the short-wave- and the FM-band. The outline of the hardware used for LOPES can be seen in the right panel of fig. 5. It samples the radio frequency signal after minimal analog treatment without the use of a local oscillator.

Antenna: The antennas for LOPES are short dipole antennas with an "inverted vee" shape (see right panel of fig. 6). The left panel of fig. 6 shows the gain pattern of a single antenna The half power beam width is ca. 85° in the direction parallel to the dipole and ca. 130° in the direction perpendicular to the dipole. The visible parts are commercial PVC pipes holding the active parts in place. The radiator consists of two copper cables extending from the top down two thirds of two opposing edges of the pyramid.

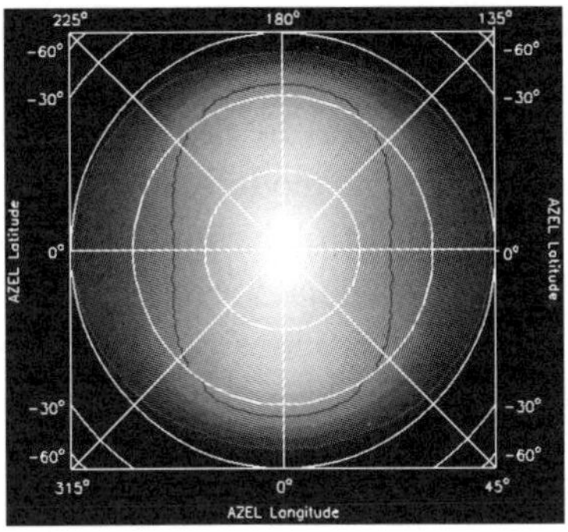

Figure 6. Left: Gain pattern of a single LOPES antenna. The vertical direction (azimuth = 0° or = 180°) is the direction perpendicular to the dipole, the horizontal direction is the one parallel to the dipole. The contours are at the 50% and 10% levels. Right: One of the LOPES antennas at the KASCADE-Grande site.

The four edges can be used for two orthogonal linear polarisations of the signal. As we expect more signal in the east–west polarisation direction this is the direction used by LOPES. Inside the container at the top resides a preamplifier. Its main functions are balanced to unbalanced conversion, amplification of the signal and transformation of the antenna impedance to the 50 Ω impedance of the cable. The PVC exterior of the antenna resides on an aluminium pedestal. This acts as a ground screen and prevents damage by the lawn mowing.

Receiver Module: In the receiver module the signal is first amplified and then fed into the anti-aliasing filter. To suppress contamination from outside our band a stopband attenuation of 60 dB is needed. Additionally the desire for high usable bandwidth makes steep edges necessary. The filter used for LOPES gives us a usable frequency band from 43 MHz to 76 MHz.

The last analog device in the signal path is the A/D-converter board. The necessary dynamic range to detect weak pulses while not saturating the ADC with radio interference is achieved by using 12-bit ADCs. We are using ADCs running at 80 MHz, thus sampling the signal in the 2nd Nyquist domain of the ADCs. Piggybacked onto the A/D-converter board is an optical transmitter board for transmission of the digital data to the backend module.

Digital Backend and Clock Module: The digital data is transferred via fibre optics to memory modules. These modules have two optical inputs.

They fit into and are read out by the front-end PCs. Each module can store up to 6.25 seconds of data from both inputs or even 12.5 seconds of data using only one input. Several of these modules can be used together by synchronising them with a common sync-signal. The modules can either start writing the data after a sync-signal or write data continuously into the memory and stop a predefined time after a sync-signal is received.
The sample clock for the A/D-converters and a synchronous clock for the memory modules is generated on a central master clock module and then distributed via slave clock modules to all receiver and memory modules. The trigger from KASCADE-Grande is first fed into a clock card to generate a time stamp for the event. The trigger is then used as the sync-signal and distributed to all memory modules via the master and slave clock modules.

2.3. *LOPES Software*

The goal of the processing of air shower events is to reconstruct the radio field strength of the pulse emitted by the air shower. Processing of air shower events proceeds in the following steps:

(1) Correction of instrumental delays from the TV-transmitter
(2) Frequency dependent gain correction
(3) Suppression of narrow band RFI
(4) Flagging of antennas with high noise
(5) Beam forming in the direction of the air shower
(6) Quantification of peak parameters
(7) Optimising the radius of curvature
(8) Identification of good events

Delay Correction: By monitoring the relative phases of a TV transmitter we can monitor the phase stability of our system and get time delay calibration values. As the position of the TV transmitter does not change, the relative delays and thus the relative phases of its signal in the different antennas remains constant. Checking the relative phases of just a single frequency cannot detect larger shifts due to the ambiguousness of the phase. But by checking at the frequencies of the picture and the two sound carriers this ambiguity can be reduced so far that shifts of an integer number of samples can be detected.
Except infrequent shifts of an integer number of samples caused by the digital electronics the delay corrections remain smaller than 0.1 sample times.

Gain Correction: The amplification or attenuation of the electronic components in the signal chain is measured in a laboratory environment. These values are then combined to a frequency dependent gain factor, that is multiplied to the data. The gain (or directivity) of the antennas for all directions is simulated from the antenna geometry. The value of the antenna gain in the direction of the air shower can then be used to calculate the field strength of the radio pulse from the ADC data.
As the input impedance of the preamplifier is not the 50 Ω used by measurement equipment, measuring its gain is difficult and the measured values have a high uncertainty. A calibration of the whole signal chain at once has recently been done[8].

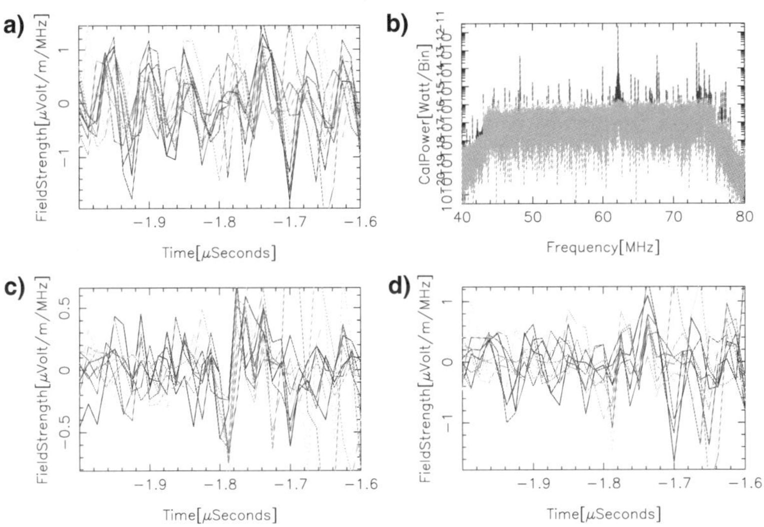

Figure 7. a) Section of the unfiltered data. The different colours show traces from different antennas. b) Gain calibrated power spectrum of one antenna with a blocksize of 65536 samples. The red spikes sticking out from the noise floor are narrow band RFI. c) Filtered data, after filtering with a blocksize of 65536 samples. A coherent pulse at $-1.78\,\mu s$ is clearly visible. d) Filtered data, but with a blocksize of 128 samples (i.e. 4 times the plotted data). In contrast to c) the coherent pulse is not easily visible.

Suppression of Narrow Band RFI: Narrow band RFI occupies only few channels in frequency space, while a short time pulse is spread over all frequency channels. So by flagging the channels with RFI one can greatly reduce the background without affecting the air shower pulse much. After gain calibration the noise floor inside our frequency band is nearly flat. So the amplitude spectrum is fitted with a line to determine the reference value at each point. All points that deviate more than $3\,\sigma$ from this are replaced by the reference value. The parts left and right outside our band are fitted extra, to remove RFI that otherwise would not be filtered. The whole process is done two times to reduce the effect that the RFI influences the fit an biases the reference values. The ratio between the filtered and unfiltered spectrum is used as a weight for the (complex) frequency space data.

Figure 7 illustrates the effect of this. Panel a) shows a section of unfiltered data with traces for all antennas. Panel b) shows the power spectrum of one antenna before and after the filtering. After filtering the RFI lines stay in the region of the noise. Panel c) shows the time domain data after the filtering. A coherent pulse at $-1.78\,\mu s$ is clearly visible (i.e. all the traces fall onto each other). Panel d) shows the effect of insufficient frequency resolution. Less frequency resolution has two effects. The first is, that with each filtered line a greater portion of the radio pulse is cut away. The second is, that RFI is more prone to be hidden in the noise and thus not filtered. In this example the second effect is dominant, the amplitude of the signals is only a little lower than in the unfiltered data.

Beam Forming: The beam forming consists of two steps: First time shifting of the data according to the given direction and then combining of the data from all antennas to calculate the resulting beam.

The geometrical delay, by which the data is shifted, is the difference of the time it takes a pulse coming from the given direction to reach the position of the corresponding antenna compared to the reference position.[a] This delay is added to the delay corrections and the shift itself is done by multiplying a phase gradient to the frequency space data before transforming it back to the time domain. Figure 8 a) and b) show the received power in single antennas of an example event before and after time shifting. The time shifting moves the peaks from the air shower pulse so that they fall onto each other. Panel c) shows that also in the electric field of the antennas the traces of the coherent pulse fall onto each other.

To form the beam from the time shifted data, the data from each unique pair of antennas is multiplied pixel by pixel, the resulting values are averaged,

---

[a] As the reference position we choose the position of the shower centre. This shifts our data, so that the pulse from the air shower is always at a certain point in our dataset.

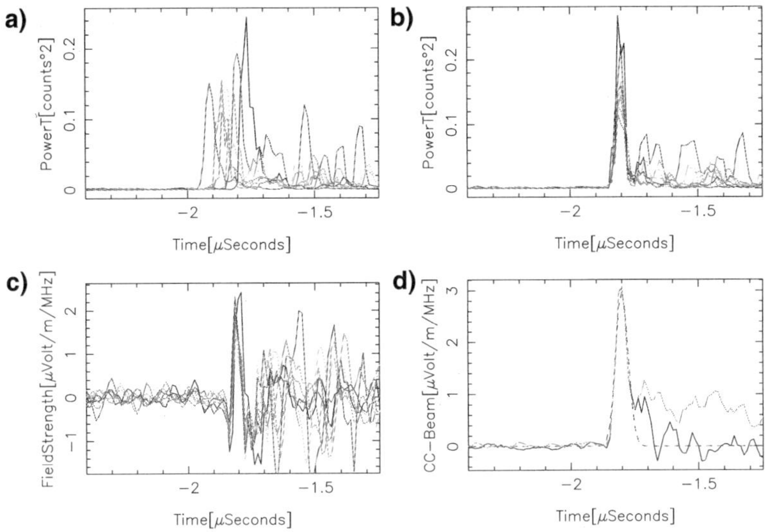

Figure 8. a) Received power in 8 antennas before beam forming. b) Received power after time shifting to form a beam in the direction of the air shower. c) The electric field of the antennas after time shifting. d) The formed (CC-)beam after combining of all antennas(dark blue), the power average of all antennas(light blue), and a Gaussian fit to the formed beam(dashed, pink). (a), b) and d) are smoothed by block averaging over 3 samples.) The coherent pulse at $-1.8\,\mu$s gives a strong peak, while the following noise is incoherent and suppressed by the beam forming.

and then the square root is taken while preserving the sign:

$$cc[t] = \pm\sqrt{\left|\frac{1}{N_{Pairs}}\sum_{i=1}^{N-1}\sum_{j>i}^{N} s_i[t]s_j[t]\right|} \qquad (1)$$

N is the number of antennas, N_{Pairs} is the number of unique antenna pairs, $s[t]$ the field strength of a single antenna, and t the time or pixel index. The negative sign is taken if the sum had a negative sign before taking the absolute values, and the positive sign otherwise. We call this the cross-correlation beam or CC-beam. In terms of radio astronomy the CC-beam is the real part of the visibility, averaged over all baselines but integrated over only one sample time. The advantage of the CC-beam is that a peak from a coherent pulse always has a positive sign. Peaks from incoherent pulses can also have a negative sign. Panel d) of figure 8 shows the CC-beam compared to the power average of all antennas. The peak from the coherent pulse at $-1.8\,\mu$s is well reproduced, while the following, incoherent noise is reduced.

Quantification of Peak Parameters: Due to the filtering of low and high frequencies, the response of the analog electronics to a short pulse is an oscillation over a short time. Sampling such a signal with an ADC gives a certain fine structure inside the pulse that is not part of the original pulse but is caused by the electronics. To suppress this fine structure the data is smoothed by block averaging over 3 samples. Although the resulting pulse shape is not really Gaussian, fitting a Gaussian to the smoothed data gives a robust value for the peak strength. So the value for the peak strength that is used in our analysis is the height of the Gaussian fitted to the CC-beam, as shown in panel d) of figure 8.

Radius of Curvature: The radio pulse of an air shower does not arrive on the ground as a plane wave, but it has some curvature. For the distances of LOPES the shape of the wavefront can be represented by a sphere with a finite radius of curvature. This radius of curvature is optimised by iterating the radius of curvature until the height of the peak is maximal.

Event Identification: Not every selected air shower is accompanied by a radio pulse that is detectable by LOPES. If there is no pulse from the air shower there can be an incoherent noise peak that is as high as an average air shower peak, even in the formed beam. So one cannot select events with air shower pulses just by the height of the fitted Gaussian, but has to classify events in an extra step. The criteria for this selection are: existence of a pulse, coherence of the pulse, position in time of the pulse, and approximately uniform pulse height in all antennas. Up to now this classification is done by hand.

2.4. *Summary*

LOPES is set up in a radio loud environment with strong interference from man made sources. The digital receiver system of LOPES allows us to do the signal processing in software. This allows us to use digital filtering and beam forming to suppress the RFI. With this LOPES is able to reliably measure the radio pulses from air showers.

References

1. G.A. Askaryan, *Soviet Phys. JETP* **14**, 441 (1962).
2. J.V. Jelley et al., *Nature* 205, 327 (1965)
3. H. Falcke & P. Gorham, *Astropart. Phys* Vol. 19, 477 (2003)

4. T. Huege & H. Falcke, *A&A*. Vol. 430, 779-798, (2005)
5. O. Ravel et al., *Nuclear Instr. & Methods in Physics A* 518, 213-215 (2004)
6. H. Falcke et al. - LOPES collab., *Nature* **435**, 313 (2005).
7. J.Petrovic et al. - LOPES collab., *Astropart. Phys.* in prep.
8. S. Nehls et al. - LOPES collab., *"Absolute Calibration of the LOPES Antenna System"*, these proceedings.

COMBINED LOPES AND KASCADE-GRANDE DATA ANALYSIS

A. HAUNGS[A], W.D. APEL[A], F. BADEA[A], L. BÄHREN[B], K. BEKK[A], A. BERCUCI[C], M. BERTAINA[D], P.L. BIERMANN[E], J. BLÜMER[A,F], H. BOZDOG[A], I.M. BRANCUS[C], M. BRÜGGEMANN[G], P. BUCHHOLZ[G], S. BUITINK[H], H. BUTCHER[B], A. CHIAVASSA[D], K. DAUMILLER[A], A.G. DE BRUYN[B], C.M. DE VOS[B], F. DI PIERRO[D], P. DOLL[A], R. ENGEL[A], H. FALCKE[B,E,H], H. GEMMEKE[I], P.L. GHIA[J], R. GLASSTETTER[K], C. GRUPEN[G], D. HECK[A], J.R. HÖRANDEL[F], A. HORNEFFER[H,E], T. HUEGE[A,E], K.-H. KAMPERT[K], G.W. KANT[B], U. KLEIN[L], Y. KOLOTAEV[G], Y. KOOPMAN[B], O. KRÖMER[I], J. KUIJPERS[H], S. LAFEBRE[H], G. MAIER[A], H.J. MATHES[A], H.J. MAYER[A], J. MILKE[A], B. MITRICA[C], C. MORELLO[J], G. NAVARRA[D], S. NEHLS[A], A. NIGL[H], R. OBENLAND[A], J. OEHLSCHLÄGER[A], S. OSTAPCHENKO[A], S. OVER[G], H.J. PEPPING[B], M. PETCU[C], J. PETROVIC[H], T. PIEROG[A], S. PLEWNIA[A], H. REBEL[A], A. RISSE[M], M. ROTH[F], H. SCHIELER[A], G. SCHOONDERBEEK[B], O. SIMA[C], M. STÜMPERT[F], G. TOMA[C], G.C. TRINCHERO[J], H. ULRICH[A], J. VAN BUREN[A], W. VAN CAPELLEN[B], W. WALKOWIAK[G], A. WEINDL[A], S. WIJNHOLDS[B], J. WOCHELE[A], J. ZABIEROWSKI[M], J.A. ZENSUS[E], D. ZIMMERMANN[G]

LOPES COLLABORATION

[A] *Institut für Kernphysik, Forschungszentrum Karlsruhe, Germany*
[B] *ASTRON Dwingeloo, The Netherlands*
[C] *NIPNE Bucharest, Romania*
[D] *Dpt di Fisica Generale dell'Università Torino, Italy*
[E] *Max-Planck-Institut für Radioastronomie, Bonn, Germany*
[F] *Institut für Experimentelle Kernphysik, Uni Karlsruhe, Germany,*
[G] *Fachbereich Physik, Universität Siegen, Germany*
[H] *Dpt of Astrophysics, Radboud Uni Nijmegen, The Netherlands*
[I] *IPE, Forschungszentrum Karlsruhe, Germany*
[J] *Ist di Fisica dello Spazio Interplanetario INAF, Torino, Italy*
[K] *Fachbereich Physik, Uni Wuppertal, Germany*
[L] *Radioastronomisches Institut der Uni Bonn, Germany*
[M] *Soltan Institute for Nuclear Studies, Lodz, Poland*

First analyses of coincident data of the LOPES (LOfar PrototypE Station) radio antennas with the particle air shower experiment KASCADE-Grande show basic correlations in the observed shower parameters, like the strength of the radio signal and the particle number, or comparing the estimated shower directions. In addition, an improvement of the experimental resolution of the shower parameters reconstructed by KASCADE-Grande can be obtained by including the data of the radio antennas. This important feature will be shown in this article explicitely by an example event.

The KASCADE[1] (KArlsruhe Shower Core and Array DEtector) experiment measures showers in a primary energy range from 100 TeV to 80 PeV and provides multi-parameter measurements on a large number of observables concerning electrons, muons at 4 different energy thresholds, and hadrons. The main detector components of KASCADE are a field array, the so called central detector and a muon tracking detector. The field array consists of 252 detector stations with shielded as well as unshielded scintillation detectors for measuring the electromagnetic and the muonic shower component.
KASCADE-Grande[2] is the extension of the multi-detector setup KASCADE to cover a primary cosmic ray energy range from 100 TeV to 1 EeV. Grande is an array of 700 x 700 m^2 equipped with 37 plastic scintillator stations sensitive to measure energy deposits and arrival times of air shower particles.
At present, LOPES[3] operates 30 dipole radio antennas (LOPES-30) positioned inside or nearby KASCADE. For the present analysis only data of LOPES-10 (10 antennas in operation), which was running for 5 months, is used. The antennas operate in the frequency range of 40-80 MHz. The radio data is collected when a "large event" trigger is received from KASCADE which translates to primary energies above 10^{16} eV.

The LOPES-10 data set is subject of various analyses using different selections: With an event sample obtained by hard cuts the proof of principle to detect air showers in the radio frequency range was given[3]. With events fallen inside KASCADE the basic correlations with shower parameters are shown[4]. Further interesting features are investigated with a sample of very inclined showers[5] and with a sample of events measured during thunderstorms[6].
Here we report results from an analysis performed by correlating the radio signals measured by LOPES-10 with EAS events reconstructed by KASCADE-Grande with remote cores included[7]. Grande is taking data

in coincidence with KASCADE and LOPES and enables to reconstruct showers with primary energies up to 10^{18} eV and with distances between shower core and the LOPES-10 antennas up to 700 m. The Grande reconstruction accuracy of shower core position and direction is in the order of 4 m (13 m) and 0.18° (0.32°) with 68% (95%) confidence level for simulated proton and iron showers at 100 PeV primary energy and 22° zenith angle[8].

A crucial element of the detection method is the digital beamforming which allows to place a narrow antenna beam in the direction of the cosmic ray event. This is possible because the phase information of the radio waves is preserved by the digital receiver and the cosmic ray produces a coherent pulse. This method is also very effective in suppressing interference from the particle detectors which radiate incoherently. The procedure of time shifting of the radio signals in the antennas is relatively safe when based on the values provided by the reconstruction for shower core and shower axis using the data of the original KASCADE field array. Due to the high granularity of the detector stations the accuracy of core and direction reconstruction is high enough to obtain a good coherence of the radio signals. The shower reconstruction using information from the Grande array is required for shower cores outside KASCADE. The Grande stations, 10 m^2 of plastic scintillator detectors each, are spaced at $\simeq$130 m and cannot assure an accuracy comparable with the original KASCADE array. So, a so-called 'optimised' beamforming is performed, which searches for a maximum coherence by varying the core and the direction around the values provided by the Grande reconstruction.

In five months of LOPES-10 data taking a sample of 862 candidate events were selected. Selection criteria were i) coincident measurement of the event by LOPES 10, KASCADE field array which have triggered LOPES, and Grande array; ii) zenith angle of the shower less than 50°; iii) a geometrical cut that the core position lies inside the Grande array (0.358 km^2); iv) to reduce the data sample additionally an energy and distance cut is applied which is motivated by Allan's fomula[9]: $\epsilon_\nu = 20 \cdot \left(\frac{E}{10^{17} eV}\right) \cdot \sin\alpha \cdot \cos\theta \cdot \exp\left(-\frac{R}{R_0(\nu,\theta)}\right) \left[\frac{\mu V}{m \cdot MHz}\right]$. The formula describes the pulse amplitude per unit bandwidth (ϵ_ν) of the radio signal induced by an EAS. Here E is the primary energy, α the angle to the geomagnetic field, θ the zenith angle, R the distance to the shower axis and R_0 the scaling radius $R_0 = 110$ m at 55 MHz; the exponential radial factor plays a significant role for remote showers. The cut has been considered as

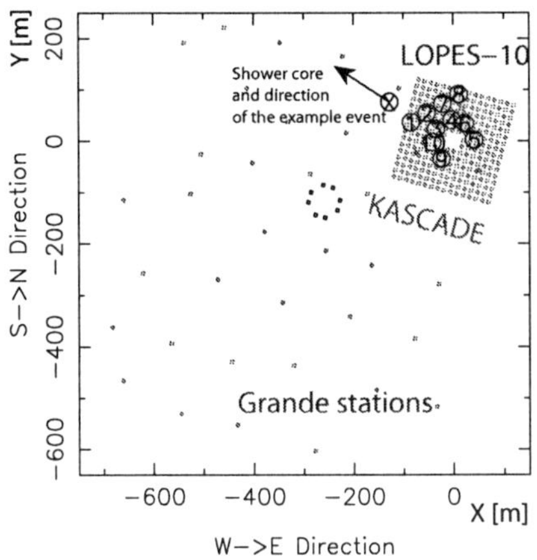

Figure 1. Sketch of the LOPES 10 layout inside KASCADE-Grande. Core position and shower direction of the candidate event 444 is also shown.

$\lg\left(\frac{E}{eV}\right) > \lg\left(\frac{E_0}{eV}\right) + 0.4343 \cdot \frac{R}{R_0}$ *OR* $\lg\left(\frac{E}{eV}\right) > 17.5$ with $E_0 = 10^{16.5}$ eV and $R_0 = 160$ m, i.e. weaker than Allan's scaling with radius.

Fig. 1 shows the layout of LOPES-10 in KASCADE-Grande including the core position and direction of an event (number 444 of 862) with a clear radio signal. Fig. 2 shows the radio signal of the event before and after the optimized beamforming. An noticeable increase is seen in the final beam signal after the optimised beamforming. Table 1 contains the values (azimuthal angle ϕ, zenith angle θ, X_{core} and Y_{core}) reconstructed by Grande and the corresponding values obtained after maximizing the radio coherence for this event; the small shifts assure an almost perfect coherence of the radio signal (lower left panel in Fig. 2).

Table 1. Shower parameters from the Grande reconstruction and after maximized radio coherence for an example event (geomagnetic angle $\alpha = 57.9°$, lg(E/eV)= 17.73).

parameter	Grande reconstruction	maximized radio coherence
ϕ	302.2°	299.3°
θ	41.0°	39.9°
X_{core}/m	-142.8	-139.2
Y_{core}/m	40.3	51.6
curvature/m	3250	4250

The maximization of the radio coherence (optimised beamforming) will be performed to all the candidate events. This will not only increase the

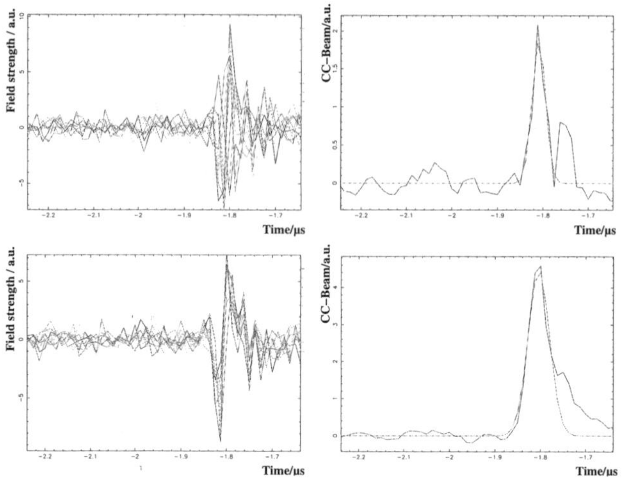

Figure 2. Radio signals in the 10 antennas and the resulting radio signal (so-called cross-correlation beam) with a gaussian fit to the signal after first beamforming (upper panels) and after *optimised* beamforming (lower panels).

efficiency ($\approx 30\%$ without optimized beamforming) for finding radio signals of air showers, but also improve the reconstruction quality on the primary energy of the cosmic rays if the information is included in the shower reconstruction of Grande.

References

1. T. Antoni et al. - KASCADE collab., *Nucl. Instr. & Meth. A* **513**, 429 (2003).
2. G. Navarra et al. - KASCADE-Grande collab., *Nucl. Instr. & Meth. A* **518**, 207 (2004).
3. H. Falcke et al. - LOPES collab., *Nature* **435**, 313 (2005).
4. A. Horneffer et al. - LOPES collab., *Detection of radio pulses from extensive air showers*, Proc. of 29th ICRC, Pune, India (2005).
5. J. Petrovic et al. - LOPES collab., *Radio emission of highly inclined cosmic ray air showers measured with LOPES*, Proc. of 29th ICRC, Pune, India (2005).
6. S. Buitink et al. - LOPES collab., *Electric field influence on the radio emission of air showers*, Proc. of 29th ICRC, Pune, India (2005);
7. A.F. Badea et al. LOPES collab., *Remote event analyses of LOPES-10*, Proc. of 29th ICRC, Pune, India (2005);
   A.F. Badea et al. LOPES collab., *First determination of the reconstruction resolution of an EAS radio detector*, Proc. of 29th ICRC, Pune, India (2005);
8. R. Glasstetter et al. - KASCADE-Grande collab., Proc. of 28th ICRC, Tsukuba, Japan, 781 (2003).
9. H.R. Allan, *Prog. in Element. Part. and Cos. Ray Phys.*, **10**, 171 (1971).

ABSOLUTE CALIBRATION OF THE LOPES ANTENNA SYSTEM

S. NEHLS[A], W. D. APEL[A], F. BADEA[A], L. BÄHREN[B], K. BEKK[A], A. BERCUCI[C], M. BERTAINA[D], P. L. BIERMANN[E], J. BLÜMER[A,F], H. BOZDOG[A], I. M. BRANCUS[C], M. BRÜGGEMANN[G], P. BUCHHOLZ[G], S. BUITINK[H], H. BUTCHER[B], A. CHIAVASSA[D], K. DAUMILLER[A], A. G. DE BRUYN[B], C. M. DE VOS[B], F. DI PIERRO[D], P. DOLL[A], R. ENGEL[A], H. FALCKE[B,E,H], H. GEMMEKE[I] , P. L. GHIA[J], R. GLASSTETTER[K], C. GRUPEN[G], A. HAKENJOS[F], A. HAUNGS[A], D. HECK[A], J. R. HÖRANDEL[F], A. HORNEFFER[H,E], T. HUEGE[A,E], K.-H. KAMPERT[K], G. W. KANT[B], U. KLEIN[L], Y. KOLOTAEV[G], Y. KOOPMAN[B], O. KRÖMER[I], J. KUIJPERS[H], S. LAFEBRE[H], G. MAIER[A], H. J. MATHES[A], H. J. MAYER[A], J. MILKE[A], B. MITRICA[C], C. MORELLO[J], G. NAVARRA[D], A. NIGL[H], R. OBENLAND,[A], J. OEHLSCHLÄGER[A], S. OSTAPCHENKO[A], S. OVER[G], H. J. PEPPING[B], M. PETCU[C], J. PETROVIC[H], T. PIEROG[A], S. PLEWNIA[A], H. REBEL[A], A. RISSE[M], M. ROTH[F], H. SCHIELER[A], G. SCHOONDERBEEK[B], O. SIMA[C], M. STÜMPERT[F], G. TOMA[C], G. C. TRINCHERO[J], H. ULRICH[A], J. VAN BUREN[A], W. VAN CAPELLEN[B], W. WALKOWIAK[G], A. WEINDL[A], S. WIJNHOLDS[B], J. WOCHELE[A], J. ZABIEROWSKI[M], J. A. ZENSUS[E], D. ZIMMERMANN[G]

[A] *Institut für Kernphysik, Forschungszentrum Karlsruhe, Germany*
[B] *ASTRON Dwingeloo, The Netherlands*
[C] *NIPNE Bucharest, Romania*
[D] *Dpt di Fisica Generale dell'Universita Torino, Italy*
[E] *Max-Planck-Institut für Radioastronomie, Bonn, Germany*
[F] *Institut für Experimentelle Kernphysik, Uni Karlsruhe, Germany*
[G] *Fachbereich Physik, Universität Siegen, Germany*
[H] *Dpt of Astrophysics, Radboud Uni Nijmegen, The Netherlands*
[I] *IPE, Forschungszentrum Karlsruhe, Germany*
[J] *Ist di Fisica dello Spazio Interplanetario INAF, Torino, Italy*
[K] *Fachbereich Physik, Uni Wuppertal, Germany*
[L] *Radioastronomisches Institut der Uni Bonn, Germany*
[M] *Soltan Institute for Nuclear Studies, Lodz, Poland*

Radio emission in extensive air showers arises from an interaction with the geomagnetic field and is subject of theoretical studies. This radio emission has advantages for the detection of high energy cosmic rays compared to secondary particle or fluorescence measurement methods. Radio antennas like the LOPES30 antenna system are suited to investigate this emission process by detecting the radio pulses. The characteristic observable parameters like electric field strength and pulse length require a calibration which was done with a reference radio source resulting in an amplification factor representing the system behavior in the environment of the KASCADE-Grande experiment. Knowing the amplification factor and the gain of the LOPES antennas LOPES30 is calibrated absolutely for systematic analyses of the radio emission.

1. Introduction

The long known radio emission in extensive cosmic ray air showers (EAS) is again under investigation with new fully digital radio antennas. Nearly 40 years ago, in the early 1960's this nano-second short weak pulses in EAS were detected and basically confirmed with theoretical predictions. With recent theoretical studies (Huege and Falcke[1]), using a more detailed Monte-Carlo technique, and a new generation of radio telescopes the comparison of predictions and measured radio emission in EAS provides us with a capable method for EAS investigation. The stochastic production process of EAS is a complicated phenomenon. Therefore as many observables as possible are needed to reconstruct the primary shower parameters correctly. The digital radio antenna field of LOPES30 placed inside the existing multiple detector-component experiment KASCADE-Grande[2] is now calibrated absolutely allowing us to measure precisely the long known radio emission in EAS and their dependencies on primary shower parameters like arrival direction, primary particle mass and energy.

2. Radio emission in EAS

Analytical calculations in the early 1970's[3] of the expected electric field strength ϵ_ν, the lateral distribution of ϵ_ν and the dependence on the shower direction predict electric field strengths at ground level in the range of $\epsilon_\nu \approx 5 - 15\ \mu\mathrm{V/m/MHz}$ ($E_\omega \approx 0.5 - 2.5\ \mu\mathrm{V/m/MHz}$) for primary energies $\sim 10^{17}$ eV. The definition of the quantity E_ω and a conversion factor for ϵ_ν can be found in [4]. On the basis of the so called geosynchroton-effect a new analytical model for the calculation of the electric field strength E_ω was developed 2003 by Huege and Falcke. The results of the simulations have been summarized with a parametrisation formula to get expected electric field strength E occurring in EAS. From this parametrisation formula one gets electric field strength at ground in the range of $E_\omega \approx 3 - 5\ \mu$V/m/MHz also for primary energies $\sim 10^{17}$ eV. There has never been a common agreement about the absolute field strength[5] and the values cited decreased over time to a tenth of a μV/m/MHz. For the absolute calibration of LOPES30 these values give a first benchmark to our detection thresholds but they do not really represent typical values of electrical field strengths occurring in model prediction of EAS.

3. Calibration setup for LOPES30

With LOPES10 the "proof of principle" in detecting radio emission from EAS was achieved by comparing relative field strengths in the antenna array [6] and comparing them with the parameters obtained by KASCADE-Grande. The analysis was done without a precise absolute calibration and therefore only a qualitative comparison with theoretical predictions was possible. An absolute comparison can be done by knowing the system (Fig. 1) response to a calibrated well defined signal where the knowledge of the voltage amplitude and signal phase is included. Due to their inverted-

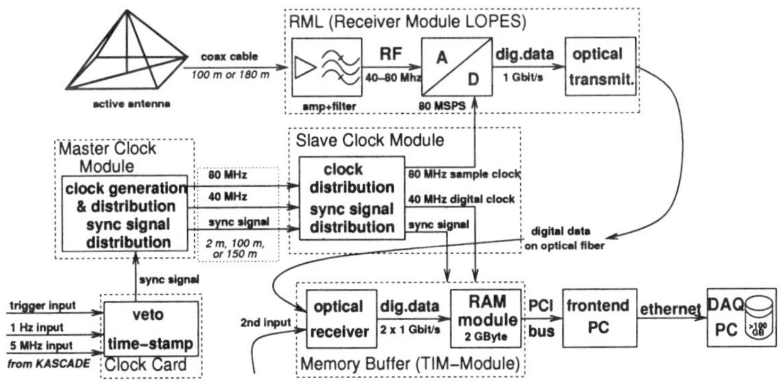

Figure 1. Scheme of LOPES30 electronic. Incoming radio pulses from EAS were detected with inverted-V-shaped antennas, transmitted over 100m to 180m coax cable to the Receiver Module (RML), digitizes, and stored.

V-shape the antennas are most sensitive to vertical EAS and less sensitive to highly inclined signals ($> 70°$) except around 60 MHz as shown in figure 2, left. A mechanically needed quadratic ground plate (2.5 x 2.5 m^2) of aluminum modifies the antenna gain, i.e. it increases the antenna gain in the range of 60 MHz towards highly inclined signals and decreases the gain for vertical signals. From a commercial reference radio source (VSQ

1000[7]) the electrical field strength E in a certain distance is known and the emitted time-continuous and frequency discrete signal is used. This means that the reference radio source emits in 1 MHz, 5 MHz, or 10 MHz steps a defined sine wave, e.g. at 55 MHz around four orders of magnitude higher in power than EAS radio emission, at 10 m distance in the main direction. In our calibration setup the radio source was placed $\approx 10\ m$ above the top of the LOPES antennas. Mounted at the end of a wooden beam fixed on an extension arm of a crane we determined for each antenna an individual frequency dependent amplification factor. These values represent the overall system behavior to the input signal emitted by the reference antenna and therefore all active and passive components in the electronic system (see figure 1) contribute with their individual gain. It is more difficult to calculate an amplification factor from a single component calibration of the full elctronical chain, because some components do not have exactly 50 Ω impedance. For the calibration the transmitted power P_t, the gain G_t of the reference radio source, and the gain G_r of the LOPES antenna correlate with the received power P_r:

$$P_r = \left(\frac{\lambda}{4d\pi}\right)^2 G_r G_t P_t cos^2(\beta) \tag{1}$$

In the temporary setup of a merely simulated LOPES antenna gain G_r the received power P_r can be determined. The polarization angle β is needed to take into account that the LOPES antennas are linearly polarized and therefore the angle between polarization axes of the emitter and polarization axes of the detecting antenna modifies the received power. For all LOPES antennas we succeeded to measure the received power in the main sensitivity direction.

4. Results

In a campaign of three days the measurements were done in the 5MHz or 1MHz step mode of the reference radio source. The fraction of transmitted power P_t to received power P_r is proportional to an amplification factor calculated from a 9.8 msec dataset for each LOPES antenna. We determined such amplification factors for all antennas which can vary from antenna to antenna by a factor of ten with a typical uncertainty of around 15%. The relatively large factor of ten between the antennas is mostly caused by the characteristics of the bandpass filter and is one of the important contributions to the amplification factor. The uncertainties at 50 MHz are larger compared to frequency ranges above and below because of amateur radio

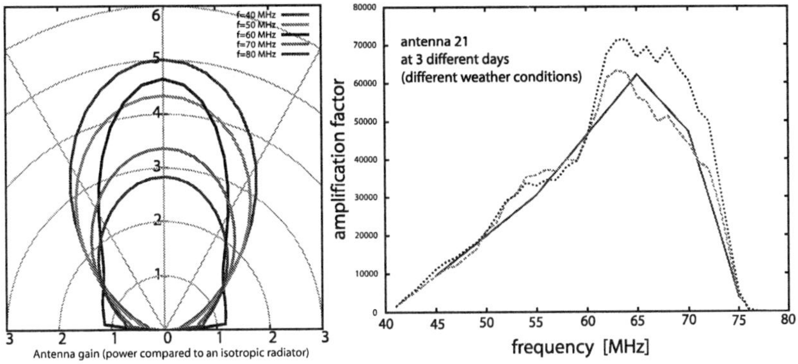

Figure 2. Left: Antenna gain from 40 to 80 MHz in a polar diagram (simulation). Right: Estimated amplification factors for one antenna. From 40 MHz to 80 MHz the influence of different conditions can be seen, especially above 60 MHz one can see deviations in the order of 25 %. Lower solid curve in 5 MHz steps for dry conditions. Middle dotted curve in 1 MHz steps wet conditions. Upper dotted curve in 1 MHz steps at rain fall.

communication occurring in this band. For one antenna we measured the received power in different weather conditions. In figure 2 three curves are shown, representing the amplification factors as a function of frequency, for very dry conditions, wet conditions, and also during rain fall. As a first result it is obvious that the conditions during the calibration measurement such as soil humidity, rain fall, or relative humidity influenced the values of the amplification factor. These first results need further detailed investigations. A weather dependent correction factor for the LOPES antenna system can minimize these calibration uncertainties. Furthermore above 60 MHz the variations in the amplification factor are much stronger than below, and there is no significant connection with the polarization axis between the two antennas or a systematic shift of the curves relative to each other. To eliminate this influence periodic calibration campaigns are needed to better understand the performance of the LOPES antenna system.

References

1. T. Huege and H. Falcke, *A&A*, **430**, 779 (2005);
2. G. Navarra et al. - KASCADE-Grande collab., *Nucl. Instr. & Meth. A* **518**, 207 (2004);
3. H.R. Allan, *Prog. in Elem. Part. and Cos. Ray Phys.*,**10**, 171 (1971);
4. T. Huege and H. Falcke, *A&A*, **412**, 19 (2003);
5. V.B. Atrashkevich et al., *Sov. J. Nucl. Phys.*, **28**, 3 (1978);
6. H. Falcke et al. - LOPES coll., *Nature*, **435**, 313 (2005);
7. Schaffner Group, *www.schaffner.com*, biconical antenna VSQ 1000

CODALEMA: A COSMIC RAY AIR SHOWER RADIO DETECTION EXPERIMENT

D. ARDOUIN[A], A. BELLETOILE[A], D. CHARRIER[A], R. DALLIER[A], L. DENIS[B], P. ESCHSTRUTH[C], T. GOUSSET[A], F. HADDAD[A], J. LAMBLIN[A], P. LAUTRIDOU[A], A. LECACHEUX[D], D. MONNIER-RAGAIGNE[C], O. RAVEL[A]

(A) SUBATECH, 4 rue Alfred Kastler, BP20722, F44307 Nantes cedex 3, France

(B) Observatoire de Paris - Station de radioastronomie, F18330 Nançay, France

(C) LAL, Université Paris-Sud, Bâtiment 200, BP 34, F91898 Orsay cedex, France

(D) LESIA, Observatoire de Paris - Section de Meudon, 5 place Jules Janssen, F92195 Meudon cedex, France

The CODALEMA experimental device currently detects and characterizes the radio contribution of cosmic ray air showers : arrival directions and electric field topologies of radio transient signals associated to cosmic rays are extracted from the antenna signals. The measured rate, about 1 event per day, corresponds to an energy threshold around 5.10^{16}eV. These results allow to determine the perspectives offered by the present experimental design for radiodetection of Ultra High Energy Cosmic Rays at a larger scale.

1. Experiment and event selection

Radio emission associated with the development of Extensive Air Showers (EAS) initiated by high energy cosmic rays was investigated in the 1960's [1]. A flurry of experiments provided initial informations about signals from 10^{17} eV cosmic rays [2], but plagued by difficulties (poor reproducibility, atmospheric effects, technical limitations) efforts almost ceased in the late 1970's to the benefit of ground particle detection and fluorescence. With the advent of cosmic ray research involving giant surface detectors as in the Auger experiment [3], the radio detection, with a potential 100% duty cycle and a sensitivity to the longitudinal development of the showers, is now reconsidered. In recent years, with the availability of new electronics, several groups have undertaken the task of reinvestigating the phenomenology of radio pulses [4,5] which are in a large extent a *terra incognita*. Using our experiment CODALEMA (COsmic ray Detection Array with Logarithmic ElectroMagnetic Antennas), located at the Nançay radio observatory, firm evidence for a radio emission counterpart of cosmic ray air showers has been obtained.

Current experimental setup has been described along with detection and analysis methods in a previous paper [6]. It uses 11 log-periodic antennas of the type constituting the Nançay DecAMetric array (DAM) [7] and 4 particle detectors originally designed as prototype detectors for the Auger array [8]. Seven antennas are equally distributed along a 600 m, East-West line, and 4 other antennas are located on a 85 m side square at the North of this line. The antenna signals are band pass filtered (24-82 MHz) and are recorded, after RF signal amplification, by LeCroy digital oscilloscopes (8-bit ADC, 500 MHz sampling frequency, 10 μs recording time). The four 2.3 m^2 particle detector modules (stations) are located on a 7.10^3 m^2 square area around the 4 northern antennas. Each station has two layers of acrylic scintillators, read out by a copper housed photomultiplier placed at the center of each sheet. The whole experiment is triggered by a fourfold coincidence from those stations in a 600 ns time window, at a rate of 0.7 event per minute. Considering the active area of the particle detector array of 7.10^3 m^2 and the arrival direction distribution of the shower, a value of 16.10^3 m^2.sr is obtained for the acceptance, which corresponds to a trigger energy threshold of about 1.10^{15} eV.

For each fourfold coincidence from the particle detectors, the 11 antenna signals are recorded. Due to the relatively low energy threshold, only a small fraction of these air shower events is expected to be accompanied by significant radio signals. The recognition of the radio transients is made during an offline analysis [6]. Radio signals are first 37-70 MHz numerically filtered to detect radio transient. The maximum voltage is searched in a given time window of 2 μs width, correlated to the trigger time, and compared to a threshold based on the noise level estimation outside this window. If the threshold condition is fulfilled, the instant of the maximum voltage sets the arrival time and the antenna is flagged. When at least 3 antennas arc flagged, a triangulation procedure calculates the arrival direction of the radio wave using a plane front fit. At this level of selection, the couting rate is about one event every two hours.

A stronger selection among these candidates is obtained using the arrival time distribution. It refers the radio wavefront arrival time at a particular point, as determined by antenna triangulation, to the particle front time extracted from the scintillator signals. This time difference distribution is

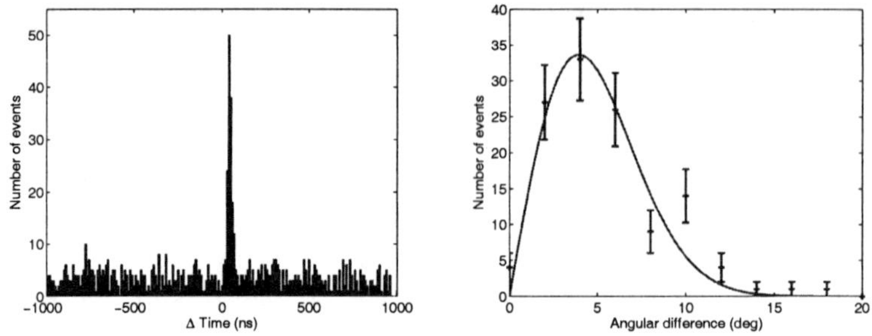

Figure 1. Left: Distribution of time delays between the radio plane front and the particle plane front. Right: Angular difference distribution with a time cut around the peak and fit by the expected analytical form (red/full line). Errors bars are calculated as the square root of the number of events in the bin.

shown in Fig. 1-left. A very sharp peak (a few tens of nanoseconds) is obtained, showing an unambiguous correlation between some radio events and the particle triggers, whereas the flat distribution corresponds to accidental radio transients which are not associated with air showers but occured in the 2 μs window where the search is conducted. Being uncorrelated to the particles, these events fill an uniform arrival time distribution. EAS events are those for which the arrival time difference between the two detector systems is within 100 ns, *i.e.* in the main peak of Fig. 1-left.

If these time-correlated events correspond to EAS, the arrival directions reconstructed from both scintillator and antenna data should be strongly correlated. Indeed, for these events arrival directions obtained by both particle and radio signals are the same within 15 degrees whereas the angle for an uncorrelated event is much bigger, its arrival direction given by the antennas being often close to the horizon. This is typical of events from radio interference due to human activity. The angular difference distribution of time-correlated events (Fig. 1-right) can be fit by a gaussian distribution centered on zero multiplied by a sine function coming from the solid angle factor. The standard deviation of the corresponding gaussian is about 4 degrees. This value includes the reconstruction accuracy of both detectors.

After this sorting, confirmed EAS radio event counting rate falls down to 1 event per day, which corresponds to an energy threshold of about 5.10^{16} eV for the antenna array with an assumed acceptance of 16.10^{3} m^2.sr. At the end of these analysis procedures, physical characteristics of the radio EAS events can be extracted.

2. EAS electric field distribution

Due to the limited extent of the North-South axis of the array, the electric field distribution has been extracted mainly along the East-West axis. Antenna responses were cross-calibrated and gains adjusted within a few %. Distributions for 4 illustrative radio EAS events are shown on the Fig. 2 (with a gaussian fit to guide the eye). A fortuitous event (triangles) and the threshold level of our setup illustrating the amount of useful signal received (circles) are also displayed. The fortuitous event was identified as resulting from an anthropic source and rejected from EAS candidate status.

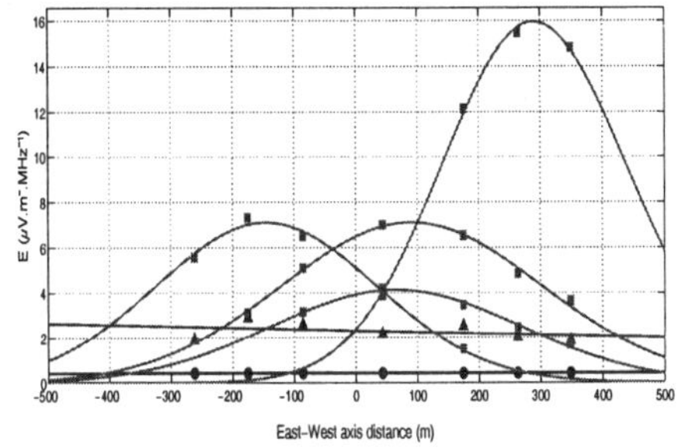

Figure 2. Profile of the maximum voltage (corrected from the cable attenuations and the amplifier gains) recorded on the antennas in the East-West direction for 4 EAS events (squares, gaussian fit: red/full line), an anthropic transient (triangles, fit: blue/dashed line) and detection threshold (circles, fit: black/dash-dotted line).

Topologies are clearly different between EAS and anthropic events. The anthropic event presents an electric field topology with a nearly linear amplitude which is not expected for an EAS candidate falling in the vicinity of the array. EAS events show highly variable field amplitudes depending on the position on the E-W axis. The widths of the distributions observed on an event by event basis is probably related to the dependence on both shower energy and zenithal angle. The projected core locations have been estimated only for events falling inside the surface delimited by the extremities of our antenna array (highly limited by the North-South extent) [9]. The differences between electric field topologies depending on its origin (EAS or anthropic) could constitute one decisive criterion to select EAS events when using a self triggered antenna system. In other words, it means that it is feasible to build a radiodetection experiment able to discriminate EAS events by itself.

3. Conclusions

Electric field transients generated by extensive air showers have been measured with CODALEMA. The current effective counting rate of 1 event/day leads to an energy threshold around 5.10^{16} eV. Electric field spread related to energy and core location of an EAS can be determined on an event by event basis. It is now possible to discriminate an EAS event from a fortuitous one using only antennas and no particle detector. This is one further step towards a stand-alone system that could be deployed over a large area. Improvements are in progress: setting up of additional scintillators will make possible an independant determination of the shower energy and core position; extension of the N-S antenna line will enable to better sample the radio signal spread; increase of the ADC dynamics using 12-bit encoding will allow to record the full 1 MHz - 100 MHz frequency band, and shower parameters could then be inferred from the full signal shape [6]. In a subsequent upgrade, it is planned to install autonomous dipoles equipped with active front-end electronics, self-triggered and self-time-tagged. This is part of current investigations on the feasibility of adding radio detection techniques to an existing surface detector such as the Pierre Auger Observatory. In the future, we expect that the radio signals should provide complementary information about the longitudinal development of the shower, as well as the ability to lower the energy threshold.

References

1. G.A. Askar'yan, Soviet Physics, J.E.T.P., 14, (2) 441 (1962)
2. H.R. Allan, in: Progress in elementary particle and cosmic ray physics, ed. by J.G. Wilson and S.A. Wouthuysen (North Holland, 1971), p. 169.
3. Auger Collaboration, Nucl. Instrum. Meth. A523, 50 (2004).
4. K. Green et al, Nucl. Instrum. Meth. A498, 256 (2003).
5. H. Falcke et al (LOPES Collaboration), Nature, May 19 (2005).
6. D. Ardouin et al (CODALEMA Collaboration), to appear in Nucl. Inst. and Meth. - A (2005), astro-ph/0504297.
7. http://www.obs-nancay.fr/
8. M. Boratav et al, The AUGER Project: First Results from the Orsay Prototype Station, Proceedings of the 24th ICRC, Rome, 954 (1995).
9. D. Ardouin et al (CODALEMA Collaboration), in Proceedings of the XXXIXth Rencontres de Moriond "Very High Energy Phenomena in the Universe", La Thuile, Italy (2005), astro-ph/0505442.

THE CONVERTED HYDROACOUSTIC ARRAY "MG-10M" – A BASIC MODULE FOR A DEEP WATER ν-TELESCOPE

DR. Y. KARLIK

Kamchatka Hydro-physical Institute,3, Viluchinsk,Kamchatka region, 684090, Russia

DR. V. SVET

N.N. Andreyev Acoustical Institute, 4 Shvernik St. 117036, Moscow Russia

The possibility of converted hydroacoustic arrays to create a module of deep water ν-Telescope is discussed. The array MG-10M is good tool to start the first real ocean experiments to investigate the problem of detecting neutrinos with energy 10^{18} eV. Technical parameters and systems which are necessary to create the module are presented. The Mediterranean Sea is an appropriate area for the first experiments due to existing of a deep underwater sound channel and stable high temperature. Some estimation of possible detecting volumes is presented. This project is continuation of works of authors in this direction, started in 1997.

Introduction

The interest to acoustic ways of registration neutrinos first of all is caused by a principal opportunity to create acoustic ν- telescopes with great volumes of detection – up to tens and even hundreds cubic kilometers, [1- 4].

Not looking at that such way of detecting has been offered more 40 years ago, in the experimental plan it has not been made practically anything lately and for quite clear reasons.

The phased hydroacoustic array is the key element of such telescopes. Existing modern hydroacoustic technologies allow making similar arrays with very big number of elementary receivers – hydrophones. Approximately the amplification array factor is simply equal to number of hydrophones, and it can make some thousands hydrophones in modern systems of underwater hydroacoustic monitoring.

Taking into account very rigid technical requirements to operation of the underwater electronic systems working on the big depths, design, making and installation of multi- element hydroacoustic array is sufficiently long-term and expensive technology. In fact such array should function not less than 20-25 years, especially, if to take into account, that detecting high-energy neutrinos is rare enough events at least on modern theoretical representations.

It is offered as an alternative variant to use already existing hydroacoustic arrays, having earlier operated in the Navy. The system MG-10M is the most suitable array to start the first ocean long-term experimental researches.

This system has sufficiently factor of amplification (more than 1700) in a frequency band of 10-25 kHz, and is capable to work on depths up to 400 meters; it is compact and has a high degree of reliability.

1. *Short review of MG-10M*

The array of MG-10M is a metal design (hollow cylinder) with diameter of 1.6 meter and height of 1 meter, Fig.1. 132 hydrophones of tube type are placed on the external surface. Internal surface of cylinder is covered by the rubber screen.

The array is designed for receive signals in range of frequencies from 0,5 KHz up to 25 KHz. The weight of array is about 1200 kg. The acoustical sensitivity of hydrophones is 170 μV/Pa. Maximum depth of operation is 400 meters. The draft of the array is shown on Fig.1.

On the technical (electro-acoustic) parameters the given array completely meets all requirements which possible to show to one acoustic module of a ν-telescope. In the further the escalating of such modules is possible. The array can be installed on special platform (tripod type).

Fig.1 Array MG10M

2. *Signaling*

The transmission of the received signals must be provided by data acquisition system and underwater cable to ground equipment. Taking in regard small amount of hydrophones and low frequencies the better technical solution is to use scheme "one hydrophone - one ADC" Such commercial "sample-hold" 14 bit ADC are available on commercial market. The sampling frequency can be not more that 100 KHz for each receiving channel.

The elementary signals must come to preliminary amplifiers with digital variation of frequency band. Amplifiers should be executed in the form of tight blocks.

3. *Information capacity*

It is better to use fiber optical serial cable to provide the transmission of all digital information to ground equipment. Throughput of the channel (information capacity) is

$$F_M \geq 2F_{\max} Nn$$

where F_M – multiplexer frequency, F_{max} – maximum frequency of signal, n – number of bits of ADC., N – number of hydrophones. For $F_{\max} = 40$ KHz, $n = 14$ and $N = 140$, $F_M = 157$ MHz that is very modest data stream for fiber optical cable with typical bandwidth of some tenths GHz.

4. *Tripod for array installation*

This platform must be supplied with special telemetry system to control the position of the array and has some additional sensors for temperature and flow measurements.

The additional system (anchor release and ballast system) must be included into the mechanical design to provide array emersion for its repair in case of need

5. *Ground equipment*

This equipment is more less typical – optical demodulator and special interface to receive incoming signals from array. It is supposed to provide signal processing on PC with sufficient power.

The basic preliminary operations are:

- Normalization of signals
- Parallel beamforming in horizontal plane (with some weighting)
- Additional filtration of signals in all beams
- Time processing and threshold detection

It is necessary to note that final detection algorithms of neutrinos signals are not clear now, because optimal detection algorithms require knowledge on space-time distribution of pulse noise of different nature. It is possible to get these data only on the base of long-term monitoring of acoustical environments in a concrete area of array installation.

6. *Calibration problem*

Calculated models of acoustical signals generated by interaction of neutrinos in sea water demonstrate that developed acoustical sources are directional. Index of directivity depends on the length of interaction and can be from 1 meter up to 50 meters. For frequencies 5-30 KHz index directivity can be some tenths degrees up to some degrees. Note that detection of signals with "narrow" directivity is a sufficiently new problem in hydro acoustics, where primary or secondary acoustical sources have very small directivity. From another side all theoretical calculations of neutrinos acoustical signals are very approximate. Therefore it is necessary to have some practical tools to understand the real situation with detection of signals in ocean environments

The design of special transducer with changeable directivity, intensity and time duration of pulse signals is the best way to start such researches.

The detection of neutrinos with high energy is rare enough event, therefore a big volume of works should be directed on studying of various pulse noise in area of array installation to develop in the future the effective algorithms minimizing the levels of false alarms.

7. *Area of installation of the array*

It is known that the effectiveness of neutrinos interaction and generation of acoustical signals depends strongly on water temperature. The conditions of sound propagation are very important too. The propagation in sound underwater channel guarantees the minimal losses. From these points of view the Mediterranean Sea is a most attractive region. In East part of Mediterranean Sea the underwater sound channel exists during year on the depth about 300 meters and the stable temperature on this depth is about 13°C. The stability of temperature is a very important factor and it is difficult to find other more suitable area in this sense.

8. *Estimations of detection volume*

The estimations of potential detection volume have been made for hydrological conditions of Mediterranean Sea. For the best conditions of underwater sound channel one array module can provide detection volume of (2-5) km^3 and even up to 10 km^3 in some cases for neutrinos with energy of 10^{18} eV. From other side the detection of signals in sound channel is correct only for sufficiently narrow angle sector. Because the direction of neutrinos propagation is unknown the real detection volume will be less and sometimes significantly less.

Gradual escalating of antenna modules is the unique decision of a problem to increase the volume of detection and all technical opportunities for this purpose now are available.

Acknowledgments

This work is supported by grants 05-02-17410 and 05-02-26620 of the Russian Foundation of Basic Researches, and by Program "Neutrino Physics" of the Presidium of Russian Academy of Sciences.

Authors are grateful to Director of INR, Academician, Dr. V.A. Matveev, for constant attention and interest to this work and to the Organizing Committee of the ARENA Workshop for support and hospitality in Zeuthen.

References

1. Karlik Ya.S., Learned J.G., Svet V.D., and Zheleznykh I.M., *Proc.32 Rencontres de Moriond*, Les Arcs, Jan.18-25, p.p.283-286 (1997).
2. Dedenko L.G., Furduev A.V., Karlik Ya.S., Learned J.G., Mironovich A.A., Svet V.D., and Zheleznykh I.M., Proc.*25th ICRC*, Durban, **7**, 89 (1997).
3. L.G. Dedenko, Ya. S. Karlik, J.G. Learned, V.D. Svet, and I.M. Zheleznykh, *AIP Conf. Proc.*, **579**, 277 (2001).
4. Capone A., Dedenko L.G., Furduev A.V., Kalenov E.N., Karaevsky S.K., Karlil Ya.S., Koske P., Learned J.G, Matveev V.A., Mironovich A.A., Smirnov E.G., Svet V.D., Tebyakin V.P., and Zheleznykh I.M., *Proc. 27th ICRC, Hamburg,* 1264 (2001).

A DEVICE FOR DETECTION OF ACOUSTIC SIGNALS FROM SUPER HIGH ENERGY NEUTRINOS *

V.M.AYNUTDINOV[A], V.A.BALKANOV[A], I.A.BELOLAPTIKOV[D], L.B.BEZRUKOV[A], D.A.BORSCHEV[A], N.M.BUDNEV[B], K.V.BURMISTROV[A], A.G.CHENSKY[B], I.A.DANILCHENKO[A], YA.I.DAVIDOV[A], A.A.DOROSHENKO[A], ZH.-A.M.DJILKIBAEV[A], G.V.DOMOGATSKY[A], A.N.DYACHOK[B], O.N.GAPONENKO[A], K.V.GOLUBKOV[A], O.A.GRESS[B], T.I.GRESS[B], O.G.GRISHIN[B], S.V.FIALKOVSKI[F], A.M.KLABUKOV[A], A.I.KLIMOV[L], A.A.KOCHANOV[B], K.V.KONISCHEV[D], A.P.KOSHECHKIN[A], VY.E.KUZNETZOV[A], V.F.KULEPOV[F], L.A.KUZMICHEV[C], B.K.LUBSANDORZHIEV[A], S.P.MIKHEYEV[A], T.MIKOLAJSKI[E], M.B.MILENIN[F], R.R.MIRGAZOV[B], E.A.OSIPOVA[C], A.I.PANFILOV[A], G.L.PAN'KOV[B], L.V.PAN'KOV[B], YU.V.PARFENOV[B], A.A.PAVLOV[B], D.P.PETUHOV[A], E.N.PLISKOVSKY[D], P.G.POKHIL[A], V.A.POLESCHUK[A], E.G.POPOVA[C], V.V.PROSIN[C], M.I.ROZANOV[G], V.YU.RUBTZOV[B], B.A.SHAIBONOV[A], A.SHIROKOV[C], CH.SPIERING[E], B.A.TARASHANSKY[B], R.V.VASILJEV[D], R.WISCHNEWSKI[E], V.A.ZHUKOV[A], I.V.YASHIN[C]

(A) Institute for Nuclear Research, Moscow, Russia
(B) Irkutsk State University, Irkutsk, Russia
(C) Skobeltsyn Institute of Nuclear Physics MSU, Moscow, Russia
(D) Joint Institute for Nuclear Research, Dubna, Russia
(E) DESY, Zeuthen, Germany
(F) Nizhni Novgorod State Technical University, Nizhni Novgorod, Russia
(G) St.Petersburg State Marine University, St.Petersburg, Russia
(L) Kurchatov Institute, Moscow, Russia

We present the design of a device for detection of acoustic signals from high energy particle showers. The module will be stationary installed above the Baikal Neutrino Telescope NT-200+.

*This work is supported by the Russian Ministry of Education and Science, the German Ministry of Education and Research and the Russian Fund of Basic Research (grants 02-02-17427, 03-02-31011, 04-02-31003, 05-02-16593).

1. Introduction

The investigation of very high energy neutrinos is one of the most interesting task for astrophysics. It requires huge arrays with effective volumes much beyond a cubic kilometre. One of the possible ways to study high energy neutrinos is to detect acoustic pulses from cascade showers [1]. The absorption length for acoustic waves in water is at least an order of magnitude larger than that of Cherenkov light, so acoustic pulses can be detected from very large distance.

However, as it was shown in [2,3], the acoustic noise in natural basins consists of pulses with a large variety in amplitude, shape and duration, including bipolar pulses. The last provide an essential background for acoustic neutrino detection. The actual energy threshold of the acoustic method depends on possibilities to increase the signal-to-noise ratio by means of both hardware and software methods. The small value of the speed of sound prevents a direct application of the proven signal-to-noise discrimination techniques known from nuclear physics. In this work we present a deep-water 4-channel digital device, intended for the search of small-pulse signals from distant, quasi-local acoustic sources. We also present the results of test measurements performed at Lake Baikal in April, 2005.

2. A device for detection of acoustic signals from high energy neutrinos

Supposing the thermo-elastic mechanism, an acoustic signal from neutrino-induced cascades is expected to peak at frequencies of 20 kHz, with calculated amplitudes for a 10 PeV cascade at 400 m distance ranging from a few μPa [4] to a few tens of μPa [1,5]. Although such a signal is close to the sensitivity of the human ear, its detection is far from being trivial since it has to be separated from various sources of noise. For this purpose a digital hydro-acoustic device with four input channels has been developed. The principal scheme of the device is shown in Fig1. Acoustic signals are recorded by four cylindrical hydrophones (Hf-1,2,3,4) with a sensitivity of about 1 mV/Pa, made from a tangentially polarized piezoceramic. The signals are further processed by preamplifiers (Preamp-1,2,3,4) with 40 dB amplification and frequency correction. In the range down from 1 kHz, the relative amplification is lowered by 20 dB per octave in order to suppress low frequency noise. High frequency noise is suppressed by discrete low-pass filters following the preamplifiers. The further processing is performed by an IC (ADM416x200), which is mounted on the base

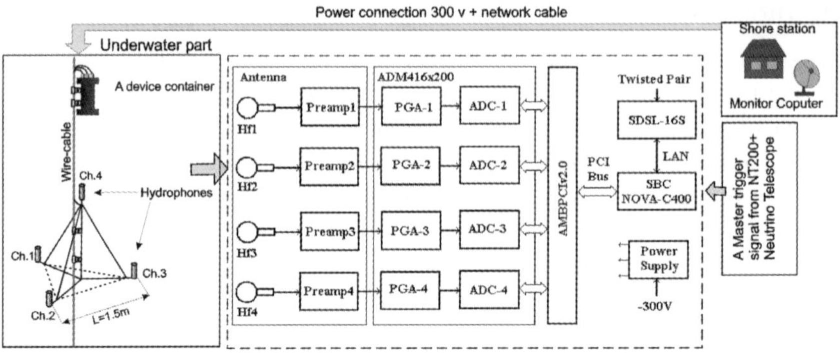

Figure 1. Block scheme of an autonomous hydro-acoustic device

board (AMBPCI v2.0) of a computer (NOVA-C400 Series) and includes four software programmable amplifiers (PGA-1,2,3,4) and the 16-bit over sampling ADC-AD7722 (ADC-1,2,3,4) with a maximum conversion rate of 0.2 Msamples/sec. A single board computer pre-processes the data and communicates with a shore computer via DSL modem. The electronics is housed by a cylindrical metallic container with 22 cm outer diameter and 40 cm height. Four hermetic connectors penetrate the upper cap of the container (1 - power connection 300 v; 2,3 - network twisted-pair cable, 4 - coaxial cable for trigger signal from NT200+) and two pairs of hydrophone connectors penetrate the left and right sides of the container.

The module is designed for operation together with the Baikal Neutrino Telescope NT200+. There are three regimes of operation of the instrument:

(1) Transmitting of a one-second sample of data from all hydrophones to the shore computer centre, after a trigger signals from NT200 or the NT200+ outer strings.
(2) Online search for short acoustic pulses of definite shape, which can be interpreted as signals from distant quasi-local sources.
(3) An autonomous analysis of acoustic background statistics.

The joined operation with NT200+ could give us an opportunity to identify the properties of acoustic emission from cascades and provide energy calibration (assuming that signal strength and flux are high enough and the energy threshold low enough to collect a usable number of true coincidences). To suppress the amount of raw information which should be transferred to the shore station the pre-processing of data will be done in situ using the algorithm described in [6]. The analysis software is able to detect an acoustic signal which exceeds the threshold and to extract the

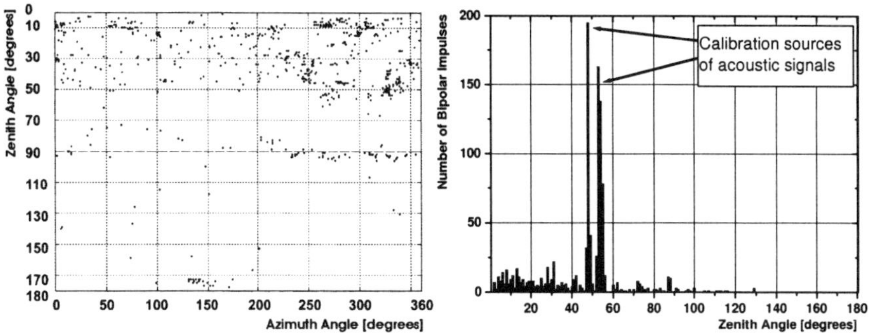

Figure 2. Left (a): The distribution of arrival of all impulses. Right (b): The distribution of bipolar impulses versus zenith angle.

parameters of the signal, such as maximum amplitude, number of periods, duration of each period and etc. Several years of study of acoustic noise in Lake Baikal [2] have shown, that most of short bipolar impulses are the result of an interference of acoustic waves from numerous distributed sources of sound. Extraction of small pulse signals from such a background requires an antenna consisting of several hydrophones. The optimum distance between the hydrophones is defined by the condition that it should clearly exceed several wave-lengths of the expected signal, but on the other hand should be not too long in order to minimize the number of background impulses captured by the 'time window'. So, we arranged the hydrophones in a pyramid-like geometry with the distance between the hydrophones of 1.5 m. Eventually we will install the device on one of mooring of underwater complex of Baikal Neutrino Telescope NT200+ and perform a long-term study of the background as well as a search for short acoustic impulses, including the regime of coincidences with NT200+.

3. Results

The combined data were taken during the test measurements in April 2005. The acoustic antenna allows us to estimate the vertical and horizontal angles of incidence of acoustic signals. The position of antenna's hydrophones in water is fixed by acoustic transponders located around NT200+. The results of a preliminary analysis show, as it was expected, that most of all impulses come from the upper hemisphere. This means that their sources are on the surface of the lake (Fig.2a). Therefore, for the next measurements we assume to place the device at small depth and to search for acoustic signals from the lower hemisphere. In Fig.2b the distribution of bipolar impulses is presented. One sees an insignificant number of bipolar impulses coming from the lower hemisphere, but with the angles not more than 20° to the horizon. These signals also could arise on the surface of the lake, but changed their direction due to refraction. The bipolar signals entering Fig.2b have been requested to have a length smaller than 50 μs and an amplitude larger than four standard deviations. These criteria have been chosen in order to select only signals which may simulate signals due to high energy particle cascades.

4. Summary and Outlook

A device for registration and preliminary analysis of acoustic signals has been constructed and tested in-situ in April, 2005. Apparently, for the search of acoustic signals from high energy cascades it is preferable to listen to water volume from top to down, that is to place the acoustic antennas at depths of the order of 100 - 200 m. One also has to reduce the sensitivity with respect to signals from above, for example, by means of caps made from a sound-absorbing material. The results of the measurement of acoustic noise in Lake Baikal show its complicated structure and its strong dependence on different factors. To study it more systematically in 2006, we intend to deploy the device for one year operation, together with the Baikal Neutrino Telescope. At the next stage we plan to use this antenna as an elementary unit of a future hydrophone array.

References

1. G.A. Askarian and B.A. Dolgoshein, ZhETF pys'ma 25 (1977).
2. V.M.Aynutdinov et al., High Frequency Noise in Lake Baikal., submitted to Acoustical Physics (Akusticheskii zhurnal)
3. J.Vandenbroucke, G.Gratta and N.Lehtinen., astro-ph/0406105
4. J.G.Learned, Phys. Rev. D19 (1979) 3293.
5. L.G.Dedenko et al., Proc 24th ICRC (1995), vol 1, 797.
6. V.M.Aynutdinov et al., Study of a possibility of acoustic detection of super high energy neutrino in Lake Baikal, HE1.5, ICRC29, 2005

ACORNE SIMULATION WORK

JONATHAN PERKIN

*for the ACoRNE Collaboration**

Department of Physics and Astronomy, University of Sheffield, Sheffield. S3 7RH, United Kingdom

A summary of the simulation studies currently underway by the UK based Acoustic Cosmic Ray Neutrino Experiment (ACoRNE) collaboration is presented. Ideas for future development are also discussed. The work described here has been developed for simulations of large scale hydrophone arrays but many of the same considerations apply for other detection techniques.

1. Introduction

All modern day particle physics experiments rely heavily on detailed simulation of their detector and the particle physics processes occurring within. Simulation requirements for the Acoustic Detection of Ultra High Energy (UHE) Neutrinos can be broken down into four key stages: the underlying neutrino interaction, the evolution of the resulting particle cascade, the formation of the acoustic signal, and finally, the propagation and detection of the signal in a detector.

2. Neutrino Interaction Simulations

Preliminary simulations of large scale acoustic UHE neutrino detectors (e.g. The ACoRNE proposal, SAUND[1]) have often assumed a constant level of inelasticity in the neutrino event. Typical energies for the hadronic component of the neutrino Deep Inelastic Scattering (DIS) interaction are $0.2 \rightarrow 0.25 \times E_\nu$. In reality this energy is not fixed but varies per interaction. There are some neutrino interaction simulators available that work at UHEs. The All Neutrino Interaction Simulation (ANIS[2]) has been developed by the AMANDA collaboration and works up to 10^{21}eV, making use of extrapolations of theory driven models for the neutrino-nucleon

*http://www.shef.ac.uk/physics/research/pppa/research/acorne.htm

cross section. PYTHIA[3] will permit simulation of neutrino events up to neutrino energies of 10^{20}eV but becomes unreliable beyond this. An event generator has been developed within the ACoRNE collaboration that can simulate events beyond 10^{21}eV. The formulae for calculating the neutrino-nucleon cross-section, and Bjorken x and y scaling variables are given in Ref.[4]. Parton Distribution Functions are provided by MRS99[5].

A comparison of the output from each generator is shown in Fig. 1. Detector simulations developed in the future might make use of such generators as a front end to their calculations.

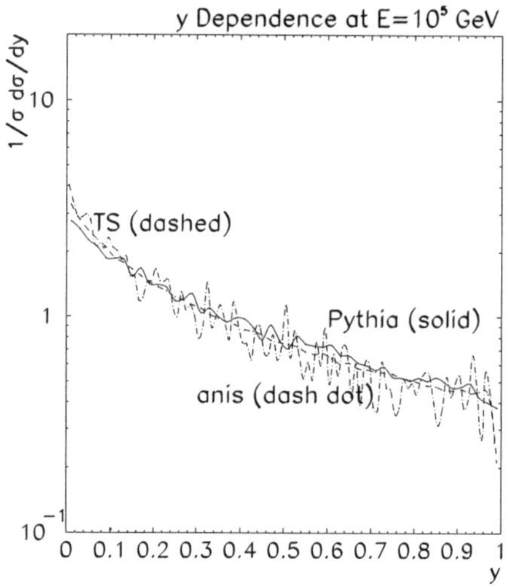

Figure 1. *Bjorken y* values from each event generator. The curve labelled TS corresponds to the ACoRNE bespoke simulation.

3. Simulating UHE Hadronic Cascades

There is no single, open source simulation toolkit available that can simulate an UHE hadronic cascade in water, ice or salt. Traditionally physicists use GEANT4 [6] for particle tracking in accelerator experiments but its range of validity ends at 10^{14}eV. CORSIKA[7] is an UHE particle shower program but it is designed for tracking of events in the atmosphere. One possible solution to the problem is a hybrid of the two. Modifications have been made to CORSIKA such that all the atmosphere parameters have been replaced by water density. A careful comparison between GEANT4 and CORSIKA is underway. If the modified version of CORSIKA can be validated by GEANT4 at energies up to 10^{14}eV, then it will be used for producing shower particle densities up to 10^{21}eV. Full implementation of the Landau-Pomeranchuk-Migdal (LPM) effect still remains in both cases.

4. Simulation of Acoustic Signal

The shape of the acoustic signal is dependent on the form of the thermal energy density of the particle cascade, initiated by the neutrino interaction. The longitudinal evolution of a shower is well described by Gamma functions up to energies of a few PeV (confirmed by Extensive Air Shower experiments) and is expected to scale with energy. The form of the transverse component however, is not so easily defined. Numerous parameterisations have been tested, from a simple Gaussian cross section, to various modified versions of the Nishimura-Kamata-Greisen (NKG)[8] distribution. If a simulation such as that described in the previous section is developed, future simulations will benefit from more accurate shower-particle distributions.

Given a form for the thermal energy density, one can compute the expected shape of the acoustic radiation field at some arbitrary location[9]. Two possible methods have been proposed by which to do this. The traditional way is to use a Monte Carlo (MC) generator in which each point in thermal energy density is of the form of a Gaussian heat source. Then one simply integrates over all points to yield the resulting pressure field. However numerical integrations over a data set of Gaussian points is computationally expensive. There is a second method, based on Signal Processing techniques, in which each point in thermal energy density is represented by a delta function (the integral of which is unity) and then convolved with a Gaussian to retrieve the signal. The second method requires approximately 5×10^6 fewer flops of calculation per integration.

5. Large Scale Detector Simulation

One important simulation goal is to be able to make a prediction for the sensitivity of the acoustic technique by way of some hypothetical large-scale array of hydrophones (or glaciophones, halophones etc). The current method by which this is achieved is to simulate hadronic cascades, initiated by a single particle that represents the excited hadronic final state of a neutrino DIS interaction. This is performed up to the maximum allowed energies for a given MC (e.g. GEANT4 =100TeV). As mentioned in Section

2, one usually assumes a constant inelasticity, y. A parameterised form of the thermal energy density resulting from such simulations is then extrapolated to UHEs. Finally, one is able to compute the expected acoustic pressure field associated with an event. Due to the long attenuation lengths of acoustic signals in water, the majority of events detected will originate from outside the instrumented volume, the acoustic pancake intersecting with an array. One is therefore required to simulate the effects of signal attenuation in order to calculate the expected far field signal. There are three factors contributing to the signal attenuation: the geometric fall off of intensity, the losses due to propagation through a dense medium, and the angular spread according to Fraunhoffer Diffraction Theory.

A threshold for detection is set depending on the Probability of False Alarm (PFA). In our study the threshold was set at 35mPa, corresponding to a PFA of one false signal in ten years due to noise, with a five-fold coincidence. We assume the signal sits on top of a flat, white noise spectrum, the level of which is calculated as a function of wind speed at the sea surface. A threshold of 35mPa indicates a noise-level of 35.9dB, the mean noise level from June to August being 35dB with an average wind speed of 5.6knots (data for Tyrrhenian Sea).

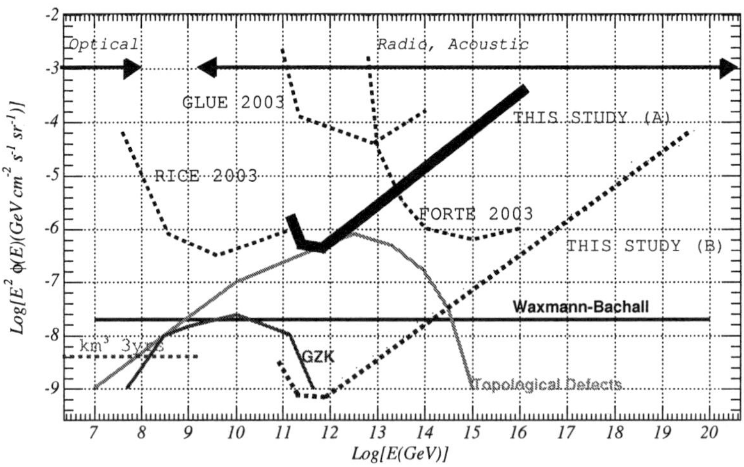

Figure 2. Various model independent flux limits. The breadth of the solid band labelled "THIS STUDY (A)" encompasses the range of performances of the array geometries under study.

The simulated sensitivities of various geometries of approximately cubic-kilometre arrays, containing typically 1000 hydrophones and running for 1 year are illustrated in Fig. 2 by the solid curve labelled "THIS STUDY (A)". The dashed curve, labelled "THIS STUDY (B)" is the predicted sensitivity for an array of 5000 hydrophones, distributed at random, in a volume with dimensions 50km×50km×1km, running for 5 years, with a threshold of 5mPa. This hypothetical array was simulated simply to illustrate a possible scenario for the investigation of sub-GZK fluxes, it is not the result of any optimisation. One calculates the expected sensitivity via a Vertex Reconstruction Algorithm based on the Time Difference Of Arrival (TDOA) of signals at hydrophones within the array. In the study described here, the effects of refraction, due to the variation of sound speed with depth have not been included. A summary of the details of this simulation can be found in Ref.[10].

6. Future Prospects

Several aspects of the simulation work now underway remain to be completed before a full detector simulation is realised. These include (but are not limited to): the variation of pulse shapes due to sound speed fluctuations, pointing and energy reconstruction. Each of these, it is hoped, will be easier to address in the presence of a suitable combination of Neutrino Event Generator and Hadronic Cascade Simulation. Furthermore it has been suggested that, in the future, hybrid simulations encompassing both acoustic and radio signal generation should be developed.

References

1. N. G. Lehtinen *et al,* Astropart.Phys. **17** (2002) 279-292 (astro-ph/0104033).
2. M. Kowalski and A. Gazizov, Proceedings of 28th ICRC (2003) 1459-1462
3. H.-U. Bengsston, Comp. Phys. Comm. **82** (1984) 323
4. R. Gandhi *et al* Astropart.Phys. **5** (1996) 81-110 (hep-ph/9512364)
5. A. D. Martin *et al* Eur.Phys.J. **C14** (2000) 133-145 (hep-ph/9907231)
6. http://geant4.web.cern.ch/geant4/
7. D. Heck *el al,* Report FZKA 6019, Institut für Kernphysik, Karlsruhe, 1998.
8. K. Greisen Progress in Cosmic Ray Physics **v3**, (1956) 1-137
9. J. G. Learned Phys. Rev. D, **v19**, No. 11 (1979).
10. J. Perkin "Simulating the Sensitivity of km^3 Hydrophone Arrays to Fluxes of UHE Neutrinos" Contribution to Proceedings of the 20th Lake Louise Winter Institute. World Scientific (2005) (article in print).

DESIGN CONSIDERATIONS AND SENSITIVITY ESTIMATES FOR AN ACOUSTIC NEUTRINO DETECTOR*

T. KARG, G. ANTON, K. GRAF, J. HÖSSL, A. KAPPES, U. KATZ, R. LAHMANN, C. NAUMANN AND K. SALOMON

Physikalisches Institut,
Friedrich-Alexander-Universität Erlangen-Nürnberg,
Erwin-Rommel-Straße 1,
91058 Erlangen, Germany
E-mail: Timo.Karg@physik.uni-erlangen.de

We present a Monte Carlo study of an underwater neutrino telescope based on the detection of acoustic signals generated by neutrino induced cascades. This provides a promising approach to instrument large detector volumes needed to detect the small flux of cosmic neutrinos at ultra-high energies ($E \gtrsim 1\,\mathrm{EeV}$). Acoustic signals are calculated based on the thermo-acoustic model. The signal is propagated to the sensors taking frequency dependent attenuation into account, and detected using a threshold trigger, where acoustic background is included as an effective detection threshold. A simple reconstruction algorithm allows for the determination of the cascade direction and energy. Various detector setups are compared regarding their effective volumes. Sensitivity estimates for the diffuse neutrino flux are presented.

1. Introduction

Very large target masses are required to detect the low neutrino fluxes predicted at ultra-high energies. Current water Čerenkov neutrino telescopes (AMANDA, BAIKAL, ANTARES, NESTOR, ...) and next-generation km^3 size detectors (IceCube, KM3NeT) do not have sufficient fiducial volume to detect, for example, GZK neutrinos. The affordable size of these detectors is limited by the attenuation length of light in water or ice which restricts the spacing between optical sensors. G.A. Askariyan described a hydrodynamic mechanism of sound generation for charged particles propagating through water[1] which can be exploited for an acoustic neutrino telescope. The thermo-acoustic model has since been verified in the laboratory several

*This work was supported by the German BMBF Grant No. 05 CN2WE1/2.

times and with high precision[2,3,4]. Utilizing the fact that, for the frequencies considered, the sonic attenuation length in water is about ten times larger than the optical attenuation length, much larger volumes could be instrumented with the same number of sensors. In the next section we describe the simulation chain used for studying acoustic neutrino telescopes. After that, sensitivity estimates for an acoustic detector are derived.

2. The simulation chain

For the simulation an isotropic flux of highest-energy neutrinos ($10^8\,\mathrm{GeV} < E_\nu < 10^{16}\,\mathrm{GeV}$) is generated. Equal numbers of neutrinos are produced in each energy bin of constant width in $\log E$, with a given energy spectrum following a power law (E^{-2}) in each E bin. It is assumed that all neutrinos from above can propagate freely down to the detector. On the other hand, the earth is assumed to be opaque for all neutrinos coming from below the horizon. The elasticity distribution of the neutrino interaction is taken from the ANIS neutrino interaction simulator[5]. For electromagnetic cascades the LPM effect, which elongates the cascade and thus reduces the energy density and the amplitude of the acoustic signal, has to be taken into account. Since there is no reliable shower simulation including the LPM effect in water so far, the leptonic branch of all neutrino interactions is discarded, even for electron-neutrino charged-current interactions.

The three-dimensional cascade development and energy deposition were studied with GEANT4 up to primary hadronic energies of 100 TeV using the QGSP interaction model. The shape and the spatial extension of the energy distribution were found to vary only slightly with the primary energy. Therefore, the spatial distribution of the energy is assumed to be the same for all energies, and the energy density scales linearly with the energy of the hadronic system. This energy distribution and the thermodynamic parameters of water are then used as an input to the thermo-acoustic model which gives the resulting bipolar acoustic signal for every sensor position. The amplitude of the bipolar pulse depends on the cascade energy only.

Sonic attenuation in sea water is strongly frequency dependent. The attenuation length for the typical signal frequency of approx. 20 kHz is 1 km (compared to 50 – 70 m optical attenuation length relevant for water Čerenkov neutrino telescopes). It is accounted for by applying a frequency filter to the acoustic signal at a given sensor position. Figure 1 shows the parameterization of the amplitude of the bipolar signal as a function of position, which is used in the simulation to determine the sensor response for a given hadronic cascade.

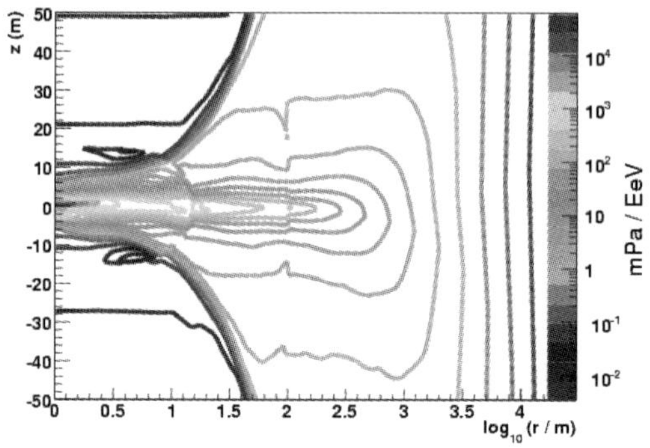

Figure 1. Parameterization of the amplitude of the sonic field for a hadronic cascade centered at the origin. The cascade has a length of approx. 15 m and develops in positive z direction.

The smallest unit of the simulated acoustic detector is an "acoustic module" (AM) which is a device that can detect bipolar acoustic signals above a given detection threshold, $p_{\rm th}$, determined by the background noise in the sea. Such an AM might be realized as a local array of hydrophones allowing the suppression of background with short correlation length. According to Ref. 6 a threshold of 35 mPa has to be used for a single hydrophone if one allows for one false signal in 10 years at a five-fold coincidence. Using AMs consisting of multiple hydrophones should allow to lower this threshold down to 5 mPa.

Our detector consists of AMs that are arranged randomly inside the instrumented volume in order to avoid geometrical effects on the sensitivity estimates. Neutrino events are generated homogeneously and with 2π sr angular distribution in a volume with a height of 2.5 km (corresponding to typical depths in the Mediterranean Sea), and a radius of 10 km; the resulting generation volume is denoted by $V_{\rm gen}$. Each AM records the arrival time and amplitude of the signal if it is above the threshold $p_{\rm th}$. An event is triggered if four or more AMs detect a signal. For our study a timing resolution of 10 μs (100 kHz sampling frequency), a positioning accuracy of 10 cm for the AMs and an amplitude resolution of 2 mPa are implied, which are all realized by Gaussian smearing.

The shower reconstruction is performed in two steps. First, the shower position is reconstructed by minimization of the residuals of the arrival times assuming an isotropic sonic point source (which is a valid assumption since the typical inter-AM distance is large compared to the shower extension). With this method the cascades center of gravity can be reconstructed with a RMS of 14 cm in each Cartesian coordinate. Based on this position and the parameterization of the sonic field (Fig. 1) the direction and energy of the cascade are reconstructed by minimizing the amplitude residuals. Without applying any selection cuts the median of the error in the direction reconstruction is 7°, where events are still included, for which the reconstruction seems to fail completely. The energy can be determined up to a factor of 3.

3. Sensitivity estimates

Based on the detector simulation chain presented above it is possible to derive sensitivity estimates for various detector configurations. We use the effective volume defined as $V_{\mathrm{eff}} = \frac{N_{\mathrm{reco}}}{N_{\mathrm{gen}}} V_{\mathrm{gen}}$ as a measure for the sensitivity of a detector, where N_{reco} is the number of reconstructed events (reconstruction fits converge) without any selection cuts obtained from N_{gen} events generated inside the volume V_{gen}. Figure 2 shows the effect of varying the instrumentation density of the detector between 50 and 800 AM/km^3. For densities much lower than approx. 200 AM/km^3 the effective volume drops dramatically at lower energies, and thus, the lower energy threshold rises.

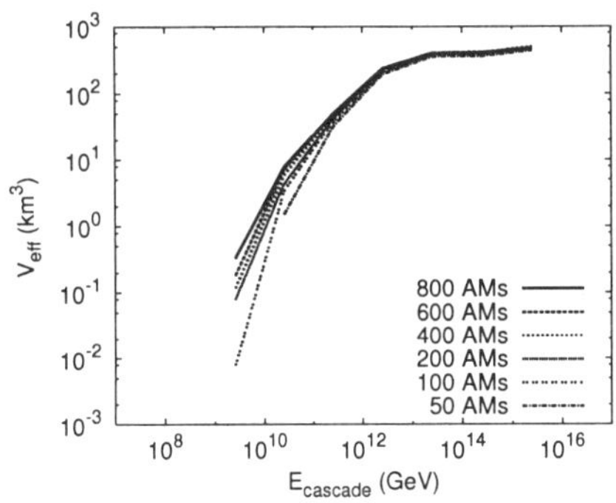

Figure 2. Effective volume for various sensor densities (instrumented volume: 1 km^3).

Further, it is essential for a future acoustic detector to have a pressure threshold p_{th} as low as possible, where the lower limit is given by the intrinsic background noise in the sea which is approx. 1 mPa (sea state 0). On the other hand, a density of only 200 AM/km^3 seems sufficient which allows to instrument very large volumes with a moderate number of DAQ channels read out at low frequencies (100 kHz), leading to manageable data rates. In figure 3 we show that, with a detector with $3 \cdot 10^5$ DAQ channels ($30 \times 50 \times 1\,\text{km}^3$, $200\,\text{AM/km}^3$, $p_{\text{th}} = 5\,\text{mPa}$), several theoretical models that predict neutrinos above 1 EeV could be verified within 5 years of runtime.

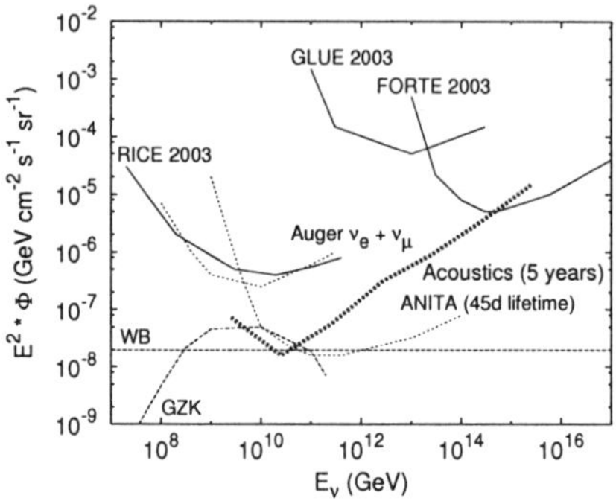

Figure 3. Neutrino flux limit derived for a $30 \times 50 \times 1\,\text{km}^3$ detector with a lifetime of 5 years. Dashed lines are theoretical models (extrapolated Waxman-Bahcall flux and GZK neutrinos). Solid lines are experimental flux limits; dotted lines are expected flux limits from future experiments.

4. Conclusions

Acoustic detection is a promising approach to detect cosmic neutrinos at highest energies. Detectors build of "acoustic modules" that can detect bipolar acoustic signals above 5 mPa are able to reconstruct neutrino-events with energies above 1 EeV with as few as $200\,\text{AM/km}^3$. This allows for the construction of a teraton detector, which is necessary to detect the small neutrino fluxes predicted by theoretical models within a reasonable time.

References

1. G. A. Askariyan, *Atomnaya Energiya* **3**, 152 (1957).
   G. A. Askariyan et al., *Nucl. Inst. Meth.* **164**, 267 (1979).
2. L. Sulak et al., *Nucl. Inst. Meth.* **161**, 203 (1979).
3. S. D. Hunter et al., *J. Acoust. Soc. Am.* **69**, 1557 (1981).
4. K. Graf et al. in these proceedings.
5. M. Kowalski and A. Gazizov, 28th ICRC, Tsukuba, 1459 (2003).
6. S. Danaher et al. in these proceedings.

STUDY OF ACOUSTIC ULTRA-HIGH ENERGY NEUTRINO DETECTION PHASE II

N. KURAHASHI*

Stanford University
Stanford, CA 94305-4060, USA
E-mail: naokok@stanford.edu

The Study of Acoustic Ultra-high energy Neutrino Detection has started its second phase (SAUND II). Although the general location of the hydrophones has not changed, SAUND II uses a new hydrophone array that uses a fiber-optic cable to connect to shore. Changes associated with the new hydrophone array as well as a new DAQ system that incorporates multiprocessor computing and accurate GPS timestamping are reported. Initial data of lightbulb calibration conducted in March 2005, and a future plan for a more accurate calibration are also presented.

1. The SAUND II Hydrophone Array

The SAUND II detector is an existing US Navy hydrophone array at the Atlantic Undersea Test and Evaluation Center (AUTEC) [1]. Since the first phase of the SAUND experiment [2], AUTEC has replaced its several-decade-old hydrophone system with a new system in which signals are digitized in water and transmitted to shore on a fiber-optic digital link. This is thought to have better immunity to transient noise which was observed in the previous copper-wire system. Signals are then converted back to analog for compatibility with the original Navy processing system. The new system has a flat response over a wider frequency range (~2 - 40kHz). The previous system had a high pass filter with a cutoff at 7.5kHz and a response that varied about 8dB across frequencies above cutoff. The sensitivity of the system has also changed from 14Pa/V to 4Pa/V. In SAUND I, the lack of gain uniformity across different channels had caused problems. In the new system, we expect channel-to-channel gain uniformity of better

*Presented by N. Kurahashi for the SAUND II collaboration: N. Kurahashi and G. Gratta, Stanford University, M. Gruell and D. Kapolka, Naval Postgraduate School, C. Galbiati, Princeton University, and J. Vandenbroucke, University of California, Berkeley.

than 1dB across the entire array. Unfortunately, the new array has hydrophones spaced at every 4km compared to the array in SAUND I where phones were spaced at 1.5km distance. SAUND II, however, will read out 50 hydrophones spanning 1000km^2.

2. New DAQ Structure and Timing Accuracy

Analog signals provided from the Navy are then fed to digitizer cards [3]. To achieve simultaneous readout of 50 hydrophones, the DAQ structure was updated to 8 PCs running in parallel; each PC processing 7 hydrophone signals. One common PC is programmed to build coincidences amongst the triggered events online. The DAQ now runs on a linux platform that uses the open source KiNOKO[a] software. The digitizer cards [3] are controlled by another open source library[b] that acts as their device driver, which allows DMA transfer from the cards to the memories of the PCs. Because the device driver is wrapped in the DAQ software, and the SAUND trigger is a digital matched filter that is a part of the DAQ software, each data sample read out passes through various buffers before it reaches the trigger. Since the amount of data in the buffers change, instead of timestamping at the time of trigger, an IRIG-B signal [4] is fed through the 8th channel of the digitizer card to resynchronize readout time. This achieves better than 20μs error in timestamping of data, an error that is sufficient for acoustic frequencies.

3. Initial Installation

In March 2005, a test system of 3 PCs reading out 14 hydrophones was installed. During this installation, we conducted a calibration run such as the one done in SAUND I in which we dropped household light bulbs with fishing weights attached. At each drop point, GPS readouts of the positions were recorded and compared to the acoustically reconstructed locations of the light bulbs imploding at their failure pressures. Fig. 1 shows the comparison. Besides errors in the reconstruction, environmental facts such as currents drifting the bulbs cause the inconsistency between the GPS readings and reconstructed locations.

---

[a] http://www.awa.tohoku.ac.jp/~naoko/kinoko-e
[b] http://www.comedi.org

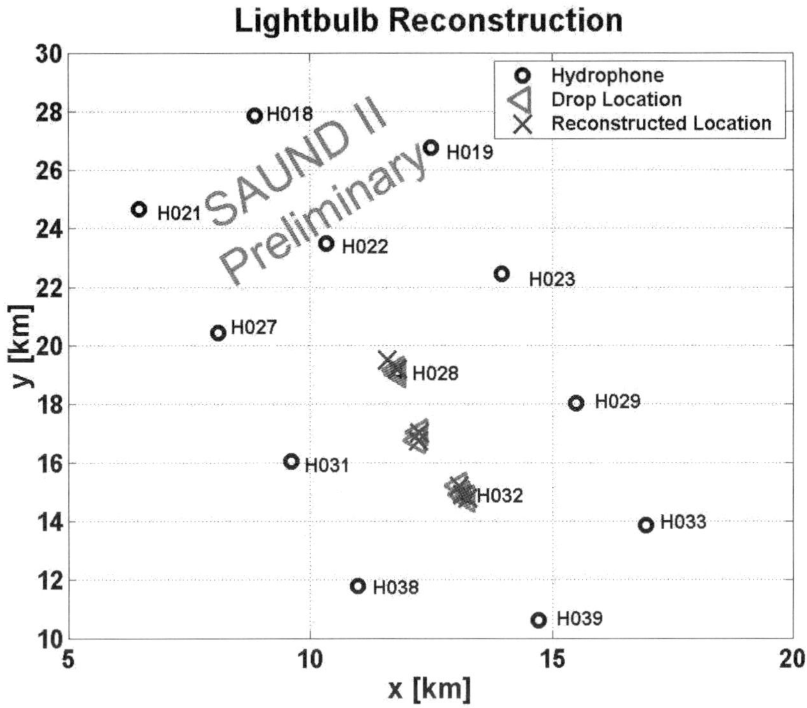

Figure 1. A preliminary comparison of GPS recorded light bulb drop location and acoustically reconstructed implosion location.

4. Future Plans

The need for better calibration of the entire system from hydrophones to the SAUND II computers is apparent. In particular, the phase response of the system is presently unknown and must be investigated for optimal digital matched filtering. This problem resulted in the largest systematic uncertainty in the analysis of the SAUND I data [2]. A plan to deploy a calibrator that produces a fast pulse of current undersea mimicking the UHE neutrino induced shower is in place. This "zapper" will consist of two electrode plates, a high voltage, fast discharge capacitor, a triggering and charging circuit, an inertial platform and data logger that will log the depth, direction and angle of tilt at the time of discharge. A high pressure float and time release ballast will allow the zapper to sink to the ~1.6km deep ocean floor and resurface autonomously. This self sufficient zapper is planned to have a discharge length of 10 to 20 meters to simulate the shower extent, and will sink carving a helicoidal path to trigger several different hydrophones during the descent. A prototype of this discharging device that is smaller in size which will operate at shallow depths tethered from a boat is scheduled to deploy before the end of calender year 2005. This prototype should provide enough information to calibrate the phase response of the system. At that time, we plan to have all 8 PCs installed, reading out 50 hydrophones.

5. Summary

The SAUND II experiment now has a system with 14 hydrophone readouts running at AUTEC. The full 50 hydrophone array is scheduled to be read out by the beginning of 2006. Our initial calibration shows consistency with SAUND I, possibly with reduced noise. A more sophisticated calibration device is being developed.

Acknowledgments

We would like to thank the Naval Undersea Warfare Center (NUWC) of the US Navy, and in particular, T. Kelly-Bissonnette, D. Deveau and others of detachment AUTEC for their continuing hospitality and help.

References

1. N.G. Lehtinen et al, *Astropart. Phys.* **17**, 279 (2002).
2. J. Vandenbroucke et al, *Astrophys. J.* **621**, 301 (2005).
3. National Instruments Model PCI-6070E
4. American Inter Range Instrumentation Group (IRIG)'s standardized time code format. http://www.jcte.jcs.mil/RCC/manuals/200-04/index.html

SPATS - A SOUTH POLE ACOUSTIC TEST SETUP

SEBASTIAN BÖSER
DESY, Zeuthen, sboeser@ifh.de

FOR THE SPATS GROUP
S.BÖSER[‡] C. BOHM[*] S. HUNDERTMARK[*] A. HALLGREN[†]
R. NAHNHAUER[‡] B. PRICE[§] J. VANDENBROUKE[§]

Due to its large Greeneisen coefficient ice is of special interest for the acoustic detection of ultra-high energetic neutrino-induced cascades. The abundant homogeneous volume and an existing neutrino observatory make the south polar ice cap a favourable location for this purpose. Theoretical calculations yield absorption lengths of $\sim 10\,\mathrm{km}$, but no measurements at all are available in the frequency range of interest. We present an experimental setup to measure the key parameters of the antarctic glacial ice.

1. Motivation

Despite its remote location, the south polar ice cap has proven to be a well-suited place for the neutrino telescopes IceCube[1] and AMANDA[2] detecting ultra-high energetic neutrinos by Cherenkov light emitted from neutrino interactions in the ice. Extensive investigations[3] have shown that the ice is very clear and homogeneous for light transmission with absorption lengths of $\sim 100\,\mathrm{m}$. Similar promising results were obtained by the RICE experiment[4] investigating the possibility of detecting neutrino induced cascades by their radio emission. Absorption lengths for radio waves were measured to be $\sim 1\,\mathrm{km}$, making this technique designated for a large-volume detector with correspondingly larger spacing and good sensitivity on the $\gtrsim$ PeV scale.
The detection of acoustic waves generated in the same neutrino interactions has only recently gained in interest again. Many km^3 of target material

[*] Universtiy Stockholm, Sweden
[†] University Uppsala, Sweden
[‡] DESY, Zeuthen,Germany
[§] University of Berkeley, California

will be needed to detect the feeble fluxes expected in the even higher energy range accesible with this technique. With larger signals than in water and expected absorption lengths of $\sim 10\,\text{km}$ [7], the south polar ice seems to be a promising medium also for an acoustic detector, allowing larger instrumented volumes and thereby suitable sensitivities on the EeV scale.
The possibility of combining the three detection methods makes the location even more favourable. A simulation of a hybrid optical – radio – acoustic array[5] yields event rates of $\gtrsim 10/\,\text{yr}$ with about 40 % percent of the events detected by at least two techniques in coincidence.
However, in contrast to optical and radio detection, the assumptions for acoustic properties of the ice are purely based on theoretical calculations[7]. For a proper evalutation of the potential of such a detector, therefore a dedicated setup will be needed to measure the four key parameters:

- **Scattering length**
  Scattering of phonons in the ice is assumed to be dominated by Raleigh scattering at the grain boundaries, and therefore dependant on crystal size a and frequency f

  $$\lambda_s \propto a^{-3} \times f^{-4} \tag{1}$$

  as shown in Fig. 1. Using an estimated grain size of 0.2 cm, values of $\lambda_s(10\,\text{kHz}) \approx 2000\,\text{km}$ and $\lambda_s(100\,\text{kHz}) \approx 0.2\,\text{km}$ are obtained[7]. Since the spectral peak frequency for a neutrino-induced acoustic wave is $\sim 50\,\text{kHz}$ in ice where $\lambda_s(50\,\text{kHz}) \approx 3.2\,\text{km}$ scattering can probably be neglected for sensor spacings closer than $\sim 1\,\text{km}$.
- **Absorption length**
  For phonon absorption, the energy loss in the relaxtion of molecular reorientations is assumed to be dominant, which is therefore also a temperature dependant effect. Using laboratory measurements on the relaxation process, a value of $\lambda_a(-51\,^\circ\text{C}) = 8.6\,\text{km}$ is predicted[7] for the coldest temperatures expected. Combined with a temperature profile, the depth dependant absorption can be determined as shown in Fig. 1[7].
- **Velocity of sound**
  Velocity of sound depends only weakly on the temperature via the elastic modulus $\bar{E}(T)$, but strongly on the density of the ice ρ

  $$\vec{v}_s = \sqrt{\frac{\bar{E}(T)\vec{e}}{\rho}} \tag{2}$$

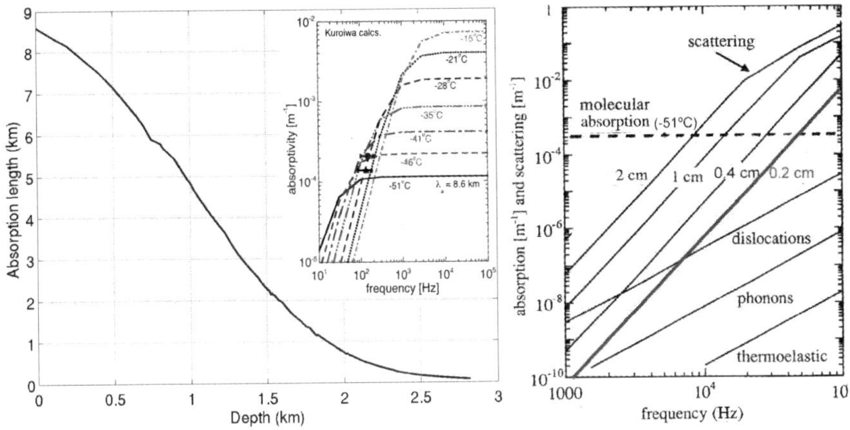

Figure 1. Absortion length vs. depth(left) and frequency(left, inset) and scattering length vs. frequency(right) in the south polar ice cap, taken from [7]

This results in a very distinct profile, with strong variation in the upper $\sim 200\,\text{m}$, where density increases strongly, and small variation below, where only temperature effects are important. Therefore, acoustic waves in the upper part will always be strongly bent towards the surface, whereas propagation will be nearly linear in the lower part.

- **Background noise**
  With only a few signal events per year, ambient noise will be of special importance for an acoustic neutrino detector. Although the south pole is known to be among the most quiet places on earth in the seismic frequency range, neither measurements nor theoretical estimates are available for the ultrasonic regime. Some possible sources may include

  – **anthropogenic** noise that should not only be damped by the firn layer, but as well be mostly refracted back to the surfaces
  – noise from **micro cracks** in the ice, similar to what is observed in the vincinty of salt mines[6]
  – noise generated in the **slip–stick motion** of the glacier over the continental bedrock

  However, it is strongly assumed that the average noise level will be well below what is observed in oceans, where not only wind and waves contribute strongly, but also natural (e.g dolphins and sperm wales) and anthropogenic (e.g ships and oil drilling platforms) sources generate many transient events.

2. SPATS

In order to access all these parameters, an experimental setup — *SPATS, the South Pole Acoustic Test Setup* — was designed. Measuring the desired quantities implies signal transmission over distances of a few hundered meters for absorption to become relevant. To confirm the assumptions on temperature and density dependance, instrumentation of the upper few hundered meters is sufficient, as variation of both is small below.

2.1. *General setup*

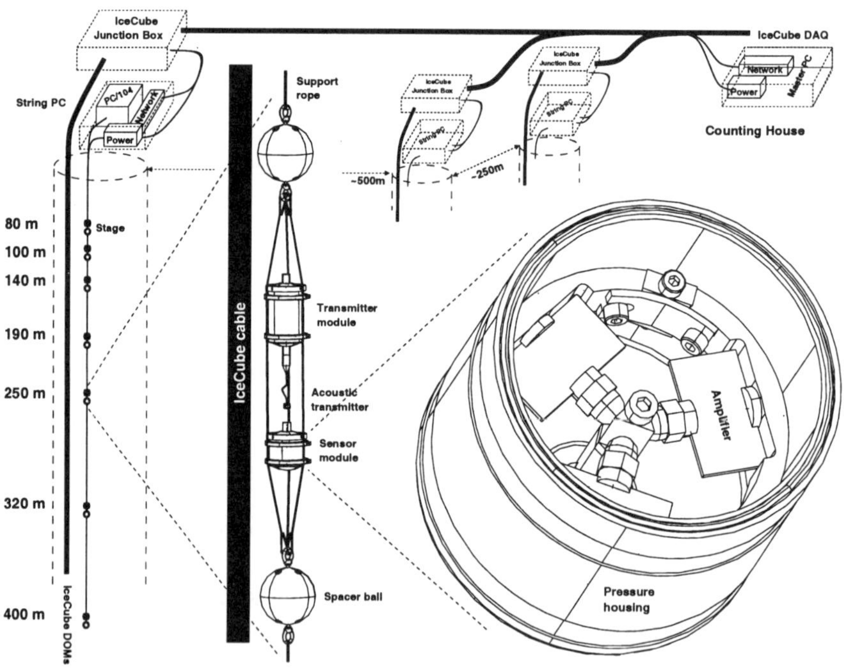

Figure 2. Schematic of the SPAT Setup(left) and acoustic sensor(right)

Therefore a transmitter–receiver array with three strings in unequal spacing of 125 m to $\sim$ 500 m is proposed, allowing redundant absorption measurements. For solving the depth-dependance, seven levels from 80 m to 400 m will be equipped with acoustic stages, each of them holding as well a sensor and a transmitter as shown in Fig. 2. With hole drilling as the major cost factor, it is suggested to use the upper part of the holes of the IceCube project, which itself instruments only the depth range of 1500 m – 2400 m. Each acoustic string is read out by a *String-PC* at the top of the hole, which then passes the data to a *Master-PC* in the counting house. The data is stored on a local disk, with a small part of it being transmitted via satellite to the northern hemisphere for immediate analysis. This same link will also allow to log on to the system from the north for control and software updates.

2.2. *Acoustic Stages*

Each acoustic stage consists of a sensor module (see Fig. 2) and a transmitter module, both of them in a custom pressure housing to withstand the static pressure and additional pressure generate in the refreezing process. Each sensor module hosts three channels arranged in a star-like pattern. While no control over the azimuthal orientation is possible in the deployment, this improves the omnidirectional sensititvity of the device, but will also allow to look for coincidences within the module and a first estimate of the direction of the incoming pressure wave.
In the transmitter housing, a HV pulse generator drives $\sim 10\,\mu s$ pulses with an externally adjustable peak voltage of up to 1 kV. A ring-shaped piezo-electric ceramic outside the pressure housing, molded into epoxy resin for stability, converts the electrical signal to an acoustic pulse. Peak currents of 8 A are reached due to the large capacitance of the ceramics. Bleeder resistors allow a read-back of the electrical signal scaled by a factor 1:100. In addition, temperature sensors are installed at each depth level but the lowest one, which contains pressure sensors for depth measurement and monitoring of the freeze-in process.

2.3. *DAQ, Communication and Timing*

The heart of the *String-PC* is an industrial standard PC/104 embedded CPU module running at 600 MHz clock frequency. Together with three 12-bit A/D sampling modules of 1.25 MHz and power supplies it is mounted in a waterproofed container, which will be buried in the snow for insulation to the strong climatic changes. The *String-PC* communicates to the *Master-PC* via a 2.2 MBit DSL connection over the $\sim$ 1 km surface cables shared with the IceCube project. DC-Power to the system is supplied via the same cable. In addition, an IRIG-B time coding sequence is used to keep the *String-PCs* synchronized on a sub-millisecond level.

3. Status

All sensor and transmitter modules for the SPATS setup were produced at DESY Zeuthen. Each of them has been individually calibrated to a reference hydrophone using a spectral comparison method [8]. The variation in sensitivity of $\sim 20\,\mathrm{dB}$ is probably a result from the mechanical setup of the sensor. Calibration of peak amplitudes and azimuthal response was also performed for all the transmitter modules. Extensive system testing has been started, with all parts that are subject to cold temperatures undergoing a freezing test at $-55°C$.
The system is planned to be installed in polar season 05/06. After a refreezing time of several weeks, first results are expected in march 2006. Once the key parameters are resolved and the suitability of glacial ice is confirmed, the ground is laid for the design and developement of a several ten km^3 array.

References

1. J. Ahrens et al., Astropart. Phys.**20**, 507-532, (2004)
2. J. Ahrens et al, Phys. Rev. D **66**, 012005 (2002) (astro-ph/0205109)
3. M. Ackermann et al, to appear in J. of Geophys. Research
4. I. Kravchenko et. al., in this proceedings
5. D. Besson et. al., Proc. 29th ICRC, Pune (2005), (astro-ph/0509330)
6. G. Manthei et. al, in this proceedings
7. B. Price, to be published in J. Geophys. Res., (astro-ph/0506648)
8. S. Böser et. al., Proc. 29th ICRC, Pune (2005) , ger-nahnhauer-R-abs1-og25-oral

INTEGRATION OF ACOUSTIC DETECTION EQUIPMENT INTO ANTARES *

R. LAHMANN, G. ANTON, K. GRAF, J. HÖSSL, A. KAPPES, T. KARG, U. KATZ, C. NAUMANN AND K. SALOMON

Physikalisches Institut,
Friedrich-Alexander-Universität Erlangen-Nürnberg,
Erwin-Rommel-Straße 1,
91058 Erlangen, Germany
E-mail: Robert.Lahmann@physik.uni-erlangen.de

The ANTARES group at the University of Erlangen is working towards the integration of a set of acoustic sensors into the ANTARES Neutrino Telescope[1]. With this setup, tests of acoustic particle detection methods and background studies shall be performed. The ANTARES Neutrino Telescope, which is currently being constructed in the Mediterranean Sea, will be equipped with the infrastructure to accommodate a 3-dimensional array of photomultipliers for the detection of Cherenkov light. Within this infrastructure, the required resources for acoustic sensors are available: Bandwidth for the transmission of the acoustic data to the shore, electrical power for the off-shore electronics and physical space to install the acoustic sensors and to route the connecting cables (transmitting signals and power) into the electronics containers. It will be explained how the integration will be performed with minimal modifications of the existing ANTARES design and which setup is foreseen for the acquisition of the acoustic data.

1. Introduction

A promising alternative to neutrino telescopes detecting Cherenkov light in a transparent medium (ice, fresh water, sea water) with optical attenuation lengths of several tens of meters arises from the fact that particle showers with energies exceeding values in the order of 100 PeV produce detectable bipolar pressure waves in water of about 50 μs length with a range of up to several km. This effect is described by the thermo-acoustic model and has been experimentally verified for proton and laser beams[2]. Acoustic neutrino telescopes therefore might allow for future giant-volume detectors

*This work was supported by the German BMBF Grant No. 05 CN2WE1/2.

in the order of $1000\,\mathrm{km}^3$.

A necessary prerequisite to develop acoustic detection methods is the detailed understanding of background conditions and the investigation of signal identification methods. In order to acquire the long-term, high-precision data needed for this purpose, it is intended to instrument a part of the ANTARES detector with acoustic sensors.

2. The ANTARES detector and its data acquisition system

The ANTARES detector is currently under construction in the Mediterranean Sea, off-shore of Toulon, and is connected to the coast by an electro-optical cable of about 40 km length. The instrumented area will range from a depth of about 2000 m to 2400 m. The detector is designed to detect the Cherenkov light from muon tracks. For this purpose, it will be equipped with a total of 900 optical modules[3] (OMs), which hold one photomultiplier tube (PMT) each inside a sphere, pointing downwards at an angle of 45°.

When finished, the detector will consist of 12 "detection lines", arranged in an octagonal shape on the seabed at a distance of about 70 m from each other (cf. Fig 1). In addition, one "instrumentation line" will hold equipment to record environmental conditions such as the current profile and the salinity of the sea water.

Each detection line comprises 25 storeys at a distance of 14.5 m from each other. Each storey consists of a mechanical support structure that holds 3 OMs and a titanium container with the required electronics ("local control module", LCM). Five storeys form a sector which constitutes one unit for purposes of data readout. It is foreseen to integrate the acoustic detection equipment into ANTARES in the form of "acoustic sectors" with acoustic sensors replacing the PMTs and using as much of the infrastructure provided by ANTARES with as little changes as possible. This will be described in detail below.

Each LCM contains a backplane that is equipped with connectors for the electronics cards and provides power and data lines to and from the connectors. In each sector, one LCM is designed as Master LCM (MLCM) which in addition to its data taking tasks manages the data transmission to the shore through single-mode optical fibres using TCP/IP. The main electronics cards of a LCM are:

- A clock card, which provides a time stamp to each recorded PMT-value with a resolution of 50 ns and a precision of about 0.5 ns[4];
- So-called ARS-boards which contain 2 Analogue Ring Sampler

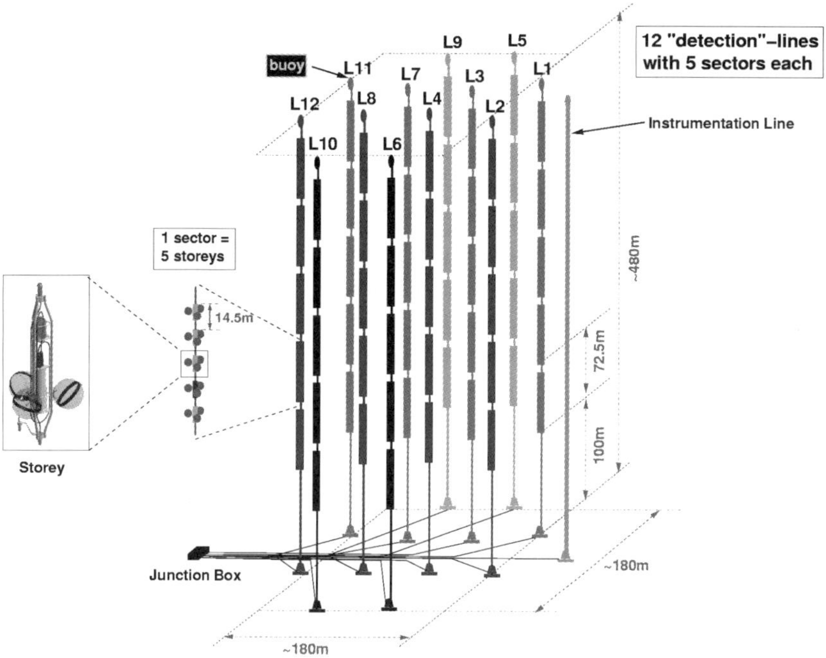

Figure 1. Schematic view of the ANTARES detector. Shown are the 5 sectors per detection line and the 5 storeys per sector. The second storey from the bottom of each sector contains the Master LCM (MLCM) for data transmission to shore (see text). Also shown is one storey with its 3 optical modules (OMs) that contain one PMT each (not shown). The connection to the shore from the Junction Box is established via a ~40 km cable comprising optical fibres and electric power lines (not shown). Equipment on the Instrumentation Line is not shown.

(ARS) ASICs[5] each, conditioning and digitising the analogue data from the PMTs. Furthermore, they subdivide the 50 ns clock signal into 256 steps. In each LCM, 3 ARS boards are installed to read out the 3 PMTs per storey.

- A data acquisition or DAQ card, which reads out the ARS boards and provides the communication to the MLCM of the sector via TCP/IP.

In the standard sampling mode, an ARS-chip is triggered by a PMT signal above a predefined threshold and then employs a pulse shape discriminator to recognise single photon events—for which the integrated analogue charge is then digitised and read out. The data rate, which is dominated by background from ^{40}K and bioluminescence, therefore can be adjusted by varying the threshold for single photon events.

The bandwidth of the complete data acquisition chain is limited by the throughput of the DAQ-boards which—as will be explained below—limits the number of acoustic sensors that can be installed per storey.

3. Integration of acoustic sensors and their readout

The acoustic sensors that are currently developed and tested by the ANTARES group at Erlangen are described in Ref. 6. The two design concepts (individual hydrophones and "acoustic modules", in which acoustic sensor elements are installed in the spheres of the optical modules instead of the PMTs) do not differ in their requirements for the electronics inside the LCM.

The fundamental guideline for the design of the acoustic detection system has been that the implementation shall be done with as few modifications to the existing ANTARES design as possible. These considerations have lead to the following layout principles:

- In order not to compromise the suitability for a deep-sea environment, no modifications must be done to the titanium container or the penetrators and connectors leading into and out of the container. Consequentially, only the 3 holes that are present in each container for the cables leading to the 3 OMs can be used to connect to the acoustic sensors.
- In the LCMs of the acoustic storeys, the ARS boards will be replaced by "Acoustic ADC boards"; no other changes to the electronics will be done. These boards will digitise the acoustic data and format them, where the format will be exactly the same as that of the optical data from the ARS boards. The acoustic data will then be read out sequentially by the DAQ board and transmitted to shore in exactly the same fashion as the PMT data.
- On-shore, the separation of acoustic and optical data will be based on their origin, i.e. on the IP address of the DAQ-board in the corresponding LCM. Acoustic data will be separated from the main data stream and processed, filtered and compressed on a dedicated PC-farm.

The number of hydrophones per storey is limited by the data throughput of the DAQ-board processor of roughly 20-25 Mb/s. Consequentially, 6 acoustic sensors per storey can be installed for a 16-bit digitisation and a

200 kHz sampling rate. The sampling rate can be further increased if an adequate down-sampling is performed on the acoustic ADC board.

Each of the 3 acoustic ADC boards per LCM will contain two 16-bit ADCs with a maximum sampling frequency of 500 kHz for the processing of two acoustic sensors. The data sampling will not be triggered but instead be continuous at an adjustable data rate of 100 kHz, 200 kHz or 400 kHz. It will be possible to individually disable each hydrophone.

In order to minimise the development time and error-proneness while maximising the flexibility of the system, no ASICs will be developed. Instead, a FPGA will be employed to process the data from the two ADCs per board and a micro controller to control the FPGA and to allow for the uploading of upgrades of the FPGA code.

The acoustic data will be provided with a time stamp derived from the standard ANTARES clock in order to allow the correlation of the data from several storeys. The electric power consumption of the design will be below 8.5 W per storey, which is the power available for acoustics per electronics container.

4. Outlook and summary

It is foreseen to install two acoustic sectors with up to 60 acoustic sensors in total into the ANTARES detector. For this goal, a conclusive concept has been devised by the Erlangen group. It is currently under discussion inside the ANTARES collaboration where to place the two acoustic sectors. In the meantime, the development of "acoustic ADC-boards" in Erlangen is progressing.

References

1. `http://antares.in2p3.fr/`
2. K. Graf et al., *Testing thermo-acoustic sound generation in water with proton and laser beams* in these proceedings and references therein.
3. P. Amram et al., *The ANTARES optical module*, Nucl. Instrum. Meth. A **484** (2002) 369 [arXiv:astro-ph/0112172].
4. F. Blanc et al., *Time calibration of the ANTARES neutrino telescope*, proceedings of 28th International Cosmic Ray Conferences (ICRC 2003), Tsukuba, Japan, 31 Jul - 7 Aug 2003.
5. F. Druillole et al., *The analogue ring sampler: An ASIC for the front-end electronics of the ANTARES neutrino telescope*, IEEE Trans. Nucl. Sci. **49** (2002) 1122.
6. C. Naumann et al., *Development of acoustic sensors for the ANTARES experiment* in these proceedings.

OVERVIEW OF THE LORD EXPERIMENT (LUNAR ORBITAL RADIO DETECTOR)

V. A. CHECHIN, E. L. FEINBERG, G. A. GUSEV, B. N. LOMONOSOV, N. G. POLUKHINA, V. A. RYABOV, V. A. TSAREV

Lebedev Physical Institute, Russian Academy of Sciences, Leninsky pr. 53, Moscow, 117924 Russia.

K. M. PICHKADZE, V. K. SYSOEV

Lavochkin Association, Leningradskoe Shosse 24, Moscow Region, Chimki, 141400 Russia.

T. SAITO

Institute for Advanced Studies, 1-29-6 Shinjyuku, Shinjyuku-ku, Tokyo 162-0022 Japan

At first, a little bit of history and brief introduction into the Lebedev Physical Institute (LPI) group activity in the radio-detection area. It is well known that the cascade-detection method, based on the use of coherent Cherenkov radio emission, was first proposed in 1961 by G. A. Askaryan [1], who worked at that time at the LPI. It is fair to say that the LPI was the place, which was perfectly adequate for inventing the radio method. Indeed, in our Institute Cherenkov radiation (called in Russia "Cherenkov—Vavilov radiation") was discovered by P. A. Cherenkov and his teacher S. I. Vavilov and explained by I. M. Frank and I. E. Tamm. Furthermore, such divisions of science as radio physics, high-energy and cosmic ray physics, and physics of cascade processes were highly evolved at the Institute at that time[*)].

As far as our LPI group is concerned, it is funny to say that the first experience of our group with radio method about 20 years ago was rather discouraging. At that time we found [2—4] that some Askaryan's predictions for radio signal from long-baseline neutrino beam in the Earth's crust [5] were too much optimistic and the really expected radio signal should be many orders of magnitude lower than that estimated by him.

Much more stable and constructive interest of our group to the radio method arose about five years ago, when we were searching for an adequate method of detecting ultrahigh-energy cosmic particles. An essential role was played also by our long-standing collaboration with Lavochkin Association (LA), the leading Russian space company, which is specialized in designing and launching automatic space instruments, and where last years the so-called "solar sail" and other deployable space constructions are under research and development. This work stimulated advancement in technology for production and deployment in space of large-area thin-film constructions, which are sometimes called "film astrophysical structures" (FAS), or "inflatable deployable space structures". Just bearing in mind such structures, a program of astrophysical studies based mainly on balloon-borne and satellite-borne FAS had been initially proposed in 2000 by the LPI-LA collaboration. The program includes the following lines of investigation: (a) Detection of Ultra-High Energy Cosmic Rays and Neutrinos (UHECR and UHEN) by radio method (here, metallized FAS could be used for constructing large-scale radio antennas); (b) Search for massive charged particles of dark matter and micrometeorite-flux monitoring using acoustic detectors (here, FAS could be used as acoustic-signal radiators); and (c) Measurement of variations of cosmic ray nuclei (in this case thin films could be used as plastic solid state detectors).

Detailed analysis of the program has been done [6—17] in the context of the Russian Federal Space Program NIR "Budushee" ("Future"), and the major results of this work with emphasis on the LORD (Lunar Orbital Radio Detector) experiment will be presented at this workshop. In my talk I'll restrict myself to principal aspects of the proposed experiment. The aim will be to estimate "the scientific potential" of the experiment, which could be obtained "in principle", without worrying about technical realization of the instrument and the background problem. More conservative estimates for some possible specific instrument configurations will be presented in the second talk by Dr. Chechin [18].

In the course of the program realization several experimental and theoretical studies have been carried out, which include: (a) Theoretical calculations and numerical modeling for various experimental configurations (ground-based, balloon-borne, and satellite-borne); (b) A series of ground-based measurements of radio pulses from EAS using one of the biggest in the world radio-astronomical instruments in the meter wavelength range, the so-called DKR-1000 radio telescope of the LPI; (c) Preliminary work for a balloon-borne experiment in a radio-quiet region, in particular near the North or South Poles. (A pilot instrument for a high-altitude balloon flight was designed (the CREED project), and some components of electronics and antenna system were constructed and tested. During negotiations between NASA and ROSAVIAKOSMOS groups in Moscow in 2003, a preliminary agreement was reached on accommodation of this instrument in an American high-altitude balloon together with the TIGER detector for flights around the North Pole. Unfortunately, this joint experiment has not been realized for some non-scientific reasons); (d) Analysis of the possibility to detect UHECR and UHEN using satellite-borne apparatus. (In particular, the estimates have been done concerning the use of a circumterrestrial satellite, which could "see" the Antarctic ice sheet and detect neutrino-induced radio pulses. Some results will be briefly shown below). (e) Finally, in recent years, our activity is concentrated mainly near the Moon. One of the motivations for that is associated with Russian Lunar Program. The LORD experiment, proposed by our collaboration, is included in the first phase of the preliminary version of this program.

It is known that the idea to use the Moon as a target for cosmic-particle detection by the radio method employing receivers on the lunar surface was originally proposed by G. A. Askaryan [1]. Next it was reanalyzed by R. D. Dagkesamansky and I. M. Zheleznykh [19], who proposed to use terrestrial ground-based radio telescopes. This approach is currently used in the KALYAZIN and GLUE experiments. A possibility to use lunar satellite-borne radio receivers was also mentioned in a few papers [8, 19, 20]. The feasibility study of the particle detection with lunar satellite-borne radio receivers carried out by the LPI group has shown high scientific potential of the proposed experiment [17]. The main merits of such an experiment are evident. First, it is huge target mass, which can be surveyed using satellite-borne antennas. Second, it is short (and variable, in principle) distance L from a few hundreds to a few thousands of km (which is much shorter than that for detection from the Earth's surface, where L is about 400000 km). Third, it is very favorable background conditions. The specific features of the Moon as compared with Earth include the absence of atmosphere and magnetic field.

I'll not dwell on the details of our calculations. They are rather standard for such kind of estimates and will be touched upon in the next talk by Dr. Chechin [18]. Some of the results are illustrated in Figs. 1 to 3.

The total aperture for CR and neutrinos may be as high as 10^5-10^6 km^2 sr. The number of events detected per year for $E > 10^{18}$ eV can reach several hundreds for GZK neutrinos and several thousands for CR. For $E > 10^{20}$ eV, the number of CR events is about 1000. It is seen that the proposed experiment can provide rather powerful constrains on the UHEN fluxes, which will allow us to check many existing models of neutrino sources.

It is apparent that the possibility to realize the potential of this experiment depends on the possibility to ensure the required parameters of the receiving system. This problem will be addressed in [18].

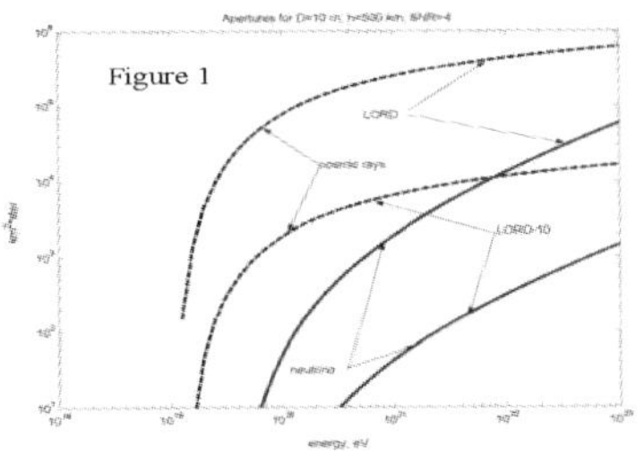

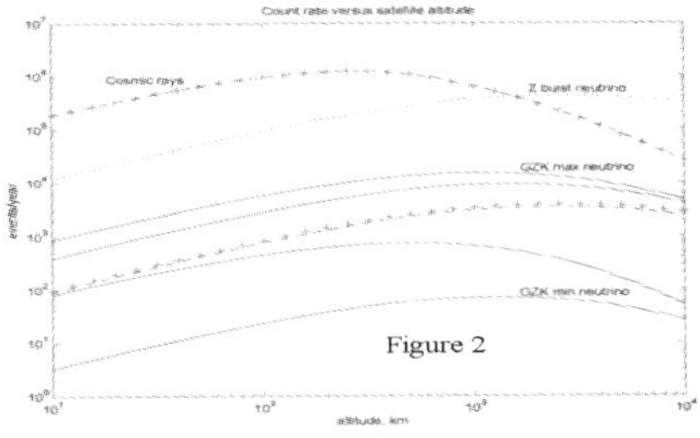

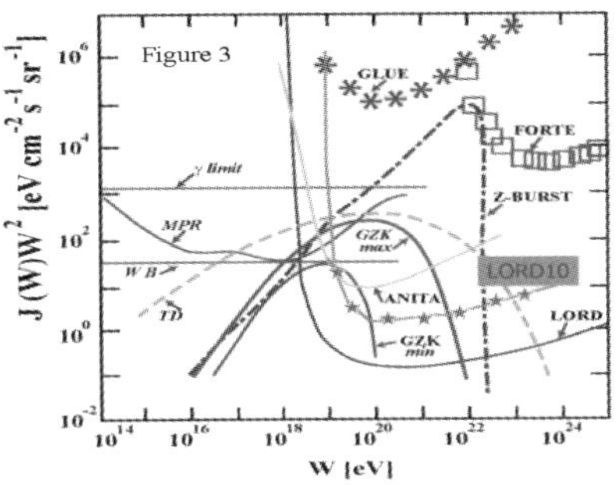

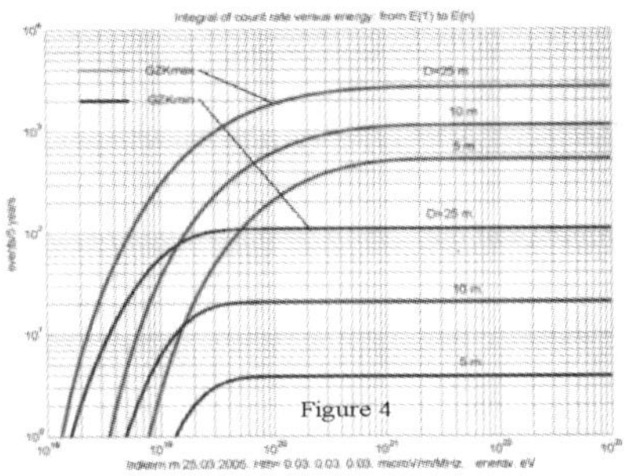

Fig. 1. Total aperture (km^2sr) for cascades from neutrinos $A_\nu(W)$ (solid lines) and from cosmic rays $A_{CR}(W)$ (dashed lines) the threshold field intensities $E_{th} = 0.03$ *μV/m/MHz*. For comparison the aperture for the LORD-10 version of the experiment is also shown (see [18]).

Fig. 2. The number of detected (per year) radio signals from cascades initiated by cosmic rays and neutrinos for various models at for $W \geq 10^{18}$ *eV* (upper curves) and $W \geq 10^{20}$ *eV* (lower curves) as a function of the satellite orbit altitude h for $E_{th} = 0.01$ *μV/m/MHz* and $f = 0.5$ *GHz*.

Fig. 3. Constrains on neutrino flux, which could be attained in the LORD and LORD-10 [18] experiments, calculated for the values $E_{th} = 0.01$ *μV/m/MHz*, $T = 1$ year, $f = 0.5$ *GHz*.

Fig. 4. The number of detected (per 5 years) radio signals from cascades initiated by cosmic neutrinos for various antenna diameters d = 5, 10, and 25 m; for $E_{th} = 0.03$ *μV/m/MHz*; $f = 0.45$ *GHz*, $\Delta f = 50$ *MHz*; polar orbit with h = 500 km, and S/N = 3.

Now, let us briefly consider an experiment with circumterrestrial satellite. In general, there are two types of trajectories, from which the satellite can "see" Antarctic either by flying it over (polar orbit), or by observing it from an equatorial orbit, which is offset by about $2 \cdot 10^4$ km from Earth. The potential of this "ROMANTICS" experiment ("Radio-wave Orbital Monitoring of Antarctic-NeuTrino Interactions from Circumterrestrial Satellite") for the case of a polar orbit with h = 500 km is shown in Fig. 4.

In conclusion we can say that for both proposed satellite experiments (LORD and ROMANTICS) the sensitivity for CR and neutrino fluxes is expected to be very high, and the experiments would be able to study the cosmic ray and neutrino fluxes in the presently controversial region and extend measurements up to higher energies.

----

*) *Personal remark of the speaker (V. A. Tsarev).* Seizing this opportunity, I would like to mention two more scientists from the LPI, who played an important role in the advent of the radio method, and whose contributions in this field are little known to the contemporary radio-detection community. The first one is Prof. E. L. Feinberg, who has gained to that time wide recognition as a specialist in radio physics, high-energy and cosmic ray physics. It has been known that G. A. Askaryan had a lot of discussions with E. L. Feinberg on radio emission from cascades, and the well-known Askaryan's papers on the radio method resulted to a great extend from these numerous and fruitful discussions. Currently E. L. Feinberg is actively working in this field, and he is in fact the scientific leader of the LORD collaboration. The other scientist from the LPI is Prof. V. I. Goldansky, who first proposed two other mechanisms of radio emission from EAS associated not with excess electrons (as in the so-called "Askaryan's mechanism") but with separation of charges by geomagnetic field and with transition radiation. These Goldansky's ideas are cited in the Askaryan's paper. Currently, it is known that this is the geomagnetic mechanism, which is mostly responsible for radio emission from air showers.

References

1. G.A. Askaryan, *JETP* **41**, 616 (1961).
2. V.A. Tsarev, V.A. Chechin, *LPI preprint,* N 248 (1984).
3. V.A. Tsarev, V.A. Chechin, *LPI preprint,* N 87 (1985).
4. P.S. Isaev, V.A. Tsarev, *Physics of Particles and Nuclei,* **20**, 997 (1989).
5. G.A. Askaryan, *JETP Lett.* **39**, 334 (1984).
6. V.A. Tsarev, Physics of Particles and Nuclei, **35**, 1 (2004).
7. V.A. Tsarev, V.A. Chechin, *Doklady RAS* **383,** 486 (2002).
8. V.A. Tsarev, Proc. of International Conference "P. A. Cherenkov and modern physics. Moscow. 22—24 June, 2004. In press": Journal of radiation physics and chemistry.
9. V.A. Tsarev, V.A. Chechin, *Kratkie Soobsh. po Fizike*, LPI **4**, 42 (2001).
10. V.A. Tsarev, V.A. Chechin, *Doklady RAS* **389**, 45 (2002).
11. K. A.Kotelnikov, N. G.Polukhina, E. L. Feinberg, *et al.*, **66**, 1638 (2002).
12. V.A. Tsarev, V.A. Chechin, *Kratkie Soobsh. po Fizike*, LPI **11,** 26 (2001).
13. V.A. Tsarev, *Kratkie Soobsh. po Fizike* , LPI **11**, 13 (2002).
14. V.A. Tsarev, V.A. Chechin, *Kratkie Soobsh. po Fizike*, LPI **11** 26(2002).
15. V.A. Chechin, , R. D.Dagkesamansky, E. L.Feinberg, et al., Proc. of the *Russian Conference "Radio Telescopes RT-2002"*, Pushchino , p. 69.
16. V.A. Chechin, E. L. Feinberg, S. M Kutuzov. *et al.*, *Proc. of the Russian Conference "Radio Telescopes RT-2002"*, Pushchino, p. 71.
17. V.A. Chechin, E. L. Feinberg, G. A.Gusev, et al., to be published in "Kosmicheskie Issledovaniya" (Space Research), (2005).
18. V.A. Chechin, *Concept of the LORD instrumen*t. Talk at this workshop.
19. R. D. Dagkesamansky, I. M. Zheleznykh, *JETP Lett.* **50**, 233 (1989)
20. A. D. Filonenko, *Uspekhi.Fiz. Nauik* **172**, 439 (2002).

CONCEPT OF THE LORD EXPERIMENT

V. A. CHECHIN, E. L. FEINBERG, G. A. GUSEV, B. N. LOMONOSOV, N. G. POLUKHINA, V. A. RYABOV, V. A. TSAREV

Lebedev Physical Institute, Russian Academy of Sciences, Leninsky pr. 53, Moscow, 117924 Russia.

K. M. PICHKADZE, V. K. SYSOEV

Lavochkin Association, Leningradskoe Shosse 24, Moscow Region, Chimki, 141400 Russia.

T. SAITO

Institute for Advanced Studies, 1-29-6 Shinjyuku, Shinjyuku-ku, Tokyo 162-0022 Japan

ANTENNA AND RECORDING SYSTEM

The LORD10 experiment is one of possible configurations of the LORD and can be considered as its first stage. In the LORD10 a parabolic antenna of diameter D≈10÷20 m is proposed to be born by an artificial lunar satellite with its orbit altitude $h \approx 10^3 \div 10^4$ km [1]. A film astrophysical structures, for example, a metallized film (or net) supported by an inflatable skeleton, could be used as such an antenna. The antenna's beams should be controlled in elevation, with the fan-shaped directional pattern being pointed at the lunar limb. The antenna can detect radio pulses from cascades produced by ultrahigh-energy cosmic rays and neutrinos in the lunar regolith. In spite of the ambient noise at circumlunar orbit being fairly low, the detection of rare weak nanosecond pulses is a highly technical task because of, in particular, hard constraints on weight and supply power of the equipment proposed.

The trigger and data recording system of the LORD10 is assumed to be fairly conventional for radio telescopes [2] (Fig.1). To broaden the antenna beamwidth in azimuth, up to three offset feeds can be used. Signals from each antenna beam are filtered to 150-200 MHz bandwidth near 400 MHz. The band is subdivided into three (or four) 50 MHz bandwidth channels with no overlap. The beam trigger condition requires that output voltages of all the channels be higher than their adjustable thresholds within a time window $\tau \approx 20$ ns. A global trigger is formed as a logical "OR" for all the beams. In this case, a 100 ns record is stored and written to a board computer.

EVENT RATE AND APERTURE

The event rate can be written out as the following integral over the shower energy W [3] :

$$\frac{\mathrm{d}N(E_{\mathrm{th}})}{\mathrm{d}t} = \int \frac{\mathrm{d}J}{\mathrm{d}W\,\mathrm{d}S\,\mathrm{d}\Omega\,\mathrm{d}t} A(W, E_{\mathrm{th}})\mathrm{d}W$$

Here, $\mathrm{d}J/(\mathrm{d}W\mathrm{d}S\mathrm{d}\Omega\mathrm{d}t)$ is the primary particle flux and $A(W, E_{\mathrm{th}})$ is the total aperture (km^2 ster) determining the rate of radio signals exceeding the threshold field strength E_{th} defined by the equation $E_{\mathrm{th}} \approx SNR\sqrt{kT_{\mathrm{sys}}Z_0/(A_{\mathrm{eff}}\Delta f)}$.

Here, A_{eff} is the affective area of the antenna, $T_{\mathrm{sys}}\approx 300$ K is the noise temperature of the system, $Z_0 = 377$ Ohm, $\Delta f \approx 50$ MHz. is the bandwidth, and SNR$\approx$3.5÷4.5 is the signal-to-noise ratio. The total aperture is defined by the integral of the angular (specific) aperture $\Delta\Omega(\theta_{\mathrm{s}},W,E_{\mathrm{th}})$ over the lunar surface visible with the antenna's beams:

$$A(W, E_{\mathrm{th}}) = R_{\mathrm{M}}^2 \int_{\delta\Omega_0} \Delta\Omega(\theta_{\mathrm{s}}, W, E_{\mathrm{th}}) \left|\frac{D(\cos\theta_{\mathrm{s}}, \varphi_{\mathrm{s}})}{D(\cos\theta_0, \varphi_0)}\right| \mathrm{d}\cos\theta_0 \mathrm{d}\varphi_0 .$$

Here, R_{M} is the Moon's radius, shower positions are given by the spherical angles $(\theta_{\mathrm{s}}, \varphi_{\mathrm{s}})$ (or (θ_0,φ_0)) with the origin at the Moon's center (or at the satellite), and the obvious Jacobian is introduced. The integration domain $\delta\Omega_0$ is determined by the directional pattern $Q(\theta_0,\varphi_0)$: $\delta\Omega_0\approx\Delta\theta_x\Delta\theta_y\approx(c/f)^2/A_{\mathrm{eff}}$ (Fig. 2).

The angular aperture $\Delta\Omega$ for CR-produced showers is determined by the following integral over the spherical angles $(\theta_{\mathrm{n}}, \varphi_{\mathrm{n}})$ of the shower axis relative to the outer normal to the lunar surface:

$$\Delta\Omega(\theta_{\mathrm{s}}, E_{\mathrm{th}}) = \int_0^{2\pi} \mathrm{d}\varphi_{\mathrm{n}} \int_0^1 \cos\theta_{\mathrm{n}} \mathrm{d}\cos\theta_{\mathrm{n}} \Theta\left[E_f(\theta_{\mathrm{n}}, \varphi_{\mathrm{n}}, \theta_{\mathrm{s}}, W) - E_{\mathrm{th}}\right],$$

where $\Theta(x)$ is the step function. This expression yields the solid angle of the primary particles that hit the surface at the point θ_{s} and produce the radiation field $E_f(\theta_{\mathrm{n}},\varphi_{\mathrm{n}},\theta_{\mathrm{s}},W)>E_{\mathrm{th}}$. The distance R_{s} from the satellite to the refraction point (coinciding with the point θ_{s}), refraction angle θ_{r} (determining the radiation direction outside the Moon), and spherical angles $(\theta_{\mathrm{a}},\varphi_{\mathrm{a}})$ should be expressed in terms of the angles $(\theta_{\mathrm{s}}, \varphi_{\mathrm{s}})$ and satellite orbit altitude h.

In the case of neutrino-produced showers the angular aperture is defined by the equation similar to (4):

$$\Delta\Omega_\nu = \int_0^{2\pi} \mathrm{d}\varphi_{\mathrm{n}} \int_{-1}^{1} \left\{ \int_0^\infty \exp\left(-\frac{l(z,\theta_{\mathrm{n}})}{L_{\nu N}}\right) \frac{\mathrm{d}z}{L_{\nu N}} \right\} \mathrm{d}\cos\theta_{\mathrm{n}} \Theta\left[E_f^{(\nu)}(\ldots, W, z) - E_{\mathrm{th}}\right].$$

The factor $\exp[-l(z,\theta_{\mathrm{n}})/L_{\nu\mathrm{N}}]$ describes the neutrino flux attenuation on the pathlength $l(z,\theta_{\mathrm{n}})$ up to the shower-production point at the depth z, $L_{\nu\mathrm{N}}$ is the neutrino interaction length taken according to [4]. The field attenuation in the lunar regolith is taken in the form

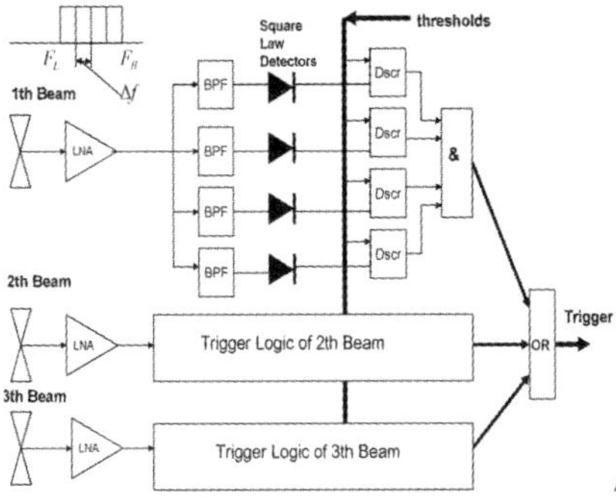

Fig. 1. LORD10 trigger and data recording system

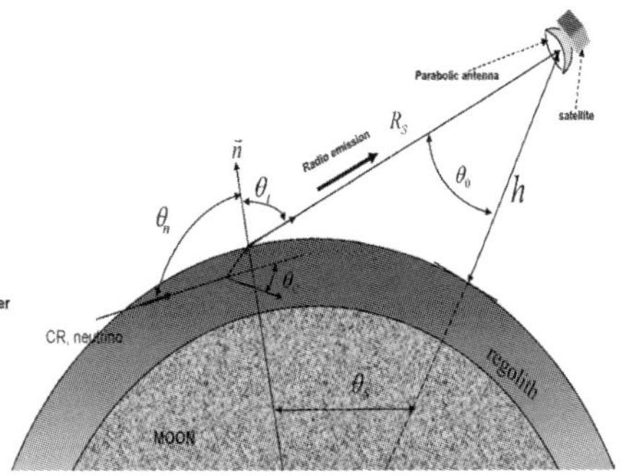

Fig. 2. Layout of the LORD10 experiment

$$E_f^{(\nu)}(\theta_n, \varphi_n, \theta_s, W, z) \approx E_f^{(\nu)}(\theta_n, \varphi_n, \theta_s, W) \exp(-z/\lambda_{abs} \cos\theta_i),$$

where $\lambda_{abs} \approx 15$ m/f(GHz) and θ_i is the incident angle of radiation ($n \sin\theta_i = \sin\theta_t$).

These formulas determine the aperture for a narrow frequency band Δf near f. In the case of the trigger system involving N_b frequency bands, the $\Theta(x)$-functions should be substituted by the products $\prod_{k=1}^{N_b} \Theta\left[E_f(\ldots) - E_{th}^k\right]$.

PARAMETERIZATION OF RADIATION FIELD STRENGTH

The calculation of the function $E_f(\theta_n, \varphi_n, \theta_s, W)$ entering into these equations is the basic problem of finding the apertures. If the shower length, wavelength, and the distance to the interface are of the same order of magnitude, the problem becomes rather complicated because in this case the processes of generation, propagation, and refraction of the radiation cannot be separated. Nevertheless, as usual, we here consider these processes singly.

Instead of performing Monte Carlo simulations, we use some known results concerning radio pulses from UHE showers [5-8]. Namely, we start from the following generalization of the parameterization proposed in [3, 7]

$$R_s E_f \approx N_f W \frac{f \exp\left[-\alpha(f, W)(\cos\theta - \cos\theta_C)^2\right]}{\left[1 + (f/f_0)^{1.44}\right]} \frac{\sin\theta}{\sin\theta_C} T_s(\theta_i, \theta_t),$$

which provides the acceptable fit of the Monte Carlo simulations of hadronic showers in ice. Here, $N_f \approx 2\ 10^{-4}$(μV/TeV/MHz^2), $f_0 \approx 0.5$ GHz, and $\cos\theta_C = 1/n \approx 0.59$. The transmission coefficient T_s with allowance for the wave divergence takes the form $T_s = 2\cos\theta_t / (n\cos\theta_t + \cos\theta_i)$ [3].

To find the function $\alpha(f,W)$ for ice, we fitted the simulation results [6] on the angular spread $\Delta\theta$ in the form $\alpha(f,W) = \log 2/(\sin\theta_C \Delta\theta)^2 \approx f^2[C_1+C\log(W)]$, which was assumed to be applicable up to the highest energies. We then used the simple scaling [8] from to ice to the lunar regolith: $N_f \propto 1/\rho$, $f_0 \propto 1/r_M$, and $\alpha \propto L_{rad}^2$, where ρ is the density, r_M is the Moliere radius, and L_{rad} is the radiation length. As a result, we found $N_f \approx 7\ 10^{-5}(\mu V/TeV/MHz^2)$, $f_0 \approx 3.3$ GHz, and $\alpha(f,W) \approx f^2(\mathrm{GHz})[70+3.3\log(W/\mathrm{TeV})]$.

This expression can be used only for hadronic showers, when the Landau-Pomeranchuk-Migdal effect weakly influences the shower development [6]. As to neutrino-produced "mixed" showers consisting of electromagnetic and hadronic parts, we allow for only the latter [3]. Indeed, because of the significant shower elongation due to the LPM effect, the contribution of the electromagnetic part to Cerenkov radiation is much less than that of the hadronic part. In this case, we evaluated the neutrino aperture for $W_{shower}=y\ W_\nu$ and then found the average with the distribution $w(y)$ defined in [4].

CALCULATION RESULTS

We calculated the angular and total apertures for a lot of possible configurations of the LORD experiment with various antenna's diameter D, orbit altitude h, bandwidth Δf, frequency range $[f_{min}, f_{max}]$, and signal-to-noise ratio SNR for each frequency channel. As the most acceptable configuration at this stage, we assumed D=10 m, h = 500 km, Δf = 50 MHz, $[f_{min}, f_{max}]$ = [350, 500] MHz, and SNR = 4.

For energies above 10^{19} eV, the cosmic ray and neutrino angular apertures $\Delta\Omega$ and $\Delta\Omega_\nu$ are both concentrated near the visible lunar horizon, with the nadir angle range for cosmic rays being noticeably less than that for neutrinos (Fig. 3). However, this effect is not pronounced enough to use it to distinguish cosmic-ray events from neutrino events.

The total aperture for cosmic ray- and neutrino-produced cascades calculated for this configuration of the LORD10 experiment is shown in Fig. 1 of Dr. V. Tsarev's report. Since the directional pattern of the LORD10 antenna is fairly narrow in azimuth ($\delta\varphi \approx 0.3$), the aperture of the LORD10 is about 20 times less than that of the maximum "2π"-LORD discussed by Dr. V. Tsarev. The corresponding constrains on neutrino flux, calculated for the LORD and LORD10 experiments (E_{th} = 0.01 μV/m/MHz, T = 1 year, f = 0.5 GHz), are shown in Fig. 3 from that report.

CONCLUSION

As is seen from the above consideration the LORD10 experiment is quite competitive with the existent and planned neutrino telescopes. The grave disadvantage of the LORD10 antenna is the narrow angle of view in azimuth. Because of this, torus parabolic antennas could be used to increase the performance of the LORD.

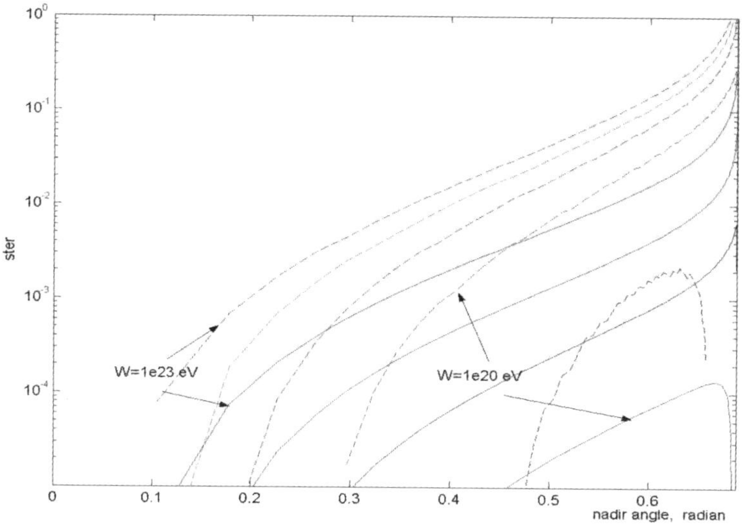

Fig. 3. The LORD10 angular apertures for (dashed) CR- and (solid) neutrino-produced cascades for various energies (f=0.5 GHz, E_{th}=0.08 μV/m/MHz).

REFERENCES

1. LORD Project (Dr. V. Tsarev's report in ARENA-2005)
2. P.W. Gorham, C.L. Hebert, K.M. Liewer, C.J. Naudet, *et al. (GLUE Experiment) arXiv: astro-ph/0310232.*
3. N.G. Lehtinen, P.W. Gorham, A.R. Jacobson, et al.,(FORTE Experiment) arXiv: astro-ph/0309656.
4. R.Gandhi, C. Quigg, M.H. Reno, et al., Phys. Rev., D **58**, 093009 (1998).
5. E. Zas, F. Halsen, and T. Stanev, *Phys. Rev. D **45**, 362 (1992).*
6. J. Alvarez-Muniz and E. Zas, Phys. Lett. B434, 396-406 (1998).
7. J. Alvarez-Muniz, R.A. Vazquez, and E. Zas, *Phys. Rev. D 62, 063001 (2000).*
8. J. Alvarez-Muniz, E. Marquez, R.A. Vazquez, and E. Zas, *in Proc. 28th ICRC, pp. 1521-1524, 2003.*

ADVANCED DETECTION METHODS OF RADIO SIGNALS FROM COSMIC RAYS FOR KASCADE GRANDE AND AUGER

H. GEMMEKE[I], W.D. APEL[A], F.A. BADEA[A], L. BÄHREN[B], K. BEKK[A], A. BERCUCI[C], M. BERTAINA[D], P.L. BIERMANN[E], J. BLÜMER[A;F], H. BOZDOG[A], I.M. BRANCUS[C], M. BRÜGGEMANN[G], P. BUCHHOLZ[G], S. BUITINK[H], H. BUTCHER[B], A. CHIAVASSA[D], K. DAUMILLER[A], A.G. DE BRUYN[B], C.M. DE VOS[B], F. DI PIERRO[D], P. DOLL[A], R. ENGEL[A], H. FALCKE[B;E;H], P.L. GHIA[J] R. GLASSTETTER[K], C. GRUPEN[G], A. HAUNGS[A], D. HECK[A], J.R. HÖRANDEL[F], A. HORNEFFER[H;E], T. HUEGE[A;E], K.-H. KAMPERT[K], G.W. KANT[B], U. KLEIN[L], Y. KOLOTAEV[G], Y. KOOPMAN[B], O. KRÖMER[I], J. KUIJPERS[H], S. LAFEBRE[H], G. MAIER[A], H.J. MATHES[A], H.J. MAYER[A], J. MILKE[A], B. MITRICA[C], C. MORELLO[J], G. NAVARRA[D], S. NEHLS[A], A. NIGL[H], R. OBENLAND[A], J. OEHLSCHLÄGER[A], S. OSTAPCHENKO[A], S. OVER[G], H.J. PEPPING[B], M. PETCU[C] J. PETROVIC[H], T. PIEROG[A], S. PLEWNIA[A], H. REBEL[A], A. RISSE[M], M. ROTH[F], H. SCHIELER[A], G. SCHOONDERBEEK[B], O. SIMA[C], M. STÜMPERT[F], G. TOMA[C], G.C. TRINCHERO[J], H. ULRICH[A], J. VAN BUREN[A], W. VAN CAPELLEN[B], W. WALKOWIAK[G], A. WEINDL[A], S. WIJNHOLDS[B], J. WOCHELE[A], J. ZABIEROWSKI[M], J.A. ZENSUS[E], D. ZIMMERMANN[G]

[A] *Institut für Kernphysik, Forschungszentrum Karlsruhe, Germany*
[B] *ASTRON Dwingeloo, The Netherlands*
[C] *National Inst. of Physics and Nuclear Engineering, Bucharest, Romania*
[D] *Dept. of General Physics, Universitá degli Studi di Torino, Italy*
[E] *Max-Planck-Institut für Radioastronomie, Bonn, Germany*
[F] *Institut für Experimentelle Kernphysik, Universität Karlsruhe, Germany,*
[G] *Fachbereich Physik, Universität Siegen, Germany*
[H] *Dept. of Astrophysics, Radboud University Nijmegen, The Netherlands*
[I] *IPE, Forschungszentrum Karlsruhe, Germany*
[J] *Institute for Interplanetary Space Physics INAF, Torino, Italy*
[K] *Fachbereich Physik, Universität Wuppertal, Germany*
[L] *Radioastronomisches Institut, Universität Bonn, Germany*
[M] *Soltan Institute for Nuclear Studies, Lodz, Poland*

The LOPES experiment (LOfar Prototype Station) has been built at the KASCADE-Grande experiment in order to test the LOFAR (LOw Frequency ARray) technology and demonstrate its capability for radio measurements in Extensive Air Showers (EAS). After the first positive results in the framework of the KASCADE-Grande experiment we developed the next generation of antennas, electronics, and trigger. The main new features are easy calibration of antennas with precise detection of polarization, and good capability of self-triggering. The results from this new design are under test in Karlsruhe. Furthermore the background situation was measured and analyzed.

1. Introduction

Geo-synchrotron radio emission by high-energy cosmic ray air showers is caused by the deflections of electrons and positrons in the earth's magnetic field. As in synchrotron radiation this radiation is narrowly beamed into the forward direction of the shower. The maximum radio emission is expected in the low frequency range (10 MHz to 200 MHz) where the atmospheric attenuation of radio signals below 10 km height can be neglected and a bolometric measure of the electromagnetic energy content of the shower is achieved. To evaluate this method a phased array of dipole antennas (LOPES) with digital RF processing has been built in co-location with the KASCADE-Grande air shower experiment to ensure the interpretation of the measured data.

The antenna array of LOPES consists of thirty short "inverted V" dipoles operating in the 40 MHz to 80 MHz range. The data readout is triggered by KASCADE-Grande events. The first results[1] verify the coherent radio emission predicted by the geo-synchrotron effect.

Parallel with these measurements we started LOPES*, an optimization of the antenna design and its electronics for the application in large cosmic ray experiments like the Pierre Auger observatory. LOPES* (LOPES^{STAR}) is the acronym for a Self Triggered Array of Radio detectors for LOPES. The main focus of LOPES* is the efficient suppression of man-made radio frequency interference (RFI), an antenna design which can be simply calibrated, with low power consumption for operating with a solar and battery supply and with the feature of self-triggering. LOPES* works in the same frequency range as LOPES.

2. Choice of Antenna for LOPES*

For the choice of antenna we considered and tested 3 types: the logarithmic periodic conical helix used in the CODALEMA[2] experiment, the inverted V-type antenna of LOPES[3] and the crossed logarithmic-periodic dipole antenna[4] (crossed LPDA). The conical helix antenna was discarded because of its large crosstalk between both polarization directions of about −6 dB, if both antennas are mounted

Figure 1. Crossed logarithmic-periodic dipole antennas of LOPES*.

on top of each other on the same pole. Furthermore it is expensive to build a heightened version protected against environmental effects at ground.

The V-type antenna is simple to construct and inexpensive. But it has a frequency dependent sensitivity and impedance with a high capacitive component, which may be partly compensated by the preamplifier. That makes calibration and data analysis difficult. The horizontal attenuation is very weak and increases the sensitivity to man-made RFI. The missing backward attenuation must be compensated by a defined ground layer otherwise the calibration of the antenna changes. The crossed LPDA in Fig.1 does not have any of these disadvantages, see Table 1, is in the overall costs for production and installation not considerably more expensive than the V-type antenna, but cheaper than the conical helix.

Table 1. Measured characteristics of the crossed LPD antennas

Characteristic	Measured values
Bandwidth:	35…90 MHz
Gain:	5.5 dBi*
Impedance:	Real, 50 Ohm
Return loss:	- 12 dB
Beam width [-3dB]:	± 50° (E-Plane), ± 70° (H-Plane)
± 90°-attenuation:	20 dB (E-Plane), 6 dB (H-Plane)
Backward attenuation:	20 dB
Polarisation isolation:	> 20 dB
Size (without pole):	4 x 4 x 3 m^3
Weight (without pole):	15 kg

The crossed LPDA can be built with any user-defined bandwidth. The directivity has a strong forward characteristic of 5.5 dBi*, a high backward attenuation and a high rejection in the horizontal direction, suppressing man-made RFI and interaction with ground and buildings efficiently. Using a crossed assembly all polarisations may be received and distinguished, therefore we choose the crossed LPDA.

3. Low power front-end electronics and measured radio background

The RF is amplified close to the antenna with a Low Noise Amplifier (LNA) with 1.8 dB noise-figure and a power consumption of 22 mW per channel. The remaining part of the analog frontend has a power consumption of 65 mW per channel. Both preamplifier and frontend fulfil our solar power budget of 100 mW per channel. The anti-aliasing band-pass filter is of 32nd order and uses

* The unit dBi is the gain of the antenna relative to an isotropic radiator.

nearly the whole 2nd Nyquist domain at 80 MHz sampling rate from 41 MHz to 79 MHz with only 3 dB passband ripple and 10 dB/MHz slope at the band limits. To reject all high-level out-of-band components (Fig. 2) the stopband attenuation is 110 dB at shortwave, 90 dB at the FM band, and 80 dB at the VHF III band.

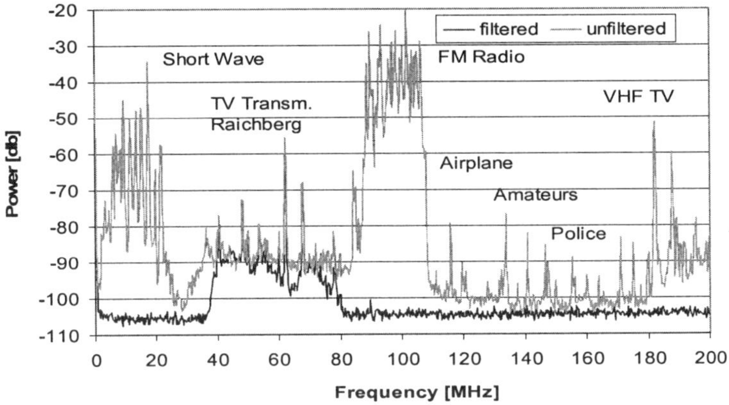

Figure 2. Measured radio background at FZ Karlsruhe and the effect of our filter.

The measured out-of-band background at Forschungszentrum Karlsruhe will be similar elsewhere and its suppression requires such strong filters. A commercial VME DAQ[5] system performs a 12 bit AD-conversion with 80 MHz sampling rate and stores the data in two ring buffers.

4. Trigger

The basic antenna array setup for self-triggering consists of three crossed LPDAs on the edges of an equilateral triangle. The triangle setup at Forschungszentrum Karlsruhe has a side length of 65 m (Fig. 3). The trigger is activated only, if all three antennas produce signals within the same 190 ns time slot given by the height of the triangle divided by the speed of light. Thereby the trigger hardware is able to distinguish most of the signals from the horizon, which are normally man made, from signals with higher elevation, which may be excited by air showers. So the trigger not only demands signals at all antennas above the thresholds, but also discriminates the elevation angle (horizontal or elevated). RFI-sources inside the trigger-triangle have to be avoided, because they can't be discriminated from real shower events by their timing.

The envelope signal (Fig. 4) is given by the RF power and produces short pulses at air showers while many man made RFI cause only slow amplitude changes, which may be easily removed by high-pass filtering. At present the first self triggered measurements have been started to investigate the dependence of trigger rates on the coincidence parameters. All events are GPS time stamped to find coincidences with KASCADE-Grande triggers.

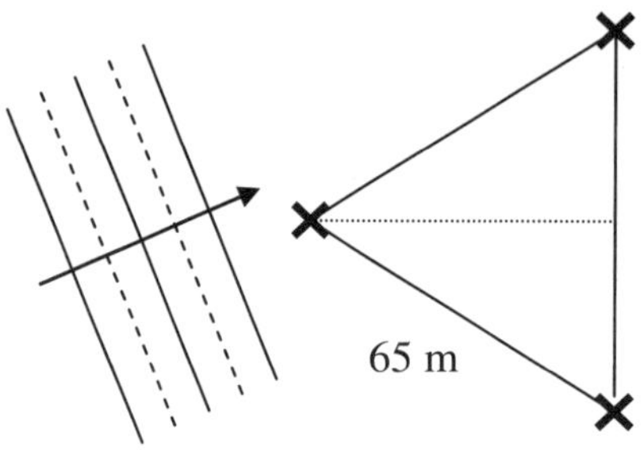

Figure 3. Triangular antenna geometry and a horizontal interference source.

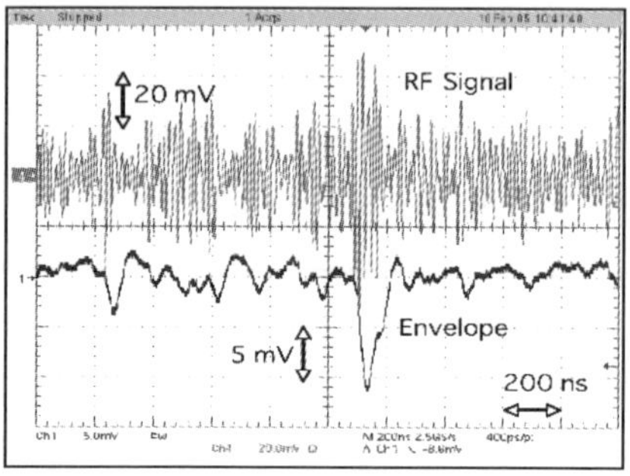

Figure 4. Triggering with envelope signals.

5. Conclusion and future steps

So far a test array of three antennas is installed in Karlsruhe and data are recorded. The expected background rejection and self-trigger features could be demonstrated. The installation is now used to measure the obtainable time resolution. The installation of several triangular setups in Karlsruhe is in preparation. A similar installation is planned on the site of the southern Pierre Auger Observatory, especially to optimize the geometry for such an array to measure energies above 10^{19} eV under realistic conditions.

References

1. H. Falcke et al., Nature 435, 313 (2005)
2. D. Ardouin et al., CODALEMA, astro-ph/0504297, submitted to NIM A
3. A. Horneffer et al., Proc. SPIE 5500-21, (2004)
4. Rothammel, 12. Auflage, 2001, S. 635 - 646
5. SIS3300, SIS GmbH, 22399 Hamburg, Germany, http://www.struck.de

NEUTRINO DETECTION IN SALT DOMES UNDER LOFAR*

A.M. VAN DEN BERG
Kernfysisch Versneller Instituut, Rijksuniversiteit Groningen, Zernikelaan 25
NL 9747 AA Groningen, the Netherlands

Large volumes of natural materials are presently under study to construct a telescope, which will be used to search for ν's at the highest energies. Although these high-energy ν's have not been discovered yet, firm predictions for their existence have been made. The very origin of these high-energy ν's remains one of the burning questions in astroparticle physics. We consider the use of huge salt domes in the shallow underground of the north-eastern part of the Netherlands as a possible site for such a high-energy ν telescope. Initial measurement of the attenuation length for radio signals at 0.3 and 1.0 GHz will be reported. The physical location of the domes and the envisioned research program has a strong overlap with those of the new radio telescope LOFAR.

1. Introduction

Although high-energy astroparticle physics has made substantial progress in recent years, the search for Ultra-High-Energy neutrinos (UHE ν's) has not yet resulted in a proof of their very existence. Presently, several experiments are finished or still running such as AMANDA [1], ANITA [2], FORTE [3], GLUE [4], and RICE [5], but only upper limits for the possible flux of these UHE ν's have been reported. The technique for the mentioned experiments is based on the detection of light or radio signals induced by the interaction of a UHE ν with solid materials like the polar icecaps or the lunar regolith. Other experiments are under study as well: the detection of extended air showers induced by UHE ν's in the Earths atmosphere [6], and the detection of radio signals inside large salt deposits located in the shallow underground [7,8]. One of the aims of these studies is to discover the sources of high-energy cosmic-ray events which have been detected with various detector systems [9,10,11], especially beyond the critical limit of 6×10^{19} eV, known as the Greisen-Zatsepin-Kuzmin (GZK) limit [12]. In case the GZK limit holds for UHE cosmic rays, as supported by the HiRes data [9], there should be a flux of UHE ν's peaking at an energy around

* This work was performed as part of the research program of the Stichting voor Fundamenteel Onderzoek (FOM) with financial support from the Nederlandse Organisatie voor Wetenschappelijk Onderzoek (NWO).

this critical limit [13]. On the other hand, in case the UHE cosmic-ray spectrum continues to follow a more or less constant power-law behavior, as suggested by the AGASA [10] and most recent Auger [11] data, it may point either to very

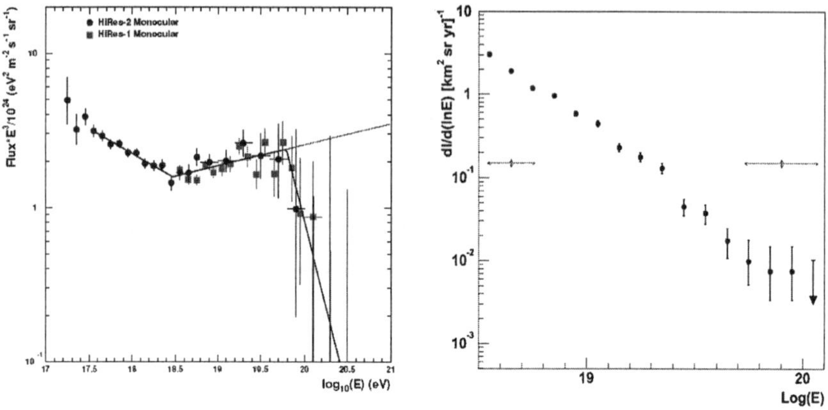

Figure 1. Flux spectra for UHE cosmic rays. The left panel shows the recent compilation from the HiRes experiments (taken from Ref. [9]); the thick solid line shows a fit to the data using a double-broken power-law spectrum, with an indication for the GZK cutoff. The right panel displays the first estimated energy spectrum from Auger (taken from Ref. [11]), where dI/dlnE = EdI/dE. The Auger data can be described by a single power-law spectrum.

nearby sources of UHE cosmic rays, to a chemical composition of these UHE cosmic rays close to iron nuclei, or to new physics at the highest energies. In this last scenario the decay of relics from the Big Bang has been proposed as the source for these UHE cosmic rays, which therefore at the same time could also be the source of UHE ν's [14]. The data from the large cosmic-ray experiments remain to contradict each other (see Figure 1) and there is a model-dependency on the calculation of the primary cosmic-ray energy as deduced from the detection of extended air showers. Therefore, an experiment based on a completely different detection technique and specifically tailored to the detection of UHE ν's in the energy domain beyond 10^{15} eV, is highly desirable. Even more so because at the highest energies, ν's are the only events which can travel over cosmological distances without being distorted by scattering or absorption processes; nor are they deflected by intergalactic or interstellar magnetic fields. Therefore, a UHE ν telescope with a good pointing accuracy provides the unique opportunity for precise backtracking to the cosmological sources. This will enable us to get a better insight into the nature of these sources, and the underlying mechanisms for the production of the UHE cosmic rays and ν's. As explained by Saltzberg [7] in more detail, the detection of UHE cosmological ν's in itself may provide new and otherwise difficult to obtain information on physics at the highest energies. The main problem for the detection of UHE ν's is the low flux predicted by several models [15], which therefore calls for very large detector systems. As was proposed by Askaryan [16], the detection of radio signals originating from the coherent Cherenkov radiation emitted by the fast moving electromagnetic cascade in a suitable dielectric medium, may offer a good opportunity to build a large ν telescope with a volume of several km^3; see also Ref. [17]. After the initial proposal of Askaryan, the first experimental proof of coherent radio Cherenkov emission was reported by Saltzberg *et al.* [18]. Presently several experiments based on this effect are running using the polar icecaps [2,5], or are envisioned using the Moon as a target [19]. Recently, rock salt is also being investigated as another suitable candidate [7]. This material has several advantages; its density is about 2.2 g/cm^3, it has a relatively long attenuation length for radio waves [8], and large sediments are found at several places in the shallow underground, which makes a telescope deployment easy.

2. Zechstein Salt in the Netherlands

Substantial salt deposits from the Zechstein period can be found in the shallow underground of the Netherlands. The four major Zechstein salt deposits are part of an extended area ranging between the United Kingdom to Poland. After the Zechstein period (in the late Perm) these salt layers were covered with several other sediments during more recent geological periods. However, because rock salt is relatively flexible and light compared to other geological deposits, domes could be formed; a few of them with an overburden of less than 100 m. Figure 2 shows the Zechstein salt layers in the underground of the Netherlands, with a concentration of domes in the north-eastern part (upper right corner). For many years, some of these domes are explored for their salt using solution mining. The storage of natural gas under a pressure of about 200 bar is a new application for these domes. In the coming years a consortium formed by several companies will make a grid of about 10 caverns separated by distances of 200 m in the

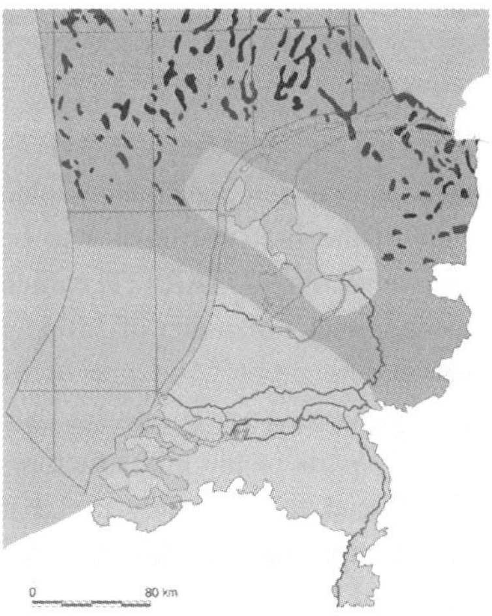

Figure 2. The Zechstein salt layers in the Netherlands; the dark spots indicate locations of salt domes (© TNO-NITG, Utrecht, the Netherlands).

Zuidwending salt dome, located near the city of Groningen. This Zuidwending dome is one the largest and shallowest in the Netherlands, actually consisting of two almost overlapping domes, the northern and the southern lobe. The depth of the dome is about 2.5 km and the surface cross section roughly 0.5 by 5 km^2. The roof of the caverns for fast gas storage will be at a depth of about 1 km; the height and diameter of these caverns are 300 m and 50 m, respectively. The site of this dome could be used for a UHE ν telescope, providing that agreements with the gas-storage consortium can made, leading to a cost-effective use of parts of the infrastructures and drilling equipment. Most importantly, however, are the dielectric properties of the rock salt; the attenuation length for radio at about 1 GHz has to be at least 250 m. In the past year, samples from the Zuidwending site were tested by Chiba *et al.* [8] using the cavity-resonator method. Initial measurements were made at 2 frequencies, resulting in rather short attenuation lengths of only (22±2) m and (77±11) m at a frequency of 0.3 and 1.0 GHz, respectively; see Ref. [8] for more details. These data deviate strongly from those determined for other sites, which might indicate that the Zuidwending dome is not very well suited for radio detection. It may also be, that there are problems with the samples themselves, which are rather old indeed; e.g. surface effects may play an important role in the determination of the bulk properties. A next stage of the project calls for an *in situ* measurement of the attenuation length, for instance in the early drilling phase of the fast gas-storage plant.

3. The connection with the LOFAR radio telescope

This project for the development of a Zechstein SAlt Neutrino Array (ZESANA) [20] in one of the Dutch salt domes fits into the recently developed strategic plan for astroparticle physics in the Netherlands. One of the first instruments to be used for this program will be the new digital radio telescope LOFAR [21], which is presently under construction. This radio telescope, with a diameter of 350 km, is centered at about 50 km south of the Zuidwending dome. And one of its spiraling arms is running very near the dome, allowing for easy access for data transmission to central computing systems. Not only parts of the infrastructures can be shared, this radio telescope itself will be used for the study of UHE cosmic rays and ν's hitting the Earths atmosphere. In this case LOFAR measures the coherent geomagnetic synchrotron emission induced by the high-energy events. The proof of principle for this technique and a first calibration of the radio signal strength versus the primary cosmic-ray energy were recently made by the LOPES collaboration [22]. In addition, the LOFAR infrastructure will also incorporate the readout of geophones [23], which will survey the shallow underground to get a better understanding of natural or induced seismic activity; e.g. caused by drilling or extraction of natural gas from underground reservoirs.

Acknowledgments

I thank J.N. Breunese, J.H. Brouwer, H. Butcher, M. Chiba, H. Falcke, M. Geluk, P.W. Gorham, T. Kamijo, R.J. de Meijer, H.F. Mijnlieff, D. Saltzberg, and O. Scholten for stimulating discussions, and TNO-NITG, Utrecht, the Netherlands, for providing the samples of rock salt from the Zuidwending dome.

References

1. K. Woschnagg for the AMANDA collaboration, *Nucl. Phys. B* **143**, 343 (2005)
2. P. Miocinovic, this conference.
3. N.G. Lehtinen *et al.*, *Phys. Rev. D* **69**, 013008 (2004).
4. P.W. Gorham *et al.*, *Phys. Rev. Lett.* **93**, 041101 (2004).
5. D. Besson, this conference.
6. K.S. Capelle *et al.*, *Astrop. Phys.* **8**, 321 (1998)
7. D. Saltzberg, this conference, and references cited therein.
8. M. Chiba *et al.*, this conference.
9. R.U. Abbasi *et al.*, *Phys. Lett. B* **619**, 271 (2005); R. Bergmann for the HiRes collaboration, 29th ICRC, Pune, India, 2005.
10. M. Takeda *et al.*, *Phys. Rev. Lett.* **81**, 1163 (1998).
11. P. Sommers, for the AUGER collaboration, 29th ICRC, Pune, India, 2005.
12. K. Greisen, *Phys. Rev. Lett.* **16**, 748 (1966); G.T. Zatsepin and V.A. Kuzmin, *Sov. Phys. JETP Lett. (Engl. Transl.)* **4**, 78 (1966).
13. D. Seckel, this conference.
14. G. Gelmini and A. Kusenko, *Phys. Rev. Lett.* **84**, 1378 (2000).
15. See for example: R. Engel, D. Seckel, and T. Stanev, *Phys. Rev. D* **64**, 093010 (2001); O.E. Kalashev *et al.*, *Phys. Rev. D* **66**, 063004 (2002).
16. G. Askaryan, *Sov. Phys. JETP* **14**, 441 (1962); *ibid* **21**, 658 (1965).
17. J. Alvarez-Muñiz, this conference.
18. D. Saltzberg *et al.*, *Phys. Rev. Lett.* **86**, 2802 (2001).
19. This conference see: J. Bacelar *et al.*, R. Dagkesamanskii *et al.*, and the LORD collaboration (contributions from V.A. Tsarev and V. Chechin).
20. See: http://www.kvi.nl/~berg/zesana/
21. See: http://www.lofar.org
22. H. Falcke for the LOPES collaboration, this conference; *Nature* **435**, 313 (2005).
23. J. Brouwer, TNO-NITG Information, May 2004.

INTRODUCTION TO THE SALSA, A SALTDOME SHOWER ARRAY AS A GZK NEUTRINO OBSERVATORY

DAVID SALTZBERG

Dept. of Physics and Astronomy, University of California, Los Angeles, 475 Portola Plaza, Los Angeles, CA 90095-1547

For the SALSA Collaboration:

Katsushi Arisaka[k], Ron Bain[a], Steven Barwick[l], James Beatty[h], David Besson[o], W. Robert Binns[t], Chien-Wen Chen[f], Pisin Chen[f], Michael Cherry[e], Amy Connolly[k], Michael DuVernois[p], Clive Field[f], Manfred Fink[q], David Goldstein[l], Peter Gorham[n], Giorgio Gratta[i], T. Gregory Guzik[e], Francis Halzen[s], Carsten Hast[f], Jay Hauser[k], Stephen Hoover[k], Charles Jui[r], Spencer Klein[d], John Learned[n], Gueylin Lin[g], Shige Matsuno[n], James Matthews[e], Radovan Milincic[n], Predrag Miocinovic[n], Rolf Nahnhauer[b], JiWoo Nam[l], Johnny Ng[f], Ryan Nichol[h], Allen Odian[f], Rene Ong[k], Buford Price[j], Kevin Reil[f], David Saltzberg[k], David Seckel[m], Pierre Sokolsky[r], Bob Svoboda[e], Ad van den Berg[c], Justin Vandenbroucke[j], Gary Varner[n], Dieter Walz[f], John Wefel[e], David Wieczorek[k], Jeffrey Wilkes[t]

[a] *ConRon Consulting; Houston TX*
[b] *DESY; Zeuthen, Germany*
[c] *Kernfysisch Versneller Instituut; Groningen, Netherlands*
[d] *Lawrence Berkeley National Lab; Berkeley, CA*
[e] *Dept. of Physics & Astronomy, Louisiana State Univ.; Baton Rouge, LA*
[f] *Stanford Linear Accelerator Center; Stanford, CA*
[g] *Inst. of Physics, National Chiao-Tung Univ.; Hsinchu, Taiwan*
[h] *Dept. of Physics, Ohio State Univ.; Columbus, OH*
[i] *Dept. of Physics, Stanford Univ.; Stanford, CA*
[j] *Dept. of Physics, Univ. of California, Berkeley; Berkeley, CA*
[k] *Dept. of Physics & Astronomy, UCLA; Los Angeles, CA*
[l] *Dept. of Physics & Astronomy, Univ. of California, Irvine; Irvine CA*

[m] *Bartol Research Inst., Univ. of Delaware; Newark, DE*
[n] *Dept. of Physics & Astronomy, Univ. of Hawaii at Manoa; Honolulu, HI*
[o] *Dept. of Physics & Astronomy, Univ. of Kansas; Lawrence, KS*
[p] *School of Physics & Astronomy, Univ. of Minnesota; Minneapolis, MN*
[q] *Dept. of Physics, Univ. of Texas at Austin; Austin, TX*
[r] *Dept. of Physics, Univ. of Utah; Salt Lake City, UT*
[s] *Dept. of Physics, Univ. of Wisconsin; Madison, WI*
[t] *Dept. of Physics, Washington Univ. in St. Louis; St. Louis, MO*

The observed spectrum of ultra-high energy cosmic rays virtually guarantees the presence of ultra-high energy neutrinos due to their interaction with the cosmic microwave background. Every one of these neutrinos will point back to its source and, unlike cosmic rays, will arrive at the Earth unattenuated, from sources perhaps as distant as z=20. The neutrino telescopes currently under construction, should discover a handful of these events, probably too few for detailed study. In this talk I will describe how an array of VHF and UHF antennas embedded in a large salt dome, SalSA (Saltdome Shower Array) promises to yield a teraton detector (> 500 km^3-sr) for contained neutrino events with energies above 10^{17} eV. Our simulations show that such a detector may observe several hundreds of these neutrinos over its lifetime. Our simulations also show how such interactions will provide high energy physicists with an energy frontier for weak interactions an order-of-magnitude larger than that of the LHC. The flavor ID capalities of SALSA, combined with the extreme L/E of these neutrinos, will provide a window on neutrino oscillations and decay times eight orders of magnitude higher than laboratory experiments. In addition to the latest simulation results, we describe progress on detectors and site selection.

NEUTRINO FLAVOR IDENTIFICATION IN SALSA

PREDRAG MIOČINOVIĆ
Department of Physics and Astronomy,
University of Hawaii at Manoa,
2505 Correa Rd,
Honolulu, HI 96822, USA
E-mail: predrag@phys.hawaii.edu

The proposed Saltdome Shower Array (SalSA) experiment will detect coherent Cherenkov radio signals from high-energy neutrino interactions in a naturally occurring salt dome. By identifying the number and the angular profile of radio emissions in any given event, distinction can be made between charged-current (CC) and neutral-current (NC) neutrino interactions. Additionally, the flavor of the neutrino can be identified in the case of charged-current interactions. Preliminary results for nominal GZK neutrino flux indicate that ~25% of all events can be correctly identified as coming from charged-current interactions of ν_μ's or ν_τ's. These charged-current initiated events can further be separated by the flavor of the original neutrino, either ν_μ or ν_τ.

1. Introduction

The Saltdome Shower Array (SalSA) experiment[1] has a potential of detecting tens to hundreds of ultra-high-energy ($E_\nu > 10^8$ GeV) cosmological neutrinos per year. Such an event rate would come from the expected flux of GZK neutrinos created by cosmic-ray interactions on cosmic microwave background. The accurate measurement of the neutrino flux would greatly aid in resolving the riddle of origin of the highest energy cosmic-rays. Additionally, being able to tell something about the flavor composition of the cosmological neutrino flux would further test models of neutrino flux generation and propagation.

This report is a preliminary study of flavor identification capability of a radio Cherenkov neutrino detector. It outlines the simulation chain used to generate events that would trigger SalSA, discusses potential ways to differentiate primary neutrino flavor causing the event trigger, and presents the initial results of the flavor identification analysis.

2. Simulation

The simulation of SalSA event detection has been done in three steps; i) generation of primary neutrino flux, its propagation through the Earth, and interaction in the detection volume, ii) tracking of secondary leptons and generation of secondary particle showers, iii) generation and detection of radio Cherenkov signal. For the first step, ANIS neutrino generator[2] is used, for the second MMC lepton propagator,[3] and for the last the custom written SalSA Monte Carlo (SMC). An isotropic GZK neutrino flux[4] in the energy range from $10^{6.5} - 10^{11}$ GeV was simulated and propagated to the detection volume. The detection volume was defined as a cylindrical salt dome with diameter of 3.75 km and extending in depth from 0.5-3.5 km, which corresponds to sizes of large salt domes found in the southern USA. The lepton resulting from the primary neutrino interaction in the salt or in the surrounding rock was propagated and all stochastic energy losses exceeding 10^6 GeV were recorded. All possible interactions, including particle decay, were taken into account.

The radio signals generated by the primary and all secondary showers were assumed to propagate without scattering in salt with a conservative attenuation length of 250 m. The signals were detected by 10x10 square array of strings, with each string consisting of 12 detection nodes. The strings were horizontally separated by 250 m, and nodes on each string were vertically separated by 182 m. The array was centered at the depth of 1750 m. Each node consisted of 12 antennas, 6 dipole and 6 slot antennas. The dipoles preferentially couple to the vertical component of the electric field, while slots to the horizontal component. Each antenna was described by a nominal gain of 2.1 over its full bandwidth (100-300 MHz) and a system temperature of 450 K.

The trigger was implemented at two levels, local and global. The local trigger was satisfied by a node if five of 12 antennas observed voltage transients exceeding 2.8 times rms voltage (2.8 σ) within the time window of 80 ns, an approximate travel time of a radio pulse across the node. The global trigger was satisfied if four nodes triggered within 4 μs, an approximate time for a radio signal to travel to the nearest neighbor nodes. If the detector triggered, all nodes that triggered were read out, and the origin of the signal exceeding the threshold at each antenna, either a specific particle shower or thermal noise, was recorded.

The key step in making neutrino flavor identification possible is an accurate description of the strength and the width of the cone of radio Cherenkov

Table 1. Particle shower types most likely expected for different neutrino flavors and primary interaction type.

	CC	NC
ν_e	hadronic and EM shower from the same vertex	single hadronic shower
ν_μ	hadronic shower and many secondary EM shower at distance	single hadronic shower
ν_τ	hadronic shower and few secondary hadronic shower at distance	single hadronic shower

radiation emitted by a particle shower. The theoretical description of differences between hadronic and electromagnetic showers, including the treatment of LPM effect, are discussed by Alvarez-Muñiz, Vázquez, and Zas.[5,6] The main points are that the energy of the shower is linearly correlated with the power of the resulting electric field, and that the width of the radio cone is inversely correlated with the length of the particle shower. Considering that at the energies below the LPM threshold hadronic showers tend to be longer than electromagnetic ones, while the opposite would be the case above the threshold, simultaneously measuring the strength and the width of the radio cone differentiates between the shower types. Combining this with the difference in probabilities that secondary leptons have to interact either electromagnetically, through bremsstrahlung or pair production, or hadronicly, through photon nuclear interaction, one can create a probability for any event to have been generated by a neutrino of a given flavor. Of course, if a neutrino interacts via a neutral-current pathway, there will be no way of telling its flavor. Table 1 summarizes the flavor identification scheme.

3. Results

The results of the simulation were analyzed considering only the number of antenna nodes triggered by any given particle shower in an event. In the absence of scattering, the timing of node triggers will, with high degree of certainty, determine whether the nodes were triggered by a single radio emission.[7] The identification strategy for separating NC from CC events required in all cases that the initial hadronic shower[a] generates a clear signature by triggering at least 4 nodes. Then, each passing event was analyzed for the presence of a secondary showers which induced one or more nodes to trigger (Figure 1). With an accurate primary vertex reconstruction, a sin-

[a]With the exception of CC ν_e where the sum of the initial hadronic and the subsequent electromagnetic shower was used.

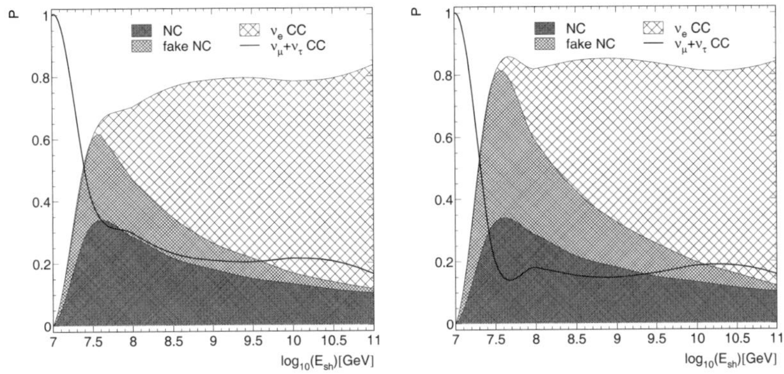

Figure 1. Probability of identifying the type of primary interaction as a function of true energy of the primary shower. The identification of ν_e CC events depends on the ability to distinguish overlapping hadronic and electromagnetic showers originating from the same location. The misidentified CC events ("fake NC") are due to partially contained events, i.e. the primary interaction occurs at or near the edge of the instrumented volume and the secondary lepton travels away from the detector center, and events where only a small fraction of ν's original energy was carried away by the secondary lepton, rendering it too weak to generate subsequent radio pulses. In the left panel it was assumed that only one secondary node is required to identify event as CC, while in the right panel three nodes were required. This three node requirement addresses the possibility that the primary vertex reconstruction produced a poor result, thus forcing a more stringent test for secondary showers.

gle subsequent node triggering at a time consistent with the reconstruction result would indicate that the event was due to CC interaction.

The events identified as charged-current interactions can be further separated by neutrino flavor using ideas summarized in Table 1. The reconstruction of the width of the strongest secondary radio cone would identify it as being of either hadronic or electromagnetic shower origin. Figure 2, left panel, shows the probability of ν_μ or ν_τ generating an event with a large, hadronic or EM, secondary shower which triggered at least three antenna nodes. It can be seen that good flavor separation exists when the strongest secondary shower is electromagnetic, but the separation is not so clear for large hadronic showers. Additional test can be made from the observation that secondary muons will produce more secondary showers (mostly through e-pair production), so that the number of secondary showers detected can serve to identify neutrino flavor (Figure 2, right panel).

By combining information from two panels of Figure 2, one can achieve good flavor separation of CC events. Further investigations, which should

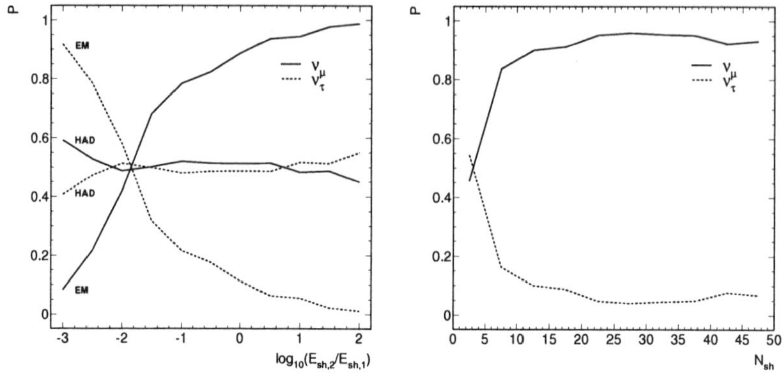

Figure 2. Left: Probability that ν_μ or ν_τ will generate an event with a large secondary shower as a function of relative energies of the secondary and the primary showers. Right: Probability that ν_μ or ν_τ will generate an event with the given number of observable secondary showers. Here, the shower was defined as detectable if it triggered two antenna nodes.

include vertex reconstruction algorithm, will better quantify the likelihood of correct flavor identification, including the identification of CC ν_e events.

Acknowledgments

This work was supported by NASA ROSS - UH grant NAG5-5387.

References

1. D. Saltzberg, these proceedings
2. A. Gazizov and M. Kowalski, DESY-04-101, astro-ph/0406439 (2004).
3. D. Chirkin and W. Rhode, hep-ph/0407075 (2004).
4. R. Engel, D. Seckel and T. Stanev, Phys. Rev. **D64**, 093010 (2001).
5. J. Alvarez-Muñiz, R. A. Vázquez, and E. Zas, Phys. Rev. **D62**, 063001 (2002).
6. J. Alvarez-Muñiz, R. A. Vázquez, and E. Zas, Phys. Rev. **D61**, 023001 (2001).
7. P. Gorham *et al.*, Phys. Rev. **D72**, 023002 (2005).

SIMULATION OF A HYBRID OPTICAL/RADIO/ACOUSTIC EXTENSION TO ICECUBE FOR EHE NEUTRINO DETECTION

J. A. VANDENBROUCKE,[*] [†] D. BESSON,[‡] S. BÖSER,[§] R. NAHNHAUER,[§] AND P. B. PRICE[†]

Astrophysical neutrinos at ~EeV energies promise to be an interesting source of information for astrophysics and particle physics. Detecting the predicted cosmogenic ("GZK") neutrinos at 10^{16} - 10^{20} eV would test models of cosmic ray production at these energies and probe particle physics at ~100 TeV center-of-mass energy. While IceCube could detect ~1 GZK event per year, it is necessary to detect 10 or more events per year in order to study temporal, angular, and spectral distributions. The IceCube observatory may be able to achieve such event rates with an extension including optical, radio, and acoustic receivers. We present results from simulating such a hybrid detector.

1. Introduction

Detecting and characterizing astrophysical neutrinos in the 10^{16} eV to 10^{20} eV range is a central goal of astro-particle physics. The more optimistic flux models in this range involve discovery physics including topological defects and relic neutrinos. Detecting the smaller flux of cosmogenic (or Greisen, Zatsepin, and Kusmin, "GZK") neutrinos produced via ultra-high energy cosmic rays interacting with the cosmic microwave background would test models of cosmic ray production and propagation and of particle physics at extreme energies. With ~100 detected events, their angular distribution would give a measurement of the total neutrino-nucleon cross section at ~100 TeV center of mass energy, probing an energy scale well beyond the reach of the LHC. Hence, as a baseline, a detector capable of detecting ~10 GZK events per year has promising basic physics potential. If any of the more exotic theories predicting greater EeV neutrino fluxes is correct, the argument in favor of such a detector is even stronger.

To detect ~10 GZK events per year, a detector with an effective volume of ~100 km^3 at EeV energies is necessary. There are three methods of ultra-high

[*]Presenting author, justin@amanda.berkeley.edu
[†]Dept. of Pysics, University of California, Berkeley, CA 94720, USA
[‡]Dept. of Physics and Astronomy, University of Kansas, Lawrence, KS 66045-2151, USA
[§]DESY, D-15738 Zeuthen, Germany

energy neutrino detection in solid media: optical, radio, and acoustic. Optical Cherenkov detection is a well-established technique that has detected atmospheric neutrinos up to 10^{14} eV and set astrophysical neutrino flux limits up to 10^{18} eV[1]. Radio efforts have produced steadily improving upper limits on neutrino fluxes from 10^{16} eV to 10^{25} eV[2]. Acoustic detection efforts are at an earlier stage, with one limit published thus far from 10^{22} to 10^{25} eV[3].

A GZK event rate of $\sim$1 per year is expected at the IceCube 1 km^3 neutrino telescope currently under construction. It is possible to extend this by adding more optical strings for a modest additional cost[4], but the increase in the sensitivity of the optical technique with energy does not compensate for the rapidly decreasing flux. The radio and acoustic methods have potentially large effective volumes at this energy, but neither has detected a neutrino. If radio experiments claim detection of a GZK signal, it will be difficult to confirm. However, it may be possible to bootstrap the large effective volumes of radio and acoustic detection the optical method by building a hybrid detector that can detect a large rate of radio or acoustic events, a fraction of which are also detected by an optical detector. A signal seen in coincidence between any two of the three methods would be convincing. The information from multiple methods can be combined for hybrid reconstruction, yielding improved angular and energy resolution.

We simulated the sensitivity of a detector that could be constructed by expanding the IceCube observatory currently under construction at the South Pole. The ice at the South Pole is likely well-suited for all three methods: Its optical clarity has been established by the AMANDA experiment[1], and its radio clarity and suitability for radio detection in the GZK energy range has been established by the RICE experiment[2]. Acoustically, the signal in ice is ten times greater than that in water. Theoretical estimates indicate low attenuation and noise[5], and efforts are planned to measure both with sensitive transducers developed for glacial ice[6]. We estimate the sensitivity of such a detector by exposing all three components to a common Monte Carlo event set and counting events detected by each method alone and by each combination of multiple methods.

2. Simulation

IceCube will have 80 strings arrayed hexagonally with a horizontal spacing of 125 m. In previous work[4], the GZK sensitivity achieved by adding more optical strings at larger distances ("IceCube-Plus") was estimated, and the possibility of also adding radio and acoustic modules was mentioned. Here we consider an IceCube-Plus configuration consisting of a "small" optical array overlapped by a "large" acoustic/radio array with a similar number of strings but larger horizontal spacing. The optimal string spacing for GZK detection was found to be ∼1 km for both radio and acoustic strings. This coincidence allows the two methods to share hole drilling and cable costs, both of which are dominant costs of such arrays.

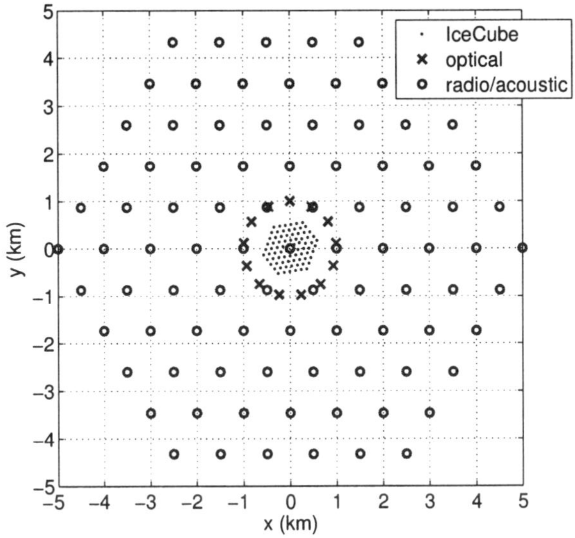

Figure 1. Geometry of the simulated hybrid array.

The geometry of the simulated array is shown in Fig. 1. We take the optical array to be IceCube as well as a ring of 13 optical strings with a 1 km radius, surrounding IceCube. All optical strings have standard IceCube geometry: 60 modules per string, spaced every 17 m, from 1.4 to 2.4 km depth. Encompassing this is a hexagonal array of 91 radio/acoustic strings with 1 km spacing. Each radio/acoustic hole has 5 radio receivers, spaced every 100 m from 200 m to 600 m depth, and 300 acoustic receivers, spaced every 5 m from 5 m to 1500 m depth. At greater depths both methods suffer increased absorption due to the warmer ice. The large acoustic density per string is necessary because the acoustic radiation pattern is thin (only ∼10 m thick) in the direction along the shower. The array geometry was designed to seek an event rate of ∼10 GZK events per year detectable with both radio and acoustic independently.

Between 10^{16} and 10^{20} eV, the neutrino interaction length ranges between 6000 and 200 km[7], so upgoing neutrinos are efficiently absorbed by the Earth and only downgoing events are detectable. Here we assume all upgoing neutrinos are absorbed before reaching the fiducial volume, and no downgoing neutrinos are;

we generate incident neutrino directions isotropically in 2π sr. Vertices are also generated uniformly in a fiducial cylinder of radius 10 km, extending from the surface to 3 km depth.

The Bjorken parameter $y = E_{had}/E_\nu$ varies somewhat with energy and from event to event, but we choose the mean value, $y = 0.2$, for simplicity. The optical method can detect both muons and showers, but here we only consider the muon channel; simulation of the shower channel is in progress. The radio and acoustic methods cannot detect muon tracks but can detect electromagnetic and hadronic showers. Under our assumptions of constant y and no event-to-event fluctuations, all flavors interacting via both CC and NC produce the same hadronic shower. Electron neutrinos interacting via the charged current also produce an electromagnetic cascade which produces radio and acoustic signals superposed on the hadronic signals. However, at the energies of interest here, electromagnetic showers are lengthened to hundreds of meters by the Landau-Pomeranchuk-Migdal effect. This weakens their radio and acoustic signals significantly, and we assume they are negligible.

For simulation of the optical response, the standard Monte Carlo chain used in current AMANDA-IceCube analyses[1] was performed. After the primary trigger requiring any 5 hits in a 2.5 μs window, a local coincidence trigger was applied: Ten local coincidences were required, where a local coincidence is at least two hits on neighboring or next-to-neighboring modules within 1 μs. Compared with a previous work[4], we used an updated ice model with increased absorption, which may account for our factor of $\sim$2 lower effective volume.

Each simulated radio "receiver" consists of two vertical half-wave dipole antennas separated vertically by 5 m to allow local rejection of down-going anthropogenic noise. We assume an effective height at the peak frequency (280 MHz in ice) equal to 10 cm, with $\pm$20% bandwidth to the –3 dB points. As currently under development for RICE-II, we assume optical fiber transport of the signal to the DAQ, with losses of 1 dB/km (measured) through the fiber. The electric field strength $E(\omega)$ is calculated from the shower according to the ZHS prescription[8,9]. Frequency-dependent ice attenuation effects are incorporated using measurements at South Pole Station[10]. The signal at the surface electronics is then transformed into the time domain, resulting in a waveform 10 ns long, sampled at 0.5 ns intervals, at each antenna. Two receivers with signals exceeding 3.5 times the estimated rms noise temperature σ_{kT} (thermal plus a system temperature of 100 K) within a time window of 30 μs are required to trigger. Four are required for high-quality vertex reconstruction.

The unattenuated acoustic pulse $P(t)$ produced at arbitrary position with respect to a hadronic cascade is calculated by integrating over the cascade energy distribution. We use the Nishimura-Kamata-Greisen cascade parametrization, with an estimated λ (longitudinal tail length) parametrization[9]. The dominant mechanism of acoustic wave absorption in South Pole ice is theorized[5] to be molecular reorientation, which increases with ice temperature. Using a temperature profile measured at the South Pole along with laboratory absorption measurements, an absorption vs. depth profile was estimated. The predicted absorption length ranges from 8.6 km at the surface to 4.8 km at 1 km depth to 0.7 km at 2 km depth. The frequency-independent absorption is integrated from source to receiver and applied in the time domain.

South Pole ice is predicted to be much quieter than ocean water at the relevant frequencies ($\sim$10-60 kHz), because there are no waves, currents, or animals. Anthropogenic surface noise will largely be waveguided back up to the surface due to the sound speed gradient in the upper 200 m of uncompactified snow ("firn"). We assume ambient noise is negligible compared to transducer self-noise. Work is underway to produce transducers with self-noise at the 2-5 mPa level[6]. For comparison, ambient noise in the ocean is $\sim$100 mPa[3]. The acoustic trigger used in this simulation required that 3 receivers detect pressure pulses above a threshold of 9 mPa.

3. Results and Conclusion

Ten-thousand events were generated at each half-decade in neutrino energy in a cylinder of volume 942 km^3. For each method and combination of methods, the number of detected events was used to calculate effective volume as a function of neutrino energy (Fig. 2). This was folded with a GZK flux model[11,12] and a cross-section parametrization[7] to estimate detectable event rates (Fig. 2). We use a flux model which assumes source evolution according to $\Omega_\Lambda = 0.7$[12]. For radio and acoustic, and their combination, all flavors and both interactions were included. For those combinations including the optical method, only the muon channel has been simulated thus far; including showers will increase event rates for these combinations.

It may be possible to build an extension like that considered here for a relatively small cost. Holes for radio antennas and acoustic transducers can be narrow and shallow. The necessarily large acoustic channel multiplicity is partially offset by the fact that the acoustic signals are slower by five orders of magnitude, making data acquisition and processing easier.

The IceCube observatory will observe the neutrino universe from 10's of GeV to 100's of PeV. Our simulations indicate that extending it with radio and acoustic strings could produce a detector competitive with other projects optimized for

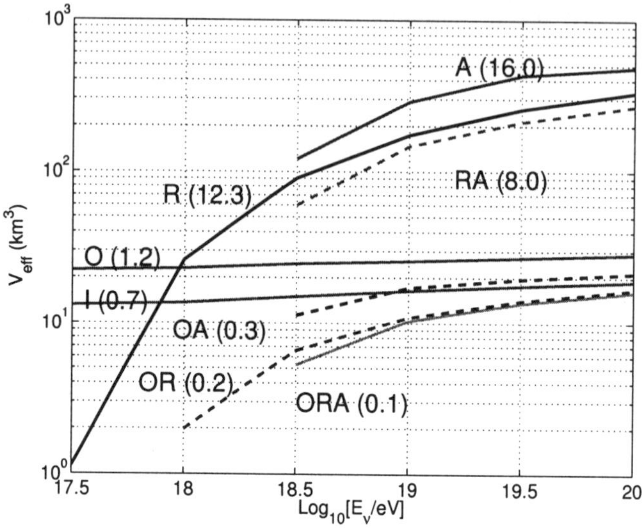

Figure 2. Effective volume for each of the seven combinations of detector components, as well as for IceCube alone ("I"). GZK event rates per year are given in parenthesis. Note that different channels were used for different combinations (see text).

high-statistics measurements of GZK neutrinos but with the unique advantage of cross-calibration via coincident optical-radio, optical-acoustic, and radio-acoustic events.

References

1. K. Woschnagg for the AMANDA Collaboration, Nucl. Phys. B 143, 343 (2005). M. Ackermann et al., Astropart. Phys. 22, 339 (2005).
2. I. Kravchenko et al., Astropart. Phys. 20, 195 (2003). P. Gorham et al., Phys. Rev. Lett. 93, 041101 (2004). N. Lehtinen et al., Phys. Rev. D 69, 013008 (2004).
3. J. Vandenbroucke, G. Gratta, and N. Lehtinen, ApJ. 621, 301 (2005).
4. F. Halzen and D. Hooper, J. Cosmol. Astropart. Phys. 01, 002 (2004).
5. P. B. Price, astro-ph/0506648.
6. S. Böser et al., these proceedings.
7. R. Gandhi et al., Phys. Rev. D 58, 093009 (1998).
8. E. Zas, F. Halzen, and T. Stanev, Phys. Lett. B 257, 432 (1991). E. Zas, F. Halzen, and T. Stanev, Phys. Rev. D 45, 362 (1992).
9. J. Alvarez-Muñiz and E. Zas, Phys. Lett. B 434, 396 (1998).
10. S. Barwick, D. Besson, P. Gorham, and D. Saltzberg, to appear in J. Glac.
11. R. Engel, D. Seckel, and T. Stanev, Phys. Rev. D 64, 093010 (2001).
12. R. Engel, D. Seckel, and T. Stanev, ftp://ftp.bartol.udel.edu/seckel/ess-gzk/

ARENA ROUND TABLE DISCUSSION
19.05.2005

SUMMARY

1. U. Katz : Introductory remarks (see also slides)

1.1.) Some Observations During the Workshop

- Radio and acoustic methods offer promising options for future high(est)-energy astroparticle detectors.
- Many efforts going on in these fields, but
  - -often in parallel, no obvious coordination between overlapping activities
  - -large synergy potential:
    - → between different acoustic scientists/groups;
    - → between different radio scientists/groups;
    - → between radio and acoustic activities.
- Path to projects unclear:
  - -funding beyond R&D ?
  - -selection of / decision on projects? Criteria?
  - -cooperative / collaborative structures?

1.2.) Example Topics for this Discussion

- Networking structures
  - -Information exchange (mailing list? common web page?)
  - -Permanent working groups?
- Tool development and dissemination
  - -simulation, filter and reconstruction code
  - -hardware: exchange of experience and developments
  - -common test environments
- Organization "on the political scene"
  - -National funding: What input is needed?
  - -Europe: FP7 activities? / US ?

1.3.) Specific project cooperations suggested

- Radio detection
  → satellite-borne CR & neutrino experiments (Tsarev)
  → telescope observation of neutrinos (Dagkesamanskii + ...)
  → radio activities in Antarctica (Dagkesamanskii, Dedenkov et al.)
- Acoustic detection
  → Kamchatka array and MG-10M antenna (Zheleznykh et al.)
- Cooperation with industry: acoustic and radio detection in salt domes (van den Berg)

2. Discussion:

Danaher:	e-mail list of workshop participants should be distributed → available already via participants list at ARENA web page
Tsarev:	Supporting contact and networking
Falcke:	European network very important, should try to use FP7, need volunteers to do the paperwork
Nahnhauer:	Common software development for processing, reconstruction and simulation would have many advantages for acoustic and radio projects
Tsarev:	Political support is important, need recommendations to Russian government
Rostovtzev:	Should have write ups of capabilities of different groups
Lyashuk:	Cooperation with seismology/ecology could be supportive
Falcke:	Do we need a road map? We are not ready for committees to judge on
Learned:	Agrees to Falcke, don't need road map, all experiments start with small groups and initiatives not with large collaborations

Seckel: No, we need resources for ice or/and salt

Sloan: Combine all salt activities in one group

Salzberg: Has happened already

Nahnhauer: SalSa is just a positive example, other topics done several times with similar results (proton beam acoustic measurements e.g.), combined efforts could improve results

Anton: Water – a lot is known already- compare background at different sites

Helbig: Start a list of working groups, circulate it

Lahmann: Not "working groups" but "News groups" are needed

v.d. Berg: Use corresponding FP6 proposals, there groups are already defined

Falcke: Need some aim, e.g. continuation of workshops, have shepherds that watch their field in between the workshops

Seckel: Working groups not an effective way for the thing we need for a 1000 km^3 detector starting in 5-10 years from now.

Anton: Not so far yet, need to evaluate problems in different working groups

Tsarev: Supportive to working groups

Budnev: Invite all to test equipment at Lake Baikal

Riccobene: Catania test site also available for tests of other groups

v.d. Berg: No competition between US/Europe in salt, choose best dome, exchange equipment

Saltzberg: Groups have largely overlapping plans

Falcke: All air-radio groups will merge in AUGER

3. Summary and conclusions

The opinion about the type of further cooperation in the field is not unique between the participants of the workshop

Most European groups support closer cooperation by networking structures and working groups. It may be advantageous to use existing activities in the frameworks FP6/FP7 of the European community.

If a collaboration has already been formed to realize a big project, which has a strong kernel and allows weakly connected satellite groups, all necessary cooperation may be organized within this structure. A positive example of this type is the SalSa project.

Self-structuring of both types will continue. The field is not ready for committees to judge on.

4. List of working/news groups proposed during the discussion

Acoustic sensors
Radio sensors
Electronics
Sensor calibration
Signal processing
Simulation

ARENA 2005 CONFERENCE SUMMARY

JOHN G. LEARNED

Department of Physics and Astronomy, University of Hawaii, 2505 Correa Road, Honolulu, Hi 96822, USA

This meeting, as the reader sees in the preceding written contributions, represented an exciting new step in a budding revolution in extremely high energy cosmic ray and neutrino physics. The focus upon acoustic and radio detection techniques, reveals a rapidly expanding interest and real progress. Most of the basic ideas have been known for many years, but it is only now that they are at last beginning to be exploited. The reasons for this are several, ranging from the advance of technology to scientific focus of the community. The former comes largely from electronics and communications technology progress. The latter comes about due to the "neutrino revolution" and the campaign to understand the highest energy (GZK) cosmic rays. At least 25 projects are in various stages of dreaming through construction. Positive detections would seem not far in the future.

1. Introduction

Two trends in particle physics and astrophysics, aided by technological advances, have driven a renewed and vigorous campaign to explore ultra-high energy cosmic rays, and neutrinos in particular. The first has been the tremendous stimulation of the discoveries in neutrino physics, triggered by the 1998 announcement by Super-Kamiokande of the discovery of neutrino oscillations and hence almost inescapably finite though small neutrino mass. The second large trend has been the claims of observation of cosmic rays of energies beyond the supposed end of the cosmic ray spectrum at about 10^{20} eV, the failure to see the GZK cutoff. The examination of the escape routes from the GZK enigma has brought about realization that neutrinos must accompany almost all cosmic ray sources solutions, in relative abundance. And even in the case that the GZK cutoff exists, we know there still must be accompanying neutrino fluxes. Measurement of these fluxes will aid in figuring out what mechanisms are able to produce these amazingly high energy cosmic rays. It remains true that 100 years after the discovery of cosmic rays we still do not know their origin at the highest energies. For these reasons and more, people have started to search for new means to study the very highest energy neutrinos.

Hence the particular focus of this meeting, examining initiatives to utilize two as yet (largely) unexploited phenomena to tag these rare but substantial depositions of energy from such neutrino interactions.

1.1 General Comments

We may well be entering heroic times for radio and acoustic detection of elementary particles, neutrinos particularly. In one way this is a renaissance, since the basic notions were examined in the 1970's and earlier. But these techniques stalled in application, mainly because the notion of such high energy neutrinos was considered to be far too speculative, and hence people (such as this author) pursued optical techniques which permitted a lower energy entrance into neutrino detection. The nice thing about Cherenkov light detection is that it works well in a transparent medium (namely water or ice) from energies of a few MeV up to the highest energies, only limited by the affordability of instrumenting huge volumes with optical detectors. The decades-long campaign of Baikal, DUMAND-US, AMANDA, NESTOR, ANTARES, and NEMO, plus ICECUBE and hopefully KM3, will we hope soon bear fruit (and indeed we had all better wish ICECUBE great success). But we can see the need for detectors with effective volumes in the range of hundreds to thousands of cubic kilometers, and this is the main chance for these new/old techniques.

Even in the mid-1970's we knew that the threshold for radio and acoustic detection in realistic media was in the range of 10 PeV or more. In that era, the idea of the existence of super-high energy neutrinos was mostly greeted with ridicule, as this author knows first hand from the reviews of his (and others') proposals for exploratory experiments thirty years ago. For these reasons, the novel methods were (mostly) put aside for nearly twenty years, and are only now being seriously pursued.

Of course we all hope that grand discoveries follow. This is an adventurous field: dangerous for careers perhaps, but providing challenges and tantalizing opportunities. It presents also a great training ground for students who have the rare opportunity to participate in the emergence of a new field.

And, it is amazing that even though we appear to be just getting started, all highest energy cosmic neutrino flux limits in the literature already come from radio and acoustic studies, as we heard at this meeting (Forte', GLUE, RICE, SAUND). It perhaps bears remarking that setting limits can be easier than making discoveries of new phenomena... knowing what you saw is vital and may require more detailed observations. We have no UHE discoveries yet, though, and not even a hints of high energy extraterrestrial neutrinos. There is a danger here. The lessons of HE and VHE Gamma Ray Astronomy are that sometimes new ventures which seem sure to succeed, based upon enthusiastic and creative theoretical models, may take decades to reach maturity. For attendees at this meeting the message is that we should keep a focus upon incorporating bread-and-butter science goals as well as the high payoff explorations into new terrain, and we should pay great attention to ties to particle physics, seek creative funding opportunities, and be ever aware that international cooperation vital and natural.

1.2 Comments on History of these Endeavors

I would like to append a comment on the history of these endeavors, adding to earlier remarks here (see Zheleznykh's talk) and at the previous meeting in this series at UCLA, RADHEP-2000 (Ref. 1). Aside from the visionaries who first proposed the mechanisms we are now hoping to exploit, I think we ought not to forget the importance of two great facilitators of neutrino physics, beyond their own direct and substantial contributions, M. A. Markov and Frederick Reines. From these far-sighted and great scientists flowed stimulation for many of the recent endeavors (and triumphs) of neutrino physics (and nucleon decay....), roles perhaps underappreciated. Great ideas are necessary, but so are the facilitators and individuals who realize those dreams.

Also, the importance of mistakes is often conveniently forgotten. Sometimes miscues are ultimately for better (eg. in the gravity wave business) in stimulating a new field, sometimes for worse (eg. cold fusion). In the area of present consideration, early overestimates of acoustic signal magnitude caused great excitement, as did claims of unexpected (non-thermal) acoustic mechanisms with large output. These inspired our (and others') initial forays to the accelerator to examine acoustic emission by a particle shower. These are clearly good things to have carried out, but had we known that there were no new mechanisms with more encouraging acoustic amplitude, our enthusiasm for the efforts might have been dampened. One should keep in mind that scientific progress is a crooked path, a matter most often neglected in the textbooks which illustrate only the "yellow brick road". It may also be worth noting that we have had some positive surprises in this thirty year endeavor towards neutrino astronomy. One of these is that the water and ice transparency turns out to be far better than was known in 1970's (when we expected 10 m absorption lengths, which turned out nearer 100 m in the deep ocean and deep ice).

1.3 Requirements for UHE Rare Particle Detection

In order to get into the business of extremely high energy neutrino detection one needs, for economic reasons, natural targets, natural radiation mechanisms, and detection at large impact parameter. We cannot afford targets which have to be processed industrially in the scale of gigatons ($\$1/l \rightarrow \$10^{12}/km^3$), and we cannot cover areas with counters in the scale of millions of square meters ($\$1000/m^2 \rightarrow \$10^9/km^2$). The basic requirement is that the event detection area be, for directional radiation,

$$A_{event} \sim (\Lambda_{transverse})^2 = (\Lambda_{atten} * \sin(\theta))^2$$

where the angle θ is the generally the Cherenkov or particle scattering angle, and the Λ_{atten} is the attenuation length for the radiation in the observation medium. This way of approaching the range of possibilities leads us to three types of radiation: optical, radio, and particles. The two electromagnetic regions are the well known regions for transparency of many materials: between roughly 300nm and 500nm optically, and between 1 cm and 10m in the radio. For secondary particle cascades spreading out laterally, as in extensive air showers (EAS) the lateral scale is of the order of 100 m growing to a few km at the highest energies.

Of course the signals grow with energy (linearly with E for particle and acoustic, as the E^2 for radio), but they also fall off with 1/distance or 1/distance^2. The backgrounds that limit detection differ as well, and it is hard to generalize. Finally, the detection elements themselves have a wide range of costs. While there have been some attempts in the literature to compare the efficacy of the various means of detection dependent upon only the physics, such efforts miss the mark of practical comparisons which are always limited by the hard facts of economics. See Figure 3 and discussion below for more upon this topic.

We know of only the following methods to realize huge high energy neutrino detectors on earth, as listed in Table 1. Note that most of these are also high energy cosmic ray detectors, with guaranteed signals.

Table 1: Various detection techniques which are operating or proposed to be used to detect natural high energy neutrinos.

Detection Technique	*Practical Energy Regime*
Particle Sampling (EAS arrays: AUGER....)	100 TeV – 1 ZeV
Fluorescence/Scintillation (Fly's Eye type detectors)	1 EeV – 1 ZeV
Optical Cherenkov in water or ice (SuperK, Baikal, AMANDA, etc.)	1 MeV – 100 PeV
Optical Cherenkov in air (Mt. Hopkins, HESS, MAGIC, NUTEL, ASHRA)	10 GeV – 10 TeV
Radio Cherenkov (GLUE, RICE, ANITA, SALSA)	1 EeV – 1 YeV
Radio Geo-Synchrotron (LOPES, etc.)	
Thermo-Acoustic Radiation (SAUND, SADCO, ACORNE,...)	>100 PeV

2. Summary of Projects

There were so many projects described at this meeting that I cannot cover all in a short summary, and for this I apologize. Please see the wonderful presentations documented earlier in this volume. Many of these projects are just at this time starting, only a few physics result are available, and as said none have yet shown indications of cosmic neutrinos. The following Table 2 is an attempt to summarize the programs in terms of mechanism and venue employed. The variety of activity is truly amazing.

2.1 Atmospheric Cherenkov experiments

In the last decade the air Cherenkov method of detecting gamma ray induced showers in the atmosphere has blossomed. There are many existing telescopes and telescope systems: CANGAROO, CAT, CLUE, HEGRA, Narrabri, PACT, Whipple Gamma-Ray Observatory on Mt. Hopkins, Arizona; plus new telescope projects: HESS, MAGIC, VERITAS, Subaru-Gamma, MACE. Also solar power facilities are being employed as light collectors: CELESTE, STACEE, CACTUS, GRAAL. There is also the MILAGRO project in New Mexico which employs Cherenkov detection in a covered water tank. None have much capability for neutrino detection, particularly at high energies, where they simply have insufficient sensitive volume. There have been projects, such as NUTEL, proposed to detect the air Cherenkov emission from tau induced showers exiting a mountain mass. The ASHRA Project, mentioned below, also aims to cover this opportunity, along with viewing the entire sky.

Table 2. Summary of projects according to detection technology and environment.

Medium/ Mechanism	***Atmos***	***Water, Ocean/Lake***	***Deep Ice***	***Salt Domes***	***Lunar Regolith***	***Bulk Earth/ Moon***
Particle Sampling	EAS Arrays… AUGER	IMB, Kam, SK, MILAGRO	Ice Top	x	??	CWI, KGF,…
Fluor/Scint	Fly's Eye, HiRes, Auger, TA, Ashra, EUSO, OWL	x	x	x?	x	x
Optical Cherenkov	HESS, MAGIC, Ashra, VERITAS, … + 17 more!	Baikal, DUMAND, NESTOR, ANTARES, NEMO	AMANDA, ICECUBE	??	x	x
Radio Cherenkov	?	x	RICE, Forte, ANITA	SALSA-R	GLUE, Kalyazin, Westerbrk LORD	x
Radio Geo-Synch	LOPES, CODALEMA AUGER+	x	x	x	x	x
Acoustic Radiation	x	SADCO, SAUND, ACORNE, ANTARES, NEMO, Baikal	ICECUBE-A	SALSA-A	x	(Seismic) ??

2.2 Atmospheric fluorescence experiments

The present or proposed air fluorescence projects of the Fly's Eye type are:

•Ashra (All-sky Survey High Resolution Air-shower detector), under construction in Hawaii

•Auger Project, Fluorescence Group in Argentina in operation.

•EUSO (Extreme Universe Space Observatory) (a proposed space experiment on ISS to observe air shower fluorescence light).

•HiRes (High Resolution Fly's Eye Cosmic Ray Detector) operating in Utah, though changing locations.

•OWL (Orbiting Wide-angle Light collectors) (a plan to build a pair of satellites for air shower detection).
•Telescope Array (Cherenkov and fluorescence light), under construction, also Utah.
All of these have the potential to observe nearly horizontal showers induced by >EeV neutrino interactions. Though the target volume is huge, the solid angle is very small, and the projected neutrino detection rates are not very encouraging. Such detectors suffer from being only able to operate on dark, clear and moonless nights, achieving livetime fractions of around 10%. It does not appear that these detectors are likely to be the first to see cosmic neutrinos, but surprises are not ruled out.

2.3 Air shower experiments with particle detectors

There have been many significant (and many more small) EAS arrays over the last 50 years, and continuing into the future: AGASA (Akeno Observatory Japan), ARGO-YBJ (new experiment under construction in Tibet), ASCE (Sydney, operational 1989-1991), Buckland Park Extensive Air Shower Array (Australia) (operational 1994-1998), CASA-MIA (operational 1990-1998), EAS-TOP (Italy, above the Gran Sasso underground laboratory, until April 2000), Haverah Park (until 1993), GREX array (Haverah Park, operational 1986-1995), HEGRA (operational 1988-2002), KASCADE (and KASCADE-GRANDE.), MILAGRO (Water Cherenkov experiment near Los Alamos), Mt. Norikura Observatory in Japan, Pierre Auger Project (South in operation, North proposed), SPASE 2 (South Pole), SUGAR (operational 1968-79), Tian-Shan Mountain Cosmic Ray Station, Tibet AS-gamma experiment (scintillation counter array 1995-present). These experiments intercept the particle cascades generated by primary cosmic rays interacting high in the atmosphere. They are thus most sensitive to vertically arriving showers, and have little ability to discriminate any neutrinos in their sample. Again, as with the fluorescence arrays, these instruments might be able to discern near horizontal showers with shower "age" too little to be due to incoming primaries. The same remarks apply as with fluorescence detectors. Of course the advantage of the EAS arrays is that they operate all the time.

3. Physics Simulations and Calibrations

In the past, there has been much uncertainty in the predicted performance of these new techniques due to the lack of sufficiently detailed and grounded simulation programs. Not only does the physics extend into terra incognita, at least for the primary interactions, but the particle numbers are so huge (1011 at

10^{20} eV) that the particle tracking routines used at lower (accelerator) energies simply cannot produce results in reasonable times. Nonetheless great progress has been made in all areas (GEANT4 + CORSIKA), and particularly in vetting these simulation codes at the dense KASCADE array. At this meeting, Dedenko reported a nice hybrid approach which should help in this area. While there has been a lot of effort on showers in the atmosphere, there is much yet to do for showers in solids, both for acoustic and radio signal production. This reviewer notes that time domain (as opposed to frequency domain) studies in the radio are needed.

We also need studies of the high energy physics potential of these research initiatives: studies of ultra high-energy neutrino cross sections, flavor separation, composition effects, etc.

There have been excellent laboratory demonstrations/calibrations for radio Cherenkov emission (at SLAC and ANL), but we need work on acoustic signals from cascades.

4. Detector Technology

Much progress has been in radio antenna design and testing. Work is ongoing, but a lot of antenna design is still an art. There are some ongoing problems due to our special need for impulse response devices. High Energy Physicists need to learn to deal with analogue signals. There are interesting new options in signal processing for everyone, as highlighted in Danaher's talk. (On the other hand engineers need to learn about non-CW signals).

For acoustic detection, fortunately the piezoelectric technology well developed by military and oil industry, and commercial devices are available and relatively inexpensive. (Such items are produced in quantities which are orders of magnitude greater than photomultiplier tubes). Particle physicists starting out in this area would be well advised to tap the existing expertise and hardware.

4.1 On Site Simulators and Calibrators

Knowing the real energy of the events one sees, to a precision of order 10% or better, is of course vital (particularly when dealing with steeply falling spectra). At lower energies the neutrino detectors can take advantage of radioactive sources and cosmic ray muons for physics calibrations at energies from ~1 MeV to ~100 GeV. Above these energies there are not many bench marks. One lovely marker would be the observation of resonant electron anti-neutrino events

(Glashow resonance) at 6.3 PeV. Otherwise we face the same problem as the EAS and fluorescence detectors, having to bootstrap from lower energies and other types of energy calibrations.

Sources of known electrical or mechanical energy can be had (pingers & collapsing bulbs) for acoustical detection, and electromagnetic transmitters (sparkers or pulse driven antennae) for radio are fairly straightforward. Shower simulation (using, say, a portable pulsed electron accelerator) is not simple in the field. Ideas have been put forward (pulse heated wire "zapper", pulsed laser) but nothing demonstrated yet.

Of course, co-location with another array would provide cross calibration, a worthy task given the current disagreement between fluorescence and ground array measurements of the 100 EeV cosmic rays.

4.2 Media Properties

There is much work needed in all areas of studying the medium properties, both for acoustics and radio. Radio wave propagation in Antarctic ice is known fairly well, but still needs work for accurate data analysis. There exists a lot of variation however, in the reported salt radio attenuation measurements, see Figure 1 below. Though (some) salt domes (salt beds generally have too many inclusions) are probably alright for experimental use, real measurements are needed *in situ*.

Acoustical attenuation is very well known in sea water, but in ice we mostly have only theory. In the ocean, work is needed to study acoustical signal propagation in the real world of temperature and salinity fluctuations, presenting a variable index of refraction, which limits the range of coherent signal detection.

Pure water absorbs wave energy due to its viscosity. In sea water, a pressure wave shifts chemical equilibrium between a molecule and ions, taking energy from wave: $B(OH)_3 = B^{3+} + 3\ (OH^-)$ (relaxation freq. ≈ 1 kHz) and $MgSO_4 = Mg^{2+} + SO_4^{2-}$ (relaxation freq. ≈ 100 kHz, as illustrated in Figure 2. For Antarctic ice we have only theory, and a feasibility demonstration needs to be made.

4.3 Comparison of Techniques

The various methods of attempting to measure ultra high energy neutrinos each have their merits, and are developed to differing degrees. This reviewer hopes that all of them will be pursued, since this is an extremely difficult business, and we know that cross checks are needed. Justin Vandenbroucke has made a comparison of optical, radio and acoustic arrays as shown in Figure 3. One sees that the crossover is in the range of 1017-18 eV, with acoustic and radio arrays exceeding optical arrays by more than an order of magnitude at the higher energies. Others (Gorham, Price) have attempted to make such plots based upon inherent signal-to-noise, which neglects the inevitable issue of practicality and cost. I like Vandenbroucke's plot because it represents an attempt to compare possible arrays in the same venue, Antarctic ice. Indeed the acoustic array he includes here seems to be much less costly than the optical array, for example, so his plot perhaps underestimates the potential for such application. The dispute about whether an acoustic array will be "better" than some similarly expensive radio array at the highest energies is not yet resolved, and will depend upon environmental measurements, and realistic costing of array designs.

5. Conclusion

We have seen from the number and quality of presentations at this workshop that there is now a wonderful level of activity in acoustic and radio detection. New ideas are being actively exploited, ranging from deep ice and salt, to the deep earth and sky. At this meeting, perhaps the most exciting news is the rapid progress in radio EAS detection, which may open channels for both neutrino and cosmic ray studies. Much work remains to do in studying media, technology, array design, and simulations. However, some projects are now close (year timescale) to having crucial results (ANITA, ICECUBE, AUGER). These previously novel research avenues seem to be opening up and we await the progress ahead with great expectations.

Acknowledgments

Many thanks to our hosts at DESY-Zeuten, conference organizer Rolf Nahnhauer, and his colleagues and helpers who made this a most intellectually stimulating and enjoyable meeting. I think everyone looks forward to next meeting in Britain.

References

1. RADIO DETECTION OF HIGH ENERGY PARTICLES: First International Workshop; RADHEP 2000, Editors David Saltzberg and Peter Gorham, *AIP Conference Proceedings* 579Los Angeles, California, USA, 16-18 November 2000, Published July 2001; ISBN 0-7354-0018-0.

2.First Informal Mini-workshop on Acoustic Cosmic Ray and Neutrino Detection, *Physics Department, Stanford University, Sept 13-14, 2003, hosted by Giorgio Gratta and Justin Vandenbroucke,* http://saund.stanford.edu/saund1/workshop/slides/index.html

3.Meeting to discuss radio neutrino detection at SLAC, February 2005, http://www.physics.ucla.edu/astroparticle/salsa/slacfeb05/talks_feb05.html

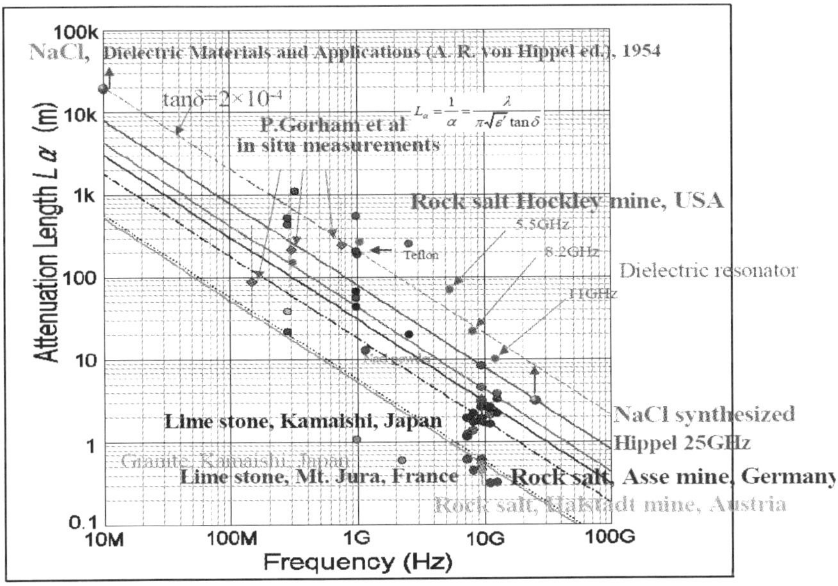

Figure 1. Radio attenuation versus frequency in salt. Note large spread of measurements. From Chiba, these Proceedings.

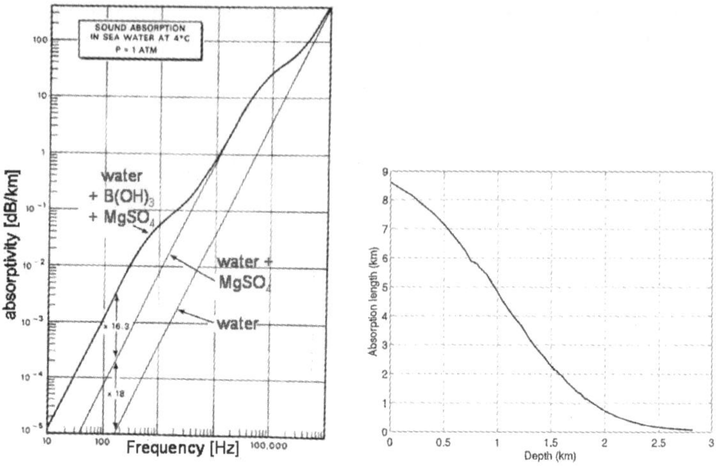

Figure 2. a) Acoustic attenuation versus frequency in seawater. b) Predicted depth (temperature)-dependent ice acoustic absorption at
~10 kHz. From Price, these Proceedings.

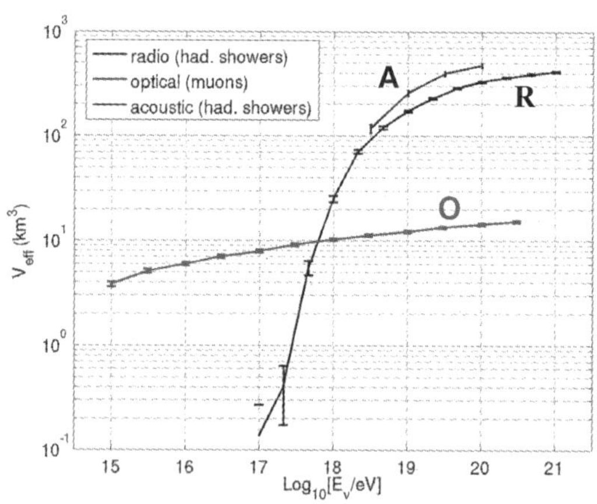

Figure 3. Rough comparison of the energy reach of radio, optical and acoustic detectors in an array at the South Pole. From Vandenbroucke, *et al*, these Proceedings.

ARENA WORKSHOP PICTURES

- Workshop participants in front of the DESY canteen
- Impressions from the workshop lectures
- Workshop excursion by pleasure boat through Berlin
- Workshop dinner on board

2
UKW
20

„Der Staat ist für die Menschen und nicht die Menschen für den Staat."
Albert Einstein
www.einsteinjahr.de

LIST OF PARTICIPANTS

Gisela Anton	University of Erlangen Nuernberg
Gerald Abich	Nautilus Marine Service GmbH
Markus Ackermann	DESY Zeuthen
Jaime Alvarez-Muniz	University of Santiago de Compostela (SPAIN)
Miguel Ardid	Universitat Politécnica de Valencia
Jose Bacelar	Kernfysisch Versneller Instituut
Julia Becker	Dortmund University
Karl-Heinz Becker	Universität Wuppertal
Elisa Bernardini	Desy-Zeuthen
Dave Besson	University of Kansas
Christian Bohm	Stockholm University
Nikolay Budnev	Irkutsk State University
Sebastian Böser	DESY, Zeuthen
Antonio Capone	Physics Department - University La Sapienza and INFN Roma

Lorenzo Cazon	Forschung Zentrum Karlsruhe. Institut für Kernphysik
Valerii Chechin	P.N.Lebedev.Physical Institute of RAS
Masami Chiba	Tokyo Metropolitan University
Amy Connolly	University of California, Los Angeles
Rustam Dagkesamanskiy	Pushchino Radio Astronomy Observatory of the Lebedev Physical Institute
Richard Dallier	SUBATECH (CNRS/IN2P3 - Universite de Nantes - Ecole des Mines de Nantes)
Sean Danaher	Northumbria University
Giulia De Bonis	Physics Department - University La Sapienza
Carlos de los Heros	Uppsala University
Leonid Dedenko	Faculty of Physics, M.V. Lomonosov Moscow State University
Jürgen Eisenblätter	Gesellschaft für Materialprüfung und Geophysik (GMuG)
Heino Falcke	ASTRON, Dwingeloo, The Netherlands
Manfred Fink	The University of Texas at Austin
Jan-Henrik Fischer	DESY, Zeuthen

Hartmut Gemmeke	Forschungszentrum Karlsruhe
Peter Gorham	University of Hawaii
Kay Graf	University of Erlangen
Allan Hallgren	Uppsala University
Andreas Haungs	Forschungszentrum Karlsruhe
Klaus Helbing	University of Erlangen
Reiner Heller	DESY, Zeuthen
Joerg Hoerandel	Universitaet Karlsruhe
Andreas Horneffer	Max-Planck-Institut fuer Radioastronomie
Tim Huege	Forschungszentrum Karlsruhe, Institut f. Kernphysik
Stephan Hundertmark	Stockholm Universitet
Jürgen Hößl	Universität Erlangen
Alexander Kappes	University Erlangen-Nuremberg
Timo Karg	University of Erlangen
Albrecht Karle	University of Wisconsin-Madison

Yakov Karlik	Kamchatka Hydro-Physics Institute
Uli Katz	Univ. Erlangen
Stefan Klepser	DESY, Zeuthen
Andrew Konstantinov	Skobeltsyn Institute of Nuclear Physics, Moscow State University
Anatoliy Kovalenko	Pushchino Radio Astronomy Observatory Lebedev Physical Institute
Naoko Kurahashi	Stanford University
Robert Lahmann	University of Erlangen
Rafael Lang	DESY, Zeuthen
John G. Learned	University of Hawaii
Vladimir Lyashuk	ITEP (Moscow)
Gerd Manthei	Gesellschaft für Materialprüfung und Geophysik (GMuG)
Hinrich Meyer	Univ. of Wuppertal
Predrag Miocinovic	University of Hawaii
Mauro Morganti	Dipartimento di Fisica and INFN of Pisa

Rolf Nahnhauer	DESY, Zeuthen
Christopher Naumann	Universität Erlangen
Steffen Nehls	Forschungszentrum Karlsruhe
Georgy Pan'kov	Irkutsk State University
Jonathan Perkin	University of Sheffield
Buford Price	University of California - Berkeley
Giorgio Riccobene	LNS-INFN
Andreas Ringwald	Deutsches Elektronen-Synchrotron DESY
Andrei Rostovtsev	ITEP
Karsten Salomon	FAU Erlangen
David Saltzberg	University of California - Los Angeles
Olaf Scholten	KVI
David Seckel	Bartol Research Institute
Terry Sloan	Lancaster University
Christian Spiering	DESY, Zeuthen

Piero Spillantini	INFN - Firenze - Italy
Victor Svet	Institute of Acoustics
Cosimo Trono	IFAC-CNR of Florence and Dipartimento di Fisica of Pisa
Vladimir Tsarev	P.N.Lebedev.Physical Institute of RAS
Huitzu Tu	University of Southern Denmark, Odense
Adriaan M van den Berg	Kernfysisch Versneller Instituut
Justin Vandenbroucke	University of California, Berkeley
Bernhard Voigt	DESY, Zeuthen
Michael Walter	DESY, Zeuthen
Yusuke Watanabe	Tokyo Metropolitan University
Christopher Wiebusch	Universität Wuppertal
Dawn Williams	Penn State University
Ralf Wischnewski	DESY, Zeuthen
Dmitry Zaborov	ITEP, Moscow, Russia
Igor Zheleznykh	Institute for Nuclear Research of the Russian Academy of Sciences